Biology in Pest and Disease Control

Biology in Pest and Disease Control

The 13th Symposium of
The British Ecological Society
Oxford, 4–7 January 1972

edited by
D. Price Jones MSc, PhD, FIBiol
Formerly of Plant Protection Ltd
Jealott's Hill Research Station
Bracknell, Berks

and

M. E. Solomon MSc, FIBiol
Long Ashton Research Station
University of Bristol

Blackwell Scientific Publications
OXFORD LONDON EDINBURGH MELBOURNE

Osney Mead, Oxford,
3 Nottingham Street, London W1,
9 Forrest Road, Edinburgh,
P.O. Box 9, North Balwyn, Victoria, Australia.

ISBN 0 632 09070 7

First published 1974

Distributed in the U.S.A. by
Halsted Press
a division of
John Wiley & Sons Inc

Printed in Great Britain by
Western Printing Services Ltd.
Bristol
and bound by
The Kemp Hall Bindery
Oxford

Contents

ix Preface

Section 1
Ecological background

3 The role of the seed predator guild in a tropical deciduous forest, with some reflections on tropical biological control
Daniel H. Janzen, Department of Biology, University of Chicago, now at Department of Zoology, University of Michigan, Ann Arbor, Michigan 48104, U.S.A.

15 Population dynamics and pest control
G. C. Varley, Hope Department of Entomology, University Museum, Oxford

28 Demographic studies on grassland weed species
José Sarukhán, School of Plant Biology, University College of North Wales, Bangor, now at Dept. de Botánica, Instituto de Biología, U.N.A.M. Apartado Postal 70–233, México 20, D.F. México

42 On the ecology of weed control
G. R. Sagar, School of Plant Biology, University College of North Wales, Bangor, Caernarvonshire

Section 2
Approaches to control

59 The use of biological methods in pest control
F. Wilson, Formerly Australian Commonwealth Scientific and Industrial Research Organisation, Sirex Unit, Imperial College Field Station, Silwood Park, Sunninghill, Ascot, Berkshire, now at Department of Zoology, Imperial College, London

73 Pest resistance in fruit breeding
The late R. L. Knight and F. H. Alston, East Malling Research Station, East Malling, Maidstone, Kent

87 Plant breeding for disease resistance
F. G. H. Lupton, Plant Breeding Institute, Cambridge

97 The biological contribution to weed control
G. W. Cussans, Agricultural Research Council, Weed Research Organization, Begbroke Hill, Oxford

106 Diversification of crop ecosystems as a means of controlling pests
J. P. Dempster, The Nature Conservancy, Monks Wood Experimental Station, Huntingdon; and T. H. Coaker, Department of Applied Biology, University of Cambridge

115 The use of pathogens in the control of insect pests
T. W. Tinsley and P. F. Entwistle, Natural Environmental Research Council, Unit of Invertebrate Virology, University of Oxford

130 Control of insects by exploiting their behaviour
J. M. Cherrett, Department of Applied Zoology, University College of North Wales, Bangor, Caernarvonshire; and T. Lewis, Department of Entomology, Rothamsted Experimental Station, Harpenden, Hertfordshire

147 Insect control by hormones
C. N. E. Ruscoe, Plant Protection Ltd., Jealott's Hill Research Station, Bracknell, Berkshire

162 Genetic control of insects
G. Davidson, Ross Institute of Tropical Hygiene, London School of Hygiene and Tropical Medicine, Keppel Street, Gower Street, London

178 Selectivity in pesticides: significance and enhancement
D. Price Jones, Formerly Plant Protection Ltd., Jealott's Hill Research Station, Bracknell, Berkshire; now at 82 Shinfield Road, Reading, Berkshire

196 Integrated control in Britain
M. J. Way, Imperial College Field Station, Silwood Park, Ascot, Berkshire

Section 3
Application of biological control

211 The use of biological methods in the control of vertebrate pests
R. K. Murton, The Nature Conservancy, Monks Wood Experimental Station, Abbots Ripton, Huntingdon

233 Biological control of water weeds
B. Stott, Salmon and Freshwater Fisheries Laboratory, Ministry of Agriculture, Fisheries and Food, London

239 Biological control of plant diseases
J. Rishbeth, Botany School, University of Cambridge

249 Control of nematode pests, background and outlook for biological control
F. G. W. Jones, Rothamsted Experimental Station, Harpenden, Hertfordshire

269 Integrated control of pests and diseases of sugar beet
R. Hull, Broom's Barn Experimental Station, Higham, Bury St. Edmunds, Suffolk

277 Role of biology in the control of pests and diseases of vegetable crops
G. A. Wheatley, National Vegetable Research Station, Wellesbourne, Warwick

294 Progress towards biological control under glass
I. J. Wyatt, Glasshouse Crops Research Institute, Littlehampton, Sussex

302 Control of forest insects: there is a porpoise close behind us
D. Bevan, Forestry Commission, Alice Holt Lodge, Farnham, Surrey

Section 4
The future of biological control in Britain

315 The future of biological control in Britain: a grower's view of the short term
R. Gair, Ministry of Agriculture, Fisheries and Food, Agricultural Development and Advisory Service, Cambridge

321 The future of biological control in Britain: a grower's view of the long term
A. H. Strickland, Ministry of Agriculture, Fisheries and Food, Plant Pathology Laboratory, Harpenden, Hertfordshire

332 The future of biological control in Britain: a research worker's view
N. W. Hussey, Glasshouse Crops Research Institute, Littlehampton, Sussex

342 The future of biological control in Britain: a manufacturer's view
W. F. Jepson, Cyanamid International, Gosport, Hampshire

349 The future of biological control in Britain: a conservationists' view
K. Mellanby, The Nature Conservancy, Monks Wood Experimental Station, Abbots Ripton, Huntingdon

Appendix:
Miscellaneous contributions

357 Introduction to a session on approaches to control
D. L. Gunn, Taylors Hill, Chilham, Canterbury, Kent; formerly Agricultural Research Council, London

357 Introduction to a session on the application of biological control
M. B. Green, Imperial Chemical Industries Limited, Mond Division, Runcorn, Cheshire

359 The development of biological control in the West Palaearctic Regional Section under the influence of OILB/IOBC
L. Brader, General Secretary, IOBC West Palaearctic Regional Section, Instituut voor Plantenziektenkundig, Onderzoek, Wageningen, the Netherlands

363 Author index

372 Subject index

Preface

This book comprises the papers presented at a Symposium at Oxford in January 1972 on the subject 'Increasing the Biological Contribution to the Control of Pests and Diseases', organized jointly by the British Ecological Society, the Association of Applied Biologists and the Society of Chemical Industry (Pesticides Group). The origins of the Symposium, and the part played by the Biological Control Subcommittee of the British National Committee for Biology, are outlined by Mr F. Wilson in the opening paragraph of his contribution, which constituted a keynote address at the Oxford meeting.

The aim of the organizers was to achieve a broad and principled account, as well as practical examples and assessments, of the role of biological factors and biological methods in the control of pests, weeds and plant pathogens. The intention was to look at the past and the present with an eye to the future.

Although pesticides have long been predominant in this field, biological factors, such as the resistance of pest to pesticide and of host-plant to pest, have of course always been relevant, and neglected only at the cost of later trouble. Biological agents, such as predators, parasitoids and pathogens of pests, are the chief means of regulation of the abundance of many pests, as has been demonstrated, for example, by the outbreaks of fruit tree red spider mite following the reduction of its natural enemies by pesticides, and by the numerous successes in pest control achieved by the introduction of predators, parasites or pathogens. A number of new methods of pest control, some entirely biological, some partly chemical, have been put forward and tested in the past ten or twenty years; these are discussed in some of the contributions to the Symposium. Some authors discuss combinations of appropriate husbandry with chemical and biological measures of control. With a few exceptions, attention is concentrated on British problems and conditions.

The plan of the book generally follows that of the Symposium meeting. A few opening contributions deal with some aspects of ecology relevant to the control of pests and weeds (Janzen, Varley, Sagar, Sarukhán). There follow eleven papers dealing with different approaches to the control of pests, weeds and plant pathogens: the principles of biological control of pests are discussed by Wilson; other contributors deal with the control of pests, chiefly insects, by the exploitation of insect pathogens (Tinsley & Entwistle), behaviour (Cherrett & Lewis), hormones (Ruscoe), genetical manipulation (Davidson), selective insecticides (Price Jones), diversification of the crop habitat (Dempster & Coaker) and integrated control (Way); others deal with plant breeding for resistance to pests (Knight & Alston) and disease (Lupton), and with biological control of weeds (Cussans).

A third section comprises estimates of the applicability of biological methods of control to vertebrates (Murton), to nematodes (Jones), to pests and diseases of sugar beet (Hull) and vegetable crops (Wheatley), to pests of glasshouse crops (Wyatt) and of forest trees (Bevan), to plant diseases (Rishbeth) and to water weeds (Stott).

The topic of the final section is 'The Future of Biological Control in Britain', from the viewpoint of the grower (Gair, Strickland), the research worker (Hussey), the manufacturer (Jepson) and the conservationist (Mellanby).

The appendix contains summaries of the opening remarks of two of the Chairmen of Sections of the Symposium, and a special contribution to the Symposium made by L. Brader on the development of biological control under the influence of the Organisation Internationale de Lutte Biologique contre les Animaux et les Plantes Nuisibles (OILB or, in its English equivalent, IOBC).

The discussions at the meeting are not included in this volume. One point, however, was raised by Professor G. C. Varley and referred to the editors for their attention, namely the possibility of confusion arising from the different meanings attached to the word 'control'. We hope that, in its context, the significance of this term will always be clear to the reader. The only example we feel may require our comment is the use of 'integrated control'. Professor M. J. Way, in his contribution to this Symposium, follows the very general usage in which integrated control means primarily the control of a pest by a combination of different methods. Dr. N. W. Hussey and Mr. I. J. Wyatt include, and indeed emphasize, the integration of control measures aimed at different pests of a single crop.

Section 1
Ecological background

The role of the seed predator guild in a tropical deciduous forest, with some reflections on tropical biological control

DANIEL H. JANZEN *Department of Biology, University of Chicago, now at Department of Zoology, University of Michigan, Ann Arbor, Michigan 48104, U.S.A.*

General theory

What effect does the array of animals that kill seeds (and seedlings), henceforth termed the seed predator guild, have on the community structure of the plants in a forest? In north temperate forests, one of the major impacts of the seed predator guild appears to be in the generation of mast year cycles in seed production; this occurs through the mechanism of natural selection against those individual trees that are not synchronized in their fruiting with the majority of their conspecifics or even other unrelated members of the community (Janzen 1971a). I have also argued that it is the generation of mast year behaviour on a population and community-wide basis that prevents the seed predator guild from lowering the density of the dominant tree species to a level where other tree species may invade and thereby increase the richness and diversity of tree species in the forest (Janzen 1970a, 1971a). These arguments, as well as those discussed in the remainder of this presentation, can be applied to foliage-eating animals as well, but will not be discussed further here in that context.

Let us now move to the other end of the gradient of climatic harshness and unpredictability, the lowland tropical evergreen and deciduous forest. Here, we enter habitats where the seed predator guild can be maximally effective at preventing population recruitment by any particular species of forest tree. The result should be that no one forest tree species may competitively displace a wide array of others. This postulation assumes that, given the absence of plant-eating animals or given immunity to them, some small subset of the trees in any complex tropical habitat will be able to multiply to where they competitively displace the others that would also grow there (Janzen 1970a).

In short, I am postulating that the population size of a tree species in a mixed forest is regulated by 'density-dependent' processes just as are animal populations. Further, I am assuming that the phytophagous animal populations in these forests are resource-limited during significant portions of their contemporary (and evolutionary) histories. Finally, I assume that the more favourable are the conditions for the seed predator guild in respect to any particular tree population, the further will that tree's population density (= proximity) have to be depressed by the seed predators before the individual seed crops are so far apart in time and space that the animals can no longer find juvenile plants at a rate that prevents an increase in the population of the tree. The obvious density-dependent mechanism need not be discussed further here. We may, however, recognize that in the case of both seed- and foliage-feeders, a major method of escape by the plants in time and space is through chemically generated inedibility (e.g. Janzen 1971b). We are then confronted with the question of why cannot the animals, and particularly the insects, 'out-run' the plant in the biochemical coevolution game, since they have so many generations during the lifetime of the tree? A formal generalized answer is still in the formulation stage (Janzen & Wolff, unpublished manuscript) but, in short, it appears that it is probably the fate of the average tree population in the lowland tropics to lose the biochemical evolutionary game in a moderately short time to the animals on a local habitat basis. This should result in a high rate of local tree species extinction (and immigration) for any particular habitat over evolutionary time (a rate which is much higher than that to be expected in temperate zone habitats of equivalent geographic size). This higher extinction and immigration rate should also be associated with a higher rate of host-switching in evolutionary time for the more host-specific insects. Incidentally, it must be remembered that all such statements are to be viewed as stochastic in nature, and thus single examples that seem contradictory may simply represent the tails of frequency distributions.

The general postulations set forth in the previous three paragraphs have been generated from a very broad miscellany of data gleaned from the literature, and from my own field work and that of my colleagues in the Central American tropics. Until very recently, almost all the relevant data were gathered for reasons other than the direct examination of these ideas and are not definitive for any given forest or set of ecological conditions. The postulations are put forth for the express purpose of encouraging other workers to gather data on seed and seedling predation, and on foliage parasitization.

In considering these general questions, I have found it always useful to start with a model of the reproduction by a single tree in a single population, and focus my ecological attention primarily at that level. In doing this, I have found a useful working model (Janzen 1970a). In most generalized terms, this model explores the idea that, on the average, an adult tree in a lowland tropical forest should be most likely to produce another adult at some intermediate distance from its crown. This follows from two observations. Reproduction close to the parent tree is very unlikely, owing to the attacks of incoming and reproducing predators on the seeds and seedlings concentrated below the parent. Reproduction is unlikely far from the parent owing to the low immigration rate of seeds. These two processes should generate an adult spacing pattern far from the contagion expected if only the seed shadow is considered. Now these considerations are of course only relevant if there is any reproduction at all. This proviso is added for the following reason. For any host-specific set of seed predators, a primary source of the seed predators at a specific crop are conspecific crops nearby (in time as well as space); there is a system of mutual infection by all the trees in the area that the insect regards as one 'species' of host. This means that, on the average, the closer a tree is to other conspecifics (with distances measured in ecological rather than metric units), the less likely it is to produce even one new adult. This type of reproductive heterogeneity may also mean that most of the adult trees of a given species may be effectively sterile (despite heavy seed production).

At this point in our brief résumé we could turn in either of two major directions. We could examine the theoretical outcomes, with particular population structures and with aggregate community structures, of modifying the behaviour or abilities of the dispersal agents, the seed predators, and/or the adult trees' reproductive behaviour. Except for a few cases, such as mangroves and some species of Dipterocarpaceae that can persist as pure stands in tropical forests, at present this would be a rather sterile exercise owing to the lack of hard data on the necessary parameters. We have just begun to build a data pool on what the seed shadow of an individual tropical tree looks like. We have almost no information on how tropical seed predators respond to an individual tree's seed crop, nor what is the effect of inter-tree spacing in time and space on tropical herbivores. Finally, we have almost no information on how individual tropical trees modulate their seed production. Some starts in these directions may be found in the following embryonic studies: Janzen (1969, 1971b, 1971c, 1971d, 1972a, 1972b), Wilson & Janzen (1972).

It is probably more profitable at this point to examine some of the more conspicuous assumptions underlying the general model. Data are now beginning to accumulate from a study of deciduous tropical forests in Costa Rica. I may add that, throughout the following discussions, it is evident that an understanding of the systems being described is relevant to the theory of applied ecology (= biological control of insect and weed pests), though not necessarily to its practice, in tropical countries; I shall, however, save discussion of this area until last.

Host specificity of seed predators

A major background assumption of this study is that the degree and kind of host specificity will influence the predator-prey interaction. Let me immediately add that it is often not clear how it will do so. I have found the seed predator guild of a deciduous forest in Guanacaste Province (Pacific coast lowlands of Costa Rica) to be subdivisible into several major groups on the basis of their type of host specificity. The most host-specific are the beetles (Bruchidae and Curculionidae) whose larvae eat the contents of seeds, usually while the seeds are still within the fruit. They may kill as much as 50 to 99% of a seed crop before it has been dispersed. To date, each of 66 of these species of seed predators (59 bruchids and 7 weevils) has been reared from only one species of host plant (in over 1000 samples of seed crops). Only five species (all bruchids) occur on two hosts (a *Mimosestes* on two *Acacia*, one *Amblycerus* on two *Cassia* and another on two *Cordia*, a *Caryedes* on two *Bauhinia*, and an undescribed bruchid genus on two *Lonchocarpus*) and none on three. These observations should be viewed in the context of a habitat containing about 300 species of woody plants, over half of which have been surveyed in the study. It appears that seed chemistry and fruit behaviour-morphology (e.g. Janzen 1969, 1971a) are the primary traits preventing broader host selection and preferences by the bruchids; it is also tempting to postulate that past millennia of plant-bruchid coevolution have resulted in rather strong character displacement among those traits that are important in defence against these insects. The plant appears to have played the evolutionary game through predator satiation of its host-specific seed predators and chemical-morphological exclusion of its potential seed predators.

There is also an array of insects that feed on fruits, immature seeds, damaged mature seeds, and even mature seeds on occasion. The amount of damage they do is unpredictable and may involve such complex things as destroying the attractiveness of a fruit to a dispersal agent, thereby causing its seeds to be found still on the tree by a seed predator at some later date. Such insects are often highly variable in their host specificity, and will not be considered further here. I should, however, point out that they may be of great importance agriculturally while having little impact on the biology of the plant.

Vertebrates may be divided into at least two groups on the basis of their host specificity. I must add early in this discussion that an individual animal may be classed in either of these two groups at different times, depending on such things as time of year, reproductive condition of the animal, and site of foraging in the habitat. The first group is made up of those vertebrates that subsist on the seeds of those tropical plants that escape through predator satiation. As in temperate forests, such trees generally produce highly edible (and often wind-dispersed) seeds during a very short period (and sometimes at long intervals, e.g. Dipterocarpaceae). Many vertebrates appear to subsist in great part on such seeds during those times of year when they are produced. Here, the vertebrate may be described as a 'polyphagous' or 'generalist' species, if we measure host specificity by the number of Latin binomials on the host list. I should point out, however, that this is quite different from the generalist that is capable of eating a wide variety of toxic plants. The epitome of escape through predator satiation is probably represented by grasses and palms, neither of which appear to have seeds toxic to seed-eating vertebrates.

On the other hand, there is another kind of polyphagy that appears to be practised by phytophagous vertebrates in tropical vegetation. A large number of tropical seeds contain high concentrations of alkaloids, uncommon amino acids and other poisonous compounds, probably as a primary result of co-evolution with insects. For example, the large seeds of the vine genus *Mucuna* contain 4–8% L-dopa and are not attacked by any insects (cf. Bell & Janzen 1971). Yet such seeds may be eaten in small amounts by small tropical rodents. For example, the seeds of the large forest vine *Dioclea megacarpa* contain 5–10% canavine (a potent argenine competitor in protein synthesis) yet are eaten by squirrels (*Sciurus variegatoides*) in quantities far below their stomach capacity. The result is that during the lifetime of a seed crop, the vine generally loses less than 10% of its seeds to this animal (Janzen 1971b). It appears that small rodents such as this eat only small amounts of a wide variety of toxic seeds during any one period, and thereby do not get enough of any one to overcome their internal detoxication system for that class of compounds (irrespective of whether detoxication is done by the gut micro-flora or their own biochemical system). This behaviour is probably functionally identical to the habit that ruminants have of browsing small amounts of foliage from each one of a wide variety of dicotyledonous plants, each of which is likely to contain a toxic dose of phenols, alkaloids, cyanogenic glucosides, etc. if eaten in large quantity.

The polyphagous phytophagous rodent that subsists on toxic seeds during all or part of its life cycle is clearly quite a different beast, from the plant's viewpoint, from the polyphagous rodent or other vertebrate that subsists primarily on seeds produced by predator satiators. The former rodent can never move into a local area and eliminate a medium-sized seed crop of toxic seeds, nor can it build up a large population on a locally abundant tree species (as could a host-specific insect). Yet such a rodent exerts a constant selective pressure against trees that find themselves evolving in the direction of producing a very small number of very large seeds. To stay in the game, such a tree would have to have ever-increasingly toxic seeds. Finally, such rodents (or even larger animals such as deer and peccaries) clearly have less potential as density-dependent population regulators of the trees than do insects or those rodents subsisting primarily on the seeds from trees on the 'predator satiation' adaptive peak.

I might add at this point that most immature fruits appear to be protected by their chemistry alone, and both the rodents and birds (especially parrots) tend to take immature tropical tree seeds in regular but small amounts. Again, this suggests a system where satiation is achieved not through filling the stomach of the predator but through nearly overloading its detoxication abilities.

Parasitization of seed predators

A second major area of background assumptions is associated with the assumption that the seed predators are for the most part resource-limited on both an evolutionary and a contemporary time scale. For the insects, this is suggested by several indirect lines of evidence from the deciduous forest study cited earlier. For the vertebrates, no direct leads have been obtained, but much anecdotal and circumstantial evidence has convinced me that they are likewise resource-limited.

One of the most conspicuous aspects of rearing bruchids and weevils from tropical seeds crops is that many appear to have no parasites. Of the 71 beetle species alluded to earlier, I have reared parasites from only 20. Further, the parasitization is at a level of 1–5% of the emerging adult beetle population, strongly suggesting that the parasites are just 'riding' their host population and having little density-dependent impact on them. A second conspicuous trait of the parasites is that they appear to be highly polyphagous in respect to bruchid species (and probably in respect to other insects as well); one is left with the distinct impression that, to survive in a community where most of the prey species are widely scattered in space and time, the parasite must be polyphagous. It is this observation that suggests why many of the bruchids lack parasites. The parasite is probably not capable of simultaneously possessing the wide array of host-finding behaviours and morphologies that are needed to attack a high number of bruchid species (each with a low density). A high number would be necessary in order to accumulate enough prey individuals to sustain a parasite population. The process is probably the same as that underlying the fact that only a small fraction of entomophagous parasites have hyperparasites in temperate zones where prey are comparatively abundant.

However, I do not intend to suggest that seed predators are unavailable to general predators. Most of the mortality in the bruchid and weevil population appears to occur when the adults are searching for hosts (and through intraspecific competition in the fruit). However, this predation is probably not density-dependent in action, as no one bruchid species constitutes a large enough fraction of the daily diet of any of the predators (birds, spiders, lizards) to generate any kind of density-dependent response from them if there is a local increase in bruchid or weevil numbers. This means that the insect population density should be most strongly related to the ecological distances in time and space between host plants (as measured in the numbers of beetles that actually pass between hosts, rather than in metric distances between hosts). These distances should in turn be long-term functions of how effective the insects are at killing seeds at any given host plant proximity in time and space.

Related to these queries is the observation that the bruchids are far more evenly distributed among their host plants than would be expected by a randomly operating process. Of the 59 host plants for the 64 bruchids referred to earlier, 48 have only one bruchid using them as a resource base, 9 have two bruchids attacking them, and 2 have three bruchids. This suggests strong competitive displacement among the bruchids in the coevolution of the system. This coevolution may operate on the level of direct interactions between two well-established species, and on the level of an occupant species preventing a second species from shifting onto the host 'island' (Janzen 1968) during an episode of evolution of host resistance by the usual host of the invading species.

Seasonal considerations

A third major area of underlying assumptions is suggested by the idea that, when hosts are not immediately available for reproduction by the seed predator, the population of seed predators is steadily declining. Since tropical vertebrates do not aestivate in any total sense, and since we are not dealing with highly migratory animals in this discussion, I shall take this assumption to be generally valid for vertebrates. Insects, on the other hand, have a general reputation for being sexually inactive and going into hiding during inimical seasons (e.g. those insects that wait in diapause between mast years of temperate zone forest trees, Janzen 1971a). However, in tropical communities, it appears that many phytophagous insects pass the dry season in reproductive diapause, but as otherwise active adults seeking those bits of moisture and nutrients adequate for survival but not for reproduction (Janzen 1972c). In the deciduous forest under scrutiny in the present study, the seeds of woody plants are for the most part produced during the dry season, and it is therefore the rainy season that is the inimical period for most bruchids. During this season, the adults of many species can be collected while they are visiting flowers for nectar or while they are resting (?) on foliage in shady places. Just by sweeping the understory of one primary deciduous forest site in July (about two months after the rains had started), 16 of the 38 bruchids known for the area (and breeding there in the dry season) were collected in 1971.

The general habitat appears to be highly heterogeneous for these 'overwintering' adults in terms of survivorship. At least one bit of evidence suggests that the 'islands' of highly favourable habitat have local carrying capacities. When one finds a local high density area for a particular bruchid species, the site exhibits the same characteristics of emigration and immigration shown by most islands. For example, at the deciduous forest site in Costa Rica, the adults of the *Acanthoscelides* bruchid that attack the seeds of the mimosaceous shrub *Mimosa pigra* feed on nectar (and pollen?) of the flowers of this plant during the early rainy season, while 'waiting' for the pods to mature in the late rainy season and early dry season. Here, the island is the source of both adult and larval food (and see Janzen 1968). In July of 1971, an isolated bush with a daily crop of about 80 to 110 inflorescences was relieved of all its beetles each morning for three consecutive days. Each day yielded about 300 beetles. The next day yielded about 25 beetles, as did the 5th to the 8th days. This suggests that the bush and its immediate environs had a pool of about 900 adult beetles, that the daily flower crop could hold only about 300 beetles at any one time, and that the immigration rate of beetles to the bush was about 25 per day. On the 9th to the 17th days, the beetles were released back onto the inflorescences after capture, with the expected result that the total numbers caught each day increased by about 25 beetles. This beetle can be obtained at a density of about 1 per 800 sweeps with a sweep net in the vegetation surrounding the *Mimosa pigra* bush; one sweep through a *M. pigra* bush generally yields 50 to 100 adult beetles.

Islands

As final comment on the background parameters of seed predator ecology, it should be mentioned that the generalities suggested in the previous paragraphs are not expected to hold on real islands in many cases. Tropical island vegetation tends to have the low plant diversity and the poverty of species characteristic of temperate zone habitats, and in my opinion for much the same reason; the seed predator guild (and foliage feeder guild) is ineffective at breaking up the pure stands of the best competitors. Major parts of the seed predator guild are absent on tropical islands through the difficulty of immigration (e.g. the absence of terrestrial vertebrates from Puerto Rico, Janzen 1972b) and through the difficulty of maintaining a population where there are few alternate hosts (e.g. to provide nectar for adults during the wet season) or alternate habitats. It is also not helpful that the host density is so low and the habitats are so monotonous that in bad years no seed may be produced by a given species of tree.

Biological control

The study of the ecology of the interactions of tropical plants and the animals that eat them is still at an embryonic level. Further, in trying to understand the relevance of field studies, such as those briefly discussed above, to the biological control of pests in agriculture, we are put in the uncomfortable position of trying to expect the unexpected, as Charles Elton has put it. These pious disclaimers notwithstanding, it may be worthwhile to enumerate a few considerations for biological control that the study of tropical seed predators and defoliators has brought to mind.

Perhaps foremost is the idea that tropical woody plants, occurring at low density as they do (on account of their herbivore load, to use Richard Root's term), have their primary external escape mechanism destroyed when planted in high-density plantation systems. Winter is on the side of the temperate zone farmer, and it is perhaps fortunate that he cannot eliminate it (Janzen 1970b) as the dryness of the desert has been eliminated with irrigation. The concentration of host plants by the tropical woody plant agriculturalist means that he is not only asking his biological control agents to depress their prey populations, but asking them to do it when the prey have suddenly had one of their main control mechanisms removed. With herbaceous (annual) crops, there is perhaps less of a problem in this area of cultural control because the farmer at least would appear to have the option of artificially creating a winter and spacing through fallow periods, crop rotation, and interplanting. However, in view of the marginal income level experienced by most tropical subsistence farmers, crop rotation and fallow periods are often not possible under contemporary human population densities and pressures for cash on an immediate basis.

If tropical pest insects turn out to have as few host-specific parasites as do the bruchids studied here, it may in general be extraordinarily difficult to keep a parasite population at a high level in the community while its prey (pest) species fluctuates with the ups and downs of its host plant during the cropping regime. Maintaining high parasite levels may be even more difficult as the natural vegetation around the field is removed, since the array of alternate hosts for the parasite to choose from will be greatly reduced by this habitat destruction. Finally, if tropical parasites turn out to be as catholic as those reared from the bruchid community, it may be difficult to keep them associated with the pest (prey) species of interest.

The evolution of resistance in crop plants cannot be ignored in any biological control scheme, since it so strongly influences the rates of increase of the pest species on a given crop. We must remember that a large number of the successful tropical crops have been tree crops of cash value to developed countries, but of little or no direct value to either the people producing them or to the majority of the animals in the field-forest community. Here, owing to the long-term nature of the crop, breeding resistant strains hardly fits with the hand-to-mouth economy of subsistence or even with more advanced agriculture. It is also not likely to be very successful in the face of a tropical insect community, the members of which have been playing the chemical coevolutionary game as their way of existence since the upper Cretaceous. With annual crops there might appear to be more hope. However, annual crops tend to be much more edible (to the animal community) than tree crops, since much of their chemical defence has been bred out of them in the process of turning them into crop plants. In breeding for resistance we are putting these compounds back into the plant (instead of on the outside, in the form of insecticides) and it seems unlikely that we can ever reconstitute an edible plant with chemical defences so effective that the plant is as lightly attacked in a field as it is when widely dispersed in a forest or other habitat.

A major component of the biological control of crop pests is the rapid build-up of the parasite or predator on a local outbreak of pests. Temperate-zone habitats appear to abound with parasites and predators that have a high reproductive rate and a behaviour to use it effectively when large amounts of prey are encountered. This is hardly surprising considering the pattern in which their normal prey species become available. However, it appears that tropical parasites and predators have been moulded by natural selection in such a manner that they put much more of their total resources into prey-finding behaviour and physiology, with the result that they are not able to consume an extraordinarily large population of items once found. Again, this is not surprising if one considers the distances that must occur between prey items in natural tropical systems if just the distribution of the prey's host plants is considered. Even if the proper parasite can be found, and even if it can be provided with a large prey population such as may be found in a field, it may be behaviourally and physiologically incapable of responding to a high host density.

As a final comment, I would point out that, over short distances, tropical phytophagous insects seem to be extraordinarily adept at locating their host plants, and at migrating back and forth between alternate habitats. This is true not only of the bruchids mentioned earlier, but also of the insect community as a whole (e.g. Janzen 1973). This means that the increased proximity of fields in large-scale agriculture is likely to accentuate the general problem (even when fields are some distance apart) of a large inoculation of pests starting off the pest population in a field well before the parasite or predator can catch up.

By way of summary, I should like to point out that the previous paragraphs unfortunately support what I consider to be an important fallacy in applied ecology. This is the idea that a general rule, which is nothing more than a statement of central tendency, is necessarily applicable to a specific case and may therefore be used to discourage experimentation in that particular case. We can all think of specific examples arguing for and against each of the general statements made in this paper. The crop plant and its pests are only one of a very small fraction of the plant–insect community whose processes are represented by my generalizations; their specific case may well lie many standard deviations from the mean. Many successful agricultural systems are very atypical of the ecological processes that generate and maintain natural communities (and often because man harvests very different things from the plant than do the other animals in the community). It seems entirely reasonable that there is only one sure way to learn which sets of plants and animals can interact for human profit in the tropics, and that is to try them. Agricultural experiment stations in the tropics are woefully understaffed, ill-equipped and out-of-date. Their personnel are generally not keen pursuers and exploiters of the special features of the tropical agricultural ecosystem. This is not surprising when we realize that their staff are generally products of the educational system of some developed country, with guiding philosophies often oriented towards agricultural ecosystems evolved to deal with the annual pulse of energy provided by the temperate-zone summer. Further, they are coevolved with respect to the economics of the countries that have long profited from that pulsed energy, both directly through yield and indirectly through the evolution of a social system to deal with pulsed yield. Finally, they are guided by governments that demand an agricultural productivity high enough to raise the standard of living for a large number of people to the level seen in the developed countries. What is missing is the active realization by the underdeveloped countries that the developed countries have probably already peaked out on their standard of living, and that one of the major routes to that peak was through exploitation of the produce of the underdeveloped countries. The question then becomes, what country is the underdeveloped country going to exploit to peak its own standard of living? It often appears that they have chosen to rape themselves. In short, the agricultural system in the tropics is being asked to produce far more than the system can produce, and the tragedy is that its inevitable failure will be blamed on such things as droughts, lack of conscientious effort by applied ecologists, and on political systems, rather than the excess of people, the lack of planning that is generating this excess, and the false expectations generated by using the developed countries as models.

References

BELL E.A. & D.H. JANZEN (1971). Medical and ecological considerations of L-dopa and 5-HTP in seeds. *Nature, Lond.* **229**, 136–7.

JANZEN D.H. (1968) Host plants as islands in evolutionary and contemporary time. *Am. Nat.* **102**, 592–5.

JANZEN D.H. (1969) Seed-eaters versus seed size, number, toxicity and dispersal. *Evolution* **23**, 1–27.

JANZEN D.H. (1970a) Herbivores and the number of tree species in tropical forests. *Am. Nat.* **104**, 501–28.

JANZEN D.H. (1970b) The unexploited tropics. *Bull. Ecol. Soc. Amer.* **51**, 4–7.

JANZEN D.H. (1971a) Seed predation by animals. *Ann. Rev. Ecol. Syst.* **2**, 501–8.

JANZEN D.H. (1971b) Escape of juvenile *Dioclea megacarpa* (Leguminosae) vines from predators in a deciduous tropical forest. *Am. Nat.* **105**, 97–112.

JANZEN D.H. (1971c) The fate of *Scheelea rostrata* fruits beneath the parent tree: predispersal attack by bruchids. *Principles* **15**, 89–101.

JANZEN D.H. (1971d) Escape of *Cassia grandis* L. beans from predators in time and space. *Ecology* **52**, 964–79.

JANZEN D.H. (1972a) Escape in space by *Sterculia apetala* seeds from the bug *Dysdercus fasciatus* in a Costa Rican deciduous forest. *Ecology* **53**, 350–61.

JANZEN D.H. (1972b) Association of a rainforest palm and seed-eating beetles in Puerto Rico. *Ecology* **53**, 258–61.

JANZEN D.H. (1973) Sweep samples of tropical foliage insects: effects of seasons, vegetation types, elevation, time of day, and insularity. *Ecology* **54** (in press).

WILSON D.E. & D.H. JANZEN (1972) Predation on *Scheelea* palm seeds by bruchid beetles: seed density and distance from the parent palm. *Ecology* **53**, 954–9.

Population dynamics and pest control

G. C. VARLEY *Hope Department of Entomology, University Museum, Oxford*

In practical pest control, ecological principles and population theory are little used at the present time, either in the planning of control measures or in the assessment of results. The applied biologists can hardly be blamed for this because ecologists have not yet reached agreement on methods nor yet made their theoretical foundations entirely secure.

My objective is to show how population models can help in the understanding of population processes. The number of possible population models is infinite and their properties are extremely diverse, so it is absolutely essential to ensure that the model chosen is realistic. The model used must be based on measurements of the population which is being studied. The measurements must be detailed enough to reveal the mechanism of existing population regulation.

After proper study when the mechanism of population regulation is understood, the effects of various possible practical strategies of pest management can be tested on the model; only those which promise success need be tried out in practice. This should materially reduce the number of failures and of unwanted 'side effects'. We shall see that the killing of species other than the target species can easily counteract the beneficial effect of killing the pest itself.

This paper is in effect a progress report which:

1 describes how we derived the model for the winter moth and its parasites and predators in an oak wood,

2 shows that the model is realistic, and

3 uses the model with various extra mortalities to mimic the possible effects of insecticides.

The final step, which is to test the effect of insecticide on the insect populations on oak trees, has not yet been taken, but is now being planned.

A detailed understanding of the mechanism of pest regulation may require 5 or even 10 years of study if a statistically significant result of high precision is needed; but the most important relationships may become clear in as little as 3 years, so that this should not be invoked as an argument against beginning the research. Many serious pests, like the codling moth, have been known for years. If a very small part of the total effort which has gone into *short-term* studies in Europe and the U.S.A. had been devoted to *long-term* studies in one place, properly planned to obtain the figures on which to base life-tables, then I think that a very small team could have reached a good understanding of the basic regulatory mechanism in less than ten years. The information now available is summarized in Chapter 4 of Clark *et al.* (1967).

Our own work, centred on the winter moth as a defoliator of oaks, has been the part time research of myself and George Gradwell; both of us have had many other duties. Our census methods have relied largely on traps of a simple kind and the trapping sequence has been illustrated (Varley 1971a). The essential point is that we have two counts of the winter moth and two counts of its larval parasites in each generation. By counting and dissecting fully grown winter moth larvae we get a count also of the ecto- and endoparasites of the larvae. Other kinds of traps operated at the right season count the adult parasites emerging from the ground under the trees. This enables us to build up life-tables for both the winter moth and its major parasites for a number of seasons. Both the winter moth and its parasites suffer very heavy mortality in the pupal stage in the soil. Tests with pupae and cocoons in the field showed a wide range of predatory insects, especially the ground beetles *Feronia* and *Abax* and the rove beetle *Philonthus* as well as shrews. Probably moles, voles and woodmice play a part in this predation (Varley 1970a). The methods we have used to analyse the census figures have been described in a number of papers (e.g. Varley & Gradwell 1970a). Essentially the process went through four major stages:

(1) *Life-tables*

The census data were restated in the form of life-tables in which the various mortality factors were treated as if they acted successively, with no overlap in time. Where, as in 'winter disappearance' (k_1) or 'pupal predation' (k_5), detailed information about causes was not available, the overall figure for the mortality during the period between census counts was used. So far as possible, similar life-tables for the commonest parasites were also prepared. Ideally, life-tables for the predators are also required; however, when our work was planned, we did not know how important predation would turn out to be and we have been unable to devise census methods to cover this.

Fortunately the overall effects of predators on winter moth were nevertheless describable in simple terms.

The life-table information is most simply expressed logarithmically, and the changes in population are measured by the difference between the logarithms; these are the *k*-values of the different successively acting mortality factors, or of the mortality periods.

(2) *Key-factor analysis*

When the *k*-values for each stage of the winter moth were plotted against generation number and compared with the changes in their sum, which is the total generation mortality K, it was clear which of the *k*-values were big enough and variable enough to be important. In particular, changes in winter disappearance, k_1, showed it to be the key-factor. It stood out because its big changes were clearly responsible for the changes in the generation mortality K.

The second point about key-factor analysis is that it shows which mortality factors other than the key-factor are also important, and which are small enough to be negligible. This part of the process directed our attention to pupal predation as being much more important than some kinds of parasitism, which in the early stages of the work had occupied a lot of attention partly because it was easy to observe.

(3) *Testing for density dependency*

By plotting individual *k*-values for a number of years against the logarithm of the population density on which they operated, we showed that pupal predation was density dependent. This relationship showed up not only when we used the mean figures for the population density for each year, but also when the figures for different trees in a single year were compared. To a first approximation, the *k*-value for pupal predation $k_5 = 0{\cdot}35 \log N$, where N is the pupal population density. This relationship (represented in the upper line of Fig. 2) is capable of stabilizing the otherwise unstable parasite-host oscillations predicted by the application of Nicholson's theory (Varley & Gradwell 1963).

(4) *Finding a sub-model for parasitism*

The first population model which we tested against winter moth census figures used Nicholson's theory. This described parasite searching behaviour by a single constant, the area of discovery *a* (Nicholson & Bailey 1935, Varley & Gradwell 1968). The value of the area of discovery was calculated from the parasite population density *P* and the *k*-value of its effect on the host, as

$$a = 2{\cdot}3\ k/P$$

Estimates for different years did not agree very well, but a mean value was used as a constant in a Nicholsonian population model, which fitted well with the census figures for the winter moth, but the predicted peaks of parasite population density were about two years too late.

Hassell & Varley (1969) proposed a new Quest Theory, which included a density dependent effect on the area of discovery, caused by mutual interference by the parasites. Measurements had shown clearly for *Nemeritis* a fall in the value of the area of discovery as in Fig. 1, where the mutual interference constant *m* determines the slope of the line. A population model for

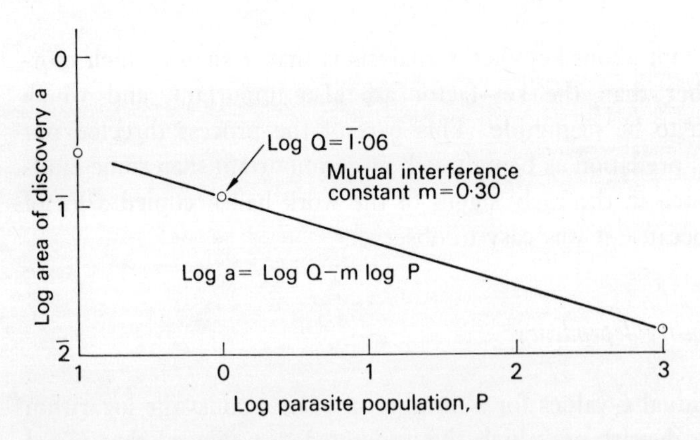

Figure 1. Parasite quest model. Quest constant $Q = 0{\cdot}115$.

winter moth and two of its parasites, whose behaviour is described by this quest theory, gives a much more satisfactory fit to the census figures for winter moth and its two main parasites, the tachinid fly *Cyzenis* and the ichneumonid *Cratichneumon* (Varley 1970b).

Model to study the possible effects of insecticides

We can now model host and parasite populations with reasonable accuracy if we use a simplified representation of the effects of predation. The model in effect has a density dependent factor acting specifically on each parasite; predation by insects and vertebrates is represented by a third factor (k_5)

dependent only on the population density of winter moth pupae in the ground. For simplicity we shall use here a model with only a single parasite species. The model has only three components:

1 The pest has a fixed annual increase × 7·5.
2 A specific parasite, whose properties are defined in Fig. 1.
3 Predation of the pupae, and of any included parasites, dependent on the pupal density as defined in Fig. 2.

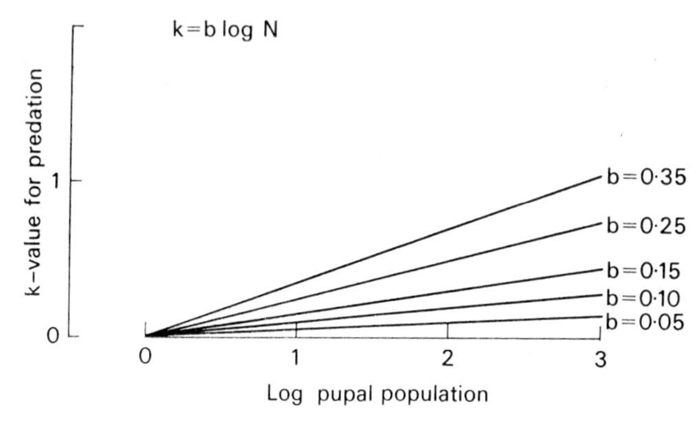

Figure 2. Predation model.

The consequences of this model (without the addition of insecticides) are given in Table 1, row A, and Fig. 3A. The model is very stable and has a larval pest population of 97 m^{-2} and adult parasites at 10 m^{-2}.

Table 1. The use of a population model to predict the effects of insecticide on the winter moth

Graph	Winter moth × 7·5 % killed	Parasite $Q = 0{\cdot}115$ $m = 0{\cdot}3$ % killed	Predation $k_5 = b \log N$ b	Steady density winter moth larvae m^{-2}
A	0	0	0·35	97
B	70	0	0·35	32
C	70	70	0·35	34
D	70	70	0·25	82
	70	70	0·15	247
E	70	70	0·10	300
	70	70	0·05	320
F	70	90	0·15	500
G	70	90	0·05	1000 (2000–400)
H	over 87	0	0	0

Model B

Table 1, row B, and Fig. 3B show the consequence of an annual destruction of 70% of the pest, with no kill of the parasites or predators by the insecticide. The pest rapidly stabilizes at a population of 32 m^{-2} and the parasite slowly drops to about one hundredth of its previous population. It is rendered ineffective, but is not exterminated.

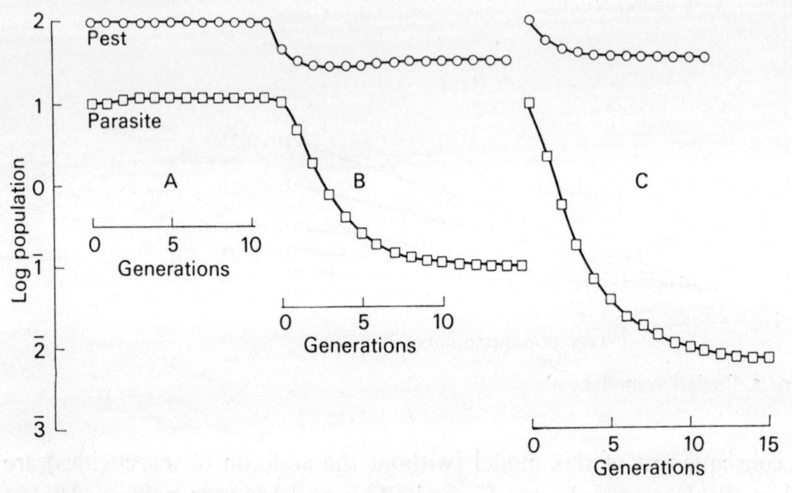

Figure 3. Predicted effect of insecticide on winter moth and its parasite.

Note that the annual 70% destruction of the pest is partly compensated by the relaxation of the density dependent pupal predation and the fall in the mortality caused by the parasite. Even the first year's 70% destruction of larvae brings the next year's initial larval population down only by 53%.

Model C

Table 1, row C, and Fig. 3C show the effect which might arise with an insecticide timed to kill the hosts after parasitism was completed, so that parasitized and unparasitized hosts were killed equally. Parasite numbers are more adversely affected but the pest stabilizes at a population of 34—almost the same as in B.

The interesting general conclusion from these models is that when there is a density dependent factor operating, which the insecticide does not influence, then the long-term effect is a reduction of the population, but the fall is much less than might be expected from the percentage kill alone.

In a routine insecticide spray programme in woodland, a high proportion of the insecticide will reach the ground layer at once or will be washed off the foliage by rain. The effects of this on insect and mammalian predators needs to be measured, but we have not done this. Let us suppose that the insecticide reduces the population of predators and test the consequences of a reduction in their effect. This is modelled by reducing the slope of the density dependent relationship in Fig. 2. Compare the models, C, D, and E in Table 1 and Figs 3 and 4.

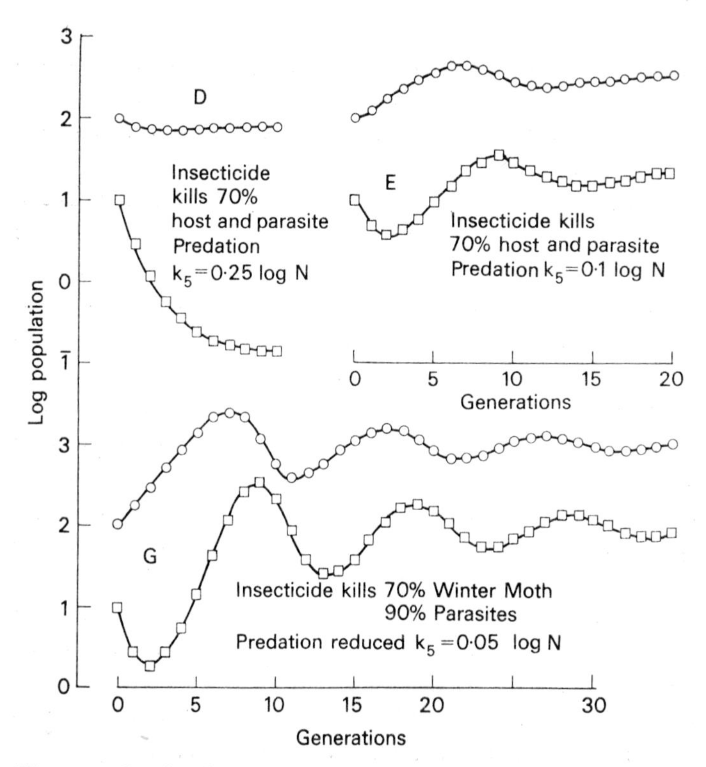

Figure 4. Predicted effects on winter moth and parasite populations when insecticide makes predators less effective.

Model D

This differs from model C only in that the density dependent pupal predation is reduced. When the *k*-value for predation, $k = 0{\cdot}25 \log N$, the steady density for winter moth rises to 82.

Model E

This with $k = 0 \cdot 10 \log N$, shows the effects of further reduction of the effectiveness of the density dependent component of the model. The population now shows a cyclic tendency, with damped oscillations, and the steady winter moth density rises to 300. Note also that, even from the first insecticide application, the pest population is rising. The adverse effect of killing the predator has been far stronger than the beneficial effect of killing the pest.

Models F and G

These (Table 1) show the predicted effects of a 90% kill of the parasite plus further reductions in the efficiency of predation. Model G (Fig. 4) is strongly cyclic, but again the population rises as a result of the insecticide effects right from the first application.

Results like those in models E, F and G are usually termed 'side effects' because they are the consequence of the mortality of species other than the target species. That they so readily appear in a realistic model is interesting. Wood (1971) gave practical examples of the upsurge of defoliators of oil palm after the use of highly toxic insecticides, but found that they could be avoided by using insecticides of lower toxicity.

This raises the question: how should we determine the percentage kill to aim at? These models suggest that for an insect such as the winter moth at Wytham, killing the target species alone reduces its population, but by less than might have been expected until the insecticide kills more than 87%. This is because the model assumes a maximum average natural rate of increase of 7·5 per generation. This figure takes into account the known egg production per female, the equal proportion of the sexes and a big early larval mortality which seems to be unavoidable. Only when insecticide kills the excess ($6 \cdot 5/7 \cdot 5 = 87\%$) can insecticides take over entirely from the biological components of predation and parasitism. It may be simple to devise an insecticide programme which will do this for one species, so that it may be easy to bring this target species down to an acceptable level of population. However, doing this will have reduced the populations of specific parasites (Models B and C) and also of the predators (Models D and G). The predators are non-specific in their feeding and their reduction by insecticide may easily allow the increase of species other than the original target species—that is to say there may be 'side effects'.

A model for the biological control of a pest

DeBach (1964, p. 121) states that to be successful in biological control a parasite needs three properties:

1 a high searching capacity,
2 specificity,
3 a high reproductive rate.

Study of population models incorporating these attributes suggests that they are not well defined. If the searching capacity is measured by the area of discovery, we have the unstable Nicholsonian system. If the parasite has a high and constant reproductive rate we have the unstable type of model considered by Thompson, where the parasite overtakes its host and exterminates it.

Quest theory worked well for winter moth and its parasites, so we have tried it in a model for biological control.

Suppose we seek to model a pest with a potential effective increase of 10-fold per generation, which we initially stabilize at 1000 m^{-2} by a density dependent factor such as competition. Suppose we introduce a specific parasite at an initial density of 1/1000 m^{-2} and that the parasite eventually regulates the pest at 1 m^{-2}. Then at this pest population density the parasite is killing 90% (k-value = 1·0) and 0·9 parasite larvae survive to become adults in the next generation. According to quest theory the killing power of the parasite population

$$k = 0{\cdot}43\, Q\, P^{1-m}$$

If $P = 0{\cdot}9$ and $m = 0$, then $Q = 2{\cdot}55$. Figure 5 uses quest theory to investi-

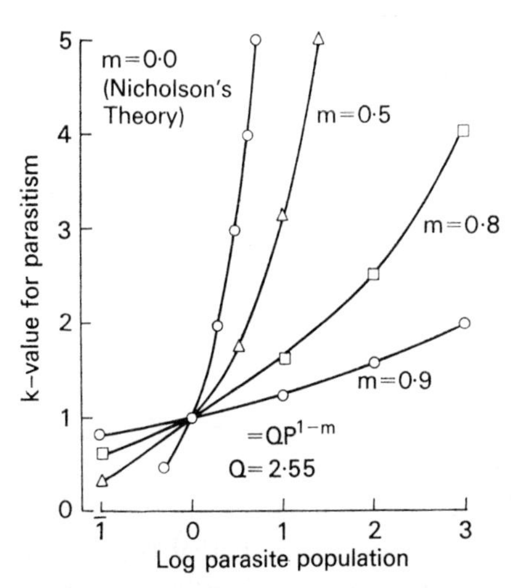

Figure 5. Parasite quest theory. Effect of mutual interference constant m on k-value caused by parasite population P.

gate the effects of parasite population density on the k-value for the pest mortality caused by the parasites. If the mutual interference constant is 0 (Nicholson's theory) the k-value is proportional to parasite density, and, on the logarithmic scale of Fig. 5, k rises very steeply. If it is to be effective, the parasite population must approach 1000 at its peak and the corresponding k-value will result in the extinction of the pest.

If the mutual interference constant $m = 0{\cdot}5$, which is near the value estimated for *Nemeritis*, the k-value still rises very high. Only when $m = 0{\cdot}8$ or $0{\cdot}9$ do we get k-values of moderate size when the parasite population is near 1000 (Fig. 5).

The course of the calculated population from the model when $m = 0{\cdot}8$ shows that high parasite population still causes an undershoot of the final equilibrium, but it is not very severe. When $m = 0{\cdot}9$ the undershoot is avoided and the population settles down steadily to $1\ m^{-2}$ in a few generations.

Is there any biological justification for assuming such high values for mutual interference? Hassell & Varley (1969) reported examples from the literature in which, in confined spaces, a maximum value of $m = 0{\cdot}54$ was reported for *Nemeritis*. Hassell (1971) found that interference arose both from encounters between adult parasites and from the reaction to the discovery of parasitized hosts. In both cases the reaction was a movement away from the encounter. If parasites can migrate away from such encounters and spread

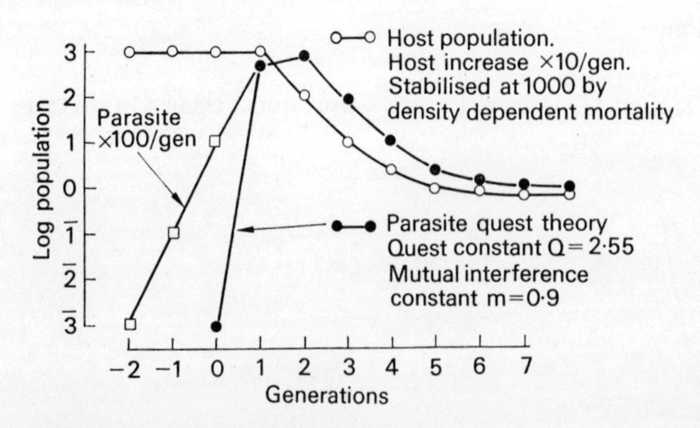

Figure 6. Parasite quest theory. Effect of mutual interference constant m on k-value caused by parasite population P.

from the original liberation point, then mutual interference will have high effective values until emigration and immigration become equal. Successful introductions of biological control agents do in fact show this rapid spread.

The main defect of these models using quest theory is that the parasite increases from a density of 10^{-3} in year 0 to one of about 10^{3} in one generation. Because quest theory neglects egg limitation, it gives an impossibly high rate of parasite increase in the first generation. Parasites must in fact have a rate of increase per generation which is limited by egg supply. Assuming that Thompson's theory holds when hosts are at very high population densities, we can model the possible effect of a parasite introduction, which in Figs 6 and 7 is now made at year −2. We can then turn to quest theory in year 1 and follow the decline of the pest population and its eventual stabilization. The picture which Varley & Gradwell (1971) produced for the decline of winter moth in Canada has a close resemblance to Fig. 7.

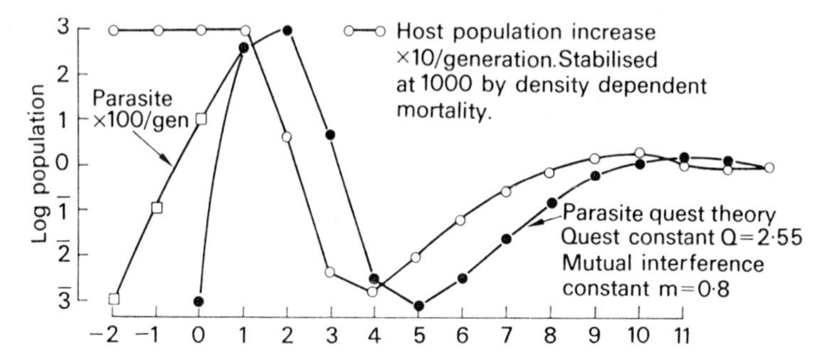

Figure 7. Parasite quest theory. Effect of mutual interference constant m on k-value caused by parasite population P.

We have yet to find a single comprehensive theory which will describe the events of biological control tidily. Watt (1959) provided a promising formulation for the attack rate of a parasite or predator, but we have failed to discover values for the constants in Watt's attack formula which would produce curves like those of Figs 6 and 7; Royama (1971) has studied this formula critically and his analysis of its faults helps to show why our attempts to use it were unsuccessful. See also Hassell & Rogers (1972).

If we seek to redefine DeBach's list of properties needed for successful biological control, we cannot simply replace '1, a high searching capacity' by 'a high Quest constant', because it is the proportion of the pest which is killed at low population densities of both species which is important, and this is affected also by the mutual interference constant. Perhaps we could say

1 the parasite must be able, when at low population density, to kill off enough of the pests to prevent pest increase when the pest is also uncommon;

2 a high maximum egg capacity may save time in the establishment of a parasite, but is not critically important to stability unless the parasites themselves suffer heavy mortality.

This study shows that a parasite requires one further property:

3 a high value for the mutual interference constant, which can be manifest by emigration when the population of the parasite is high.

At the present time population models are already helping to clarify ideas about the ways in which species interact, but no satisfactory general theory has emerged. The practical use of any modelling methods needs careful field measurements to determine the type of model to be used and the values to put into it.

Summary and conclusions

The effects of insecticide application and of schemes for integrated control will be understood and predictable only if adequate population models have been worked out. Without this, we are interfering with a complex system which we do not understand.

The effects of parasites and predators on winter moth near Oxford can be simply described in a model which fits fairly well with observation. We have therefore used this model to predict the effects of insecticide application. If the predators are adversely affected by insecticide, the winter moth population may easily increase to a new high level. This has not yet been tested in the field.

The theoretical basis for biological control is re-examined. DeBach's view that the main attributes of an effective natural enemy are a high searching capacity, specificity and a high potential rate of increase, are modified after studying these properties in simple population models. An attribute not mentioned by DeBach—strong mutual interference between the parasites—is also important if they are to avoid extinction and achieve rapid spread.

Acknowledgement

I am grateful to Dr M. P. Hassell of Imperial College Field Station, Silwood Park, who used the facilities kindly provided by the Ford Foundation to make the computations for Figs 3–7.

References

CLARK L.R., P.W. GEIER, R.D. HUGHES & R.F. MORRIS (1967) *The Ecology of Insect Populations in Theory and Practice*. Methuen, London.

DEBACH P. (1964) *Biological Control of Insect Pests and Weeds*, ed. P. DeBach. Chapman & Hall, London.

HASSELL M.P. (1971) Mutual interference between searching insect parasites. *J. Anim. Ecol.* 40, 473–86.

HASSELL M.P. & D.J. ROGERS (1972) Insect parasite responses in the development of population models. *J. Anim. Ecol.* 41, 661–76.

HASSELL M.P. & G.C. VARLEY (1959) New inductive population model for insect parasites and its bearing on biological control. *Nature, Lond.* 223, 1133–7.

NICHOLSON A.J. & V.A. BAILEY (1935) The balance of animal populations. Part 1, *Proc. zool. Soc. Lond.* 551–98.

ROYAMA T. (1971) A comparative study of models for predation and parasitism. *Res. Pop. Ecol.* Sept. 1971, suppl. 1, 91 pp.

THOMPSON W.R. (1939) Biological control and the theories of the interaction of populations. *Parasitology* 31, 299–388.

VARLEY G.C. (1970a) The concept of energy flow applied to a woodland community. Pp. 389–405 in *Animal Populations in Relation to Their Food Resources*, ed. A. Watson, British Ecol. Soc. Symp. No. 10. Blackwell, Oxford.

VARLEY G.C. (1970b) The need for life tables for parasites and predators. Pp. 59–70 in *Concepts of Pest Management*, eds. Rabb, R.L. and F.E. Guthrie. University of North Carolina Press, Chapel Hill.

VARLEY G.C. (1971a) The effects of natural predators and parasites on winter moth populations in England. Proc. Tall Timbers Conference on Ecological Animal Control by Habitat Management No. 2, 103–16.

VARLEY G.C. & G.R. GRADWELL (1963) The interpretation of insect population changes. *Proc. Ceylon Assoc. Advan. Sci.* 18, 142–56.

VARLEY G.C. & G.R. GRADWELL (1968) Population models for the winter moth. In *Insect Abundance*, Symp. Roy. Entomol. Soc. London, No. 4, 132–42, ed. T.R.E. Southwood. Blackwell, Oxford.

VARLEY G.C. & G.R. GRADWELL (1970) Recent advances in insect population dynamics. In *A. Rev. Ent.* Vol. 15, pp. 1–24.

VARLEY G.C. & G.R. GRADWELL (1971) The use of models and life tables in assessing the role of natural enemies. In *Biological Control*, ed. C.B. Huffaker. Plenum Press, New York & London.

WATT K.E.F. (1959) A mathematical model for the effect of densities of attacked and attacking species on the number attacked. *Can. Ent.* 91, 129–44.

WOOD B.J. (1971) Development of integrated control programs for pests of tropical perennial crops in Malaysia. In *Biological Control*, ed. C.B. Huffaker. Plenum Press, New York & London.

Demographic studies on grassland weed species

JOSÉ SARUKHÁN *School of Plant Biology, University College of North Wales, Bangor, Caernarvonshire, now at Dept. de Botánica, Instituto de Biología, U.N.A.M. Apartado Postal 70–233, México 20, D.F. México*

Demography and population dynamics have been classically the realm of zoologists, and have constituted the backbone of studies on mechanisms of population regulation and consequently of the methods of biological control of many animal species which constitute serious threats to the economy of human populations (see G. C. Varley 1973, this volume).

Only a handful of studies provide actuarial data about populations of plants; these are mainly of herbaceous species typical of grassland communities (Tamm 1948, 1956 and analyses of his data by Harper 1967; Canfield 1957; Rabotnov 1956, 1958; Sagar 1959, 1970; Antonovics 1966, 1972; Williams 1970).

It is a tenet of population theory that species rarely if ever achieve their reproductive potential because numerous biotic agents (generally along a food chain), or ultimately lack of resources, thwart the intrinsic rate of increase r of the population and bring its size within limits in the community in which it lives. However, when a species' population is freed from the controlling agents, or these become insufficient to regulate the population, such a species may for a time attain a rate of increase nearer to r and may become a 'pest'. The spread of *Opuntia* following its introduction in Australia in a totally new environment is a clear example of an almost unimpeded attainment of r.

Plants become 'weeds' when the natural regulation of their populations occurs at levels which allow rates of increase and population densities that hinder crop production. Most of these weeds have traditionally been dealt with by man by the application of agricultural practices such as tillage or the use of a number of highly efficient herbicides (see G. R. Sagar 1973, this volume).

A worked example

The present paper illustrates the type of approach that might be followed before an informed attempt is made to apply methods of biological control to weed populations. The study was carried out with the three common species of 'buttercup' (*Ranunculus repens*, *R. bulbosus* and *R. acris*) in a lowland grassland at College Farm, Aber, Caerns., which has been under a continuous traditional sheep-cattle grazing regime for over 50 years. The three species, which are undesirable elements of grasslands, have close taxonomic relationships, are commonly found growing abundantly in the same field, and possess widely different reproductive and energetic strategies.

A pantograph was used to produce maps of permanent sites recording the location of each individual buttercup plant and seedling present, in order to measure in detail the changes in the number of individuals with time and to be able to identify each component of the population as a separate unit. Such maps were subsequently superimposed and the number of plants and the fate and reproductive performance of each individual were recorded.

The data discussed in this paper refer only to *R. repens*. Analyses and a comparative study of the three species have been presented elsewhere (Sarukhán 1971). The three buttercups overwinter as rosettes of 2–3 cm in diameter with 2 to 4 leaves. *R. repens* produces epigeal stolons with 3 or 4 internodes of variable length; the nodes bear 1 or 2 leaves and a pair of root initials that grow downwards, anchoring the stolon to the ground. Stolon activity reaches a maximum by mid-July and can extend well into October during mild weather. Once a node is successfully rooted, additional leaves and roots are produced. By the end of August the stolons connecting the daughter plants to the parent start to senesce and wither. A large proportion of the parent plants die after vegetative reproduction. At this time the newly established daughters are in all respects similar to the surviving older plants. From this time, the whole population starts transforming to the winter rosette habit. Establishment by seed in *R. repens* is a much rarer event than vegetative reproduction, which constitutes the main source of plant recruitment.

Population flux and density dependent regulation

In agreement with animal populations, density fluctuations in populations of *R. repens* were remarkably small when compared to the number of plants which are recruited and lost from the populations. The observed population flux for an 'average 1 m^2 population' in an 'average year' has been represented in the model in Fig. 1.

The mature population in the Growing Fraction showed a steady decrease from the start of the year until the end of September, when it was only 40% of its original size. At this time the recruitment from the newly independent daughter plants brought the population size above its level at the start of the year. On average, each plant present at the beginning of August produced one vegetative offspring, so that total replacement of the population was ensured.

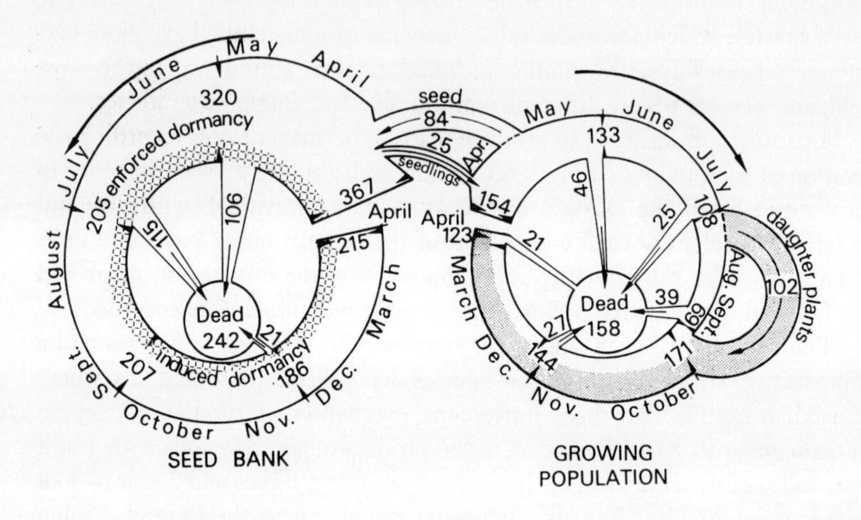

Figure 1. Population flux diagram for *R. repens.* Data for the growing populations are the averages of all the permanent sites for the two years of study. The shaded area represents populations of vegetative offspring. Data for the original seed population in the soil bank were obtained by samples near the permanent sites in March, 1970. The data about the fate of the seed populations come from the results of an experiment on seed decay. The radii are drawn to scale and represent population sizes. All data are plants or seeds/m².

By April of the following year the newly recruited vegetative offspring constituted two-thirds of the plants present. The difference between the 'starting' and 'final' populations was only 31 plants, while the total turnover observed was of 285 plants/m². The total time for a complete turnover was calculated to be between 2·7 and 2·9 years: owing to this high rate of physiologic renewal, an observer sporadically surveying the vegetation would see very much the same numbers of plants in the same site from year to year, but he would be counting mainly different plants.

Mortality of mature plants occurred in an exponential fashion, i.e. the risks of mortality were rather constant and independent of the age of the plants. However, it was possible to detect an element of seasonality in the intensity of the mortality process (Fig. 2). The severest mortality was found to occur in the spring, at the time at which plants in the sward initiate active growth, while the lowest mortality occurred in the harsher winter months.

A clear relationship between plant growth and population regulation can be seen in Fig. 3 where the total biomass of plants of *R. repens* is compared with their rate of death per week in populations of the same species. Most of the mortality among plants of *R. repens* occurred when the survivors were growing fastest. The maximum mortality occurred slightly earlier than the peak growth rate of survivors, but it is not clear whether this is a cause and effect relationship.

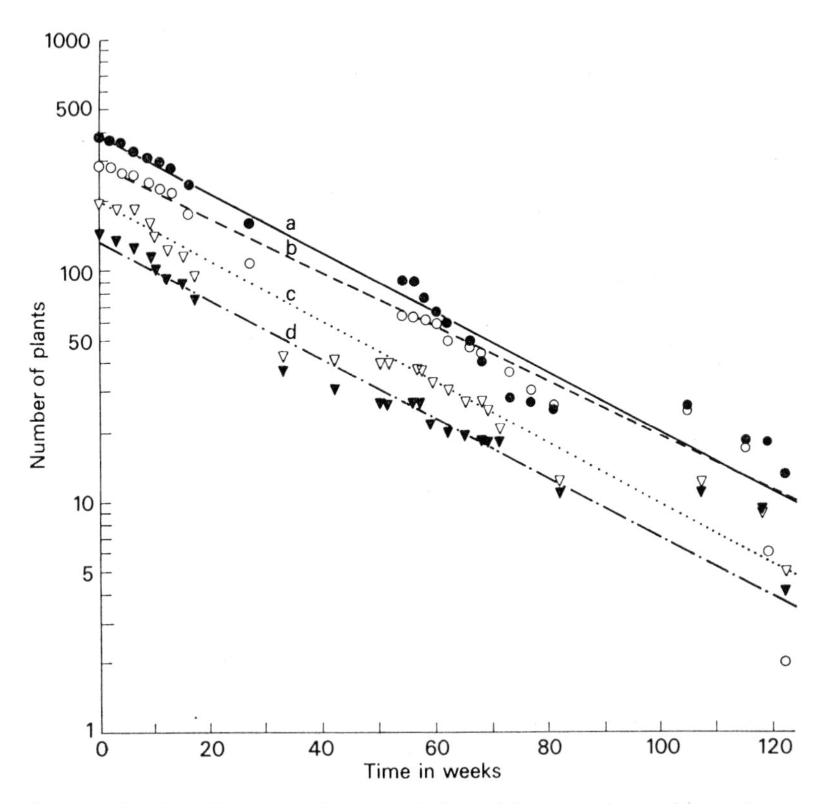

Figure 2. Survivorship lines of four populations of *R. repens* observed in 1 m² permanent sites from April 1969 to August 1971. Time in weeks after first observation. Sites, A1(a), A2(b), C1(c), C2(d).

The ecologist concerned with the adaptation of plants to the physical environment commonly lays much stress on the harshest experiences of the year—the coldest weeks of the winter and the driest days of summer. It is of the greatest interest that this species faces the greater mortality risks close to the period of most rapid growth. It may indicate that the greatest hazard to an individual is its rapidly growing neighbours! Many other component species of the sward show very similar seasonal patterns of individual development to that of *R. repens* (e.g. Anslow & Green 1967, Alcock *et al.* 1968) and probably contribute to increasing the simultaneous demand on resources in the environment and therefore to the population stresses acting on this species.

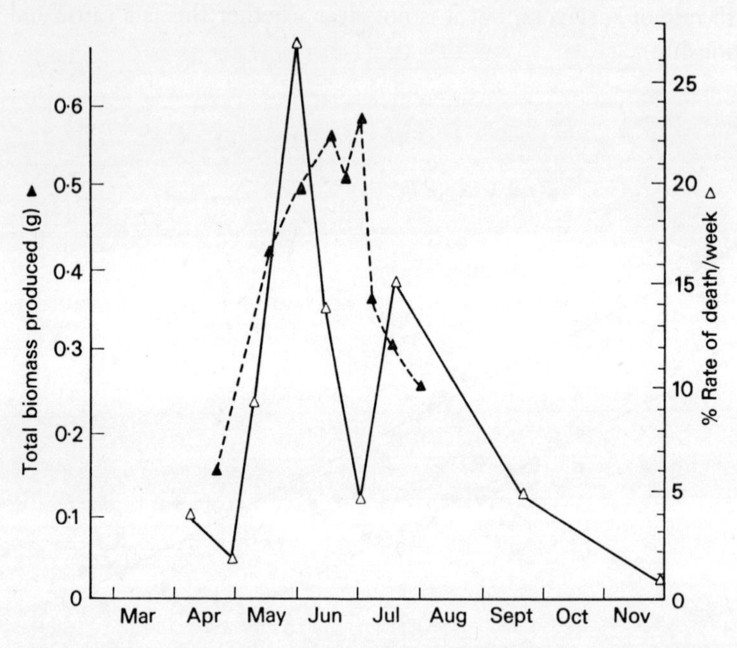

Figure 3. The relationships between the total biomass produced by plants of *R. repens* and the mortality observed in the 1 m² permanent plots. Data from the growing cycle of 1969.

Competition for resources in an area constitutes a typical density regulating factor. Such density dependent control is apparent in populations of *R. repens* through the variation of the risks of mortality at different plant densities. Figure 4 shows that vegetative propagules of *R. repens* live significantly longer in sites of low density of their own species than in high-density sites. Another interesting fact is that plants derived from seed have much lower chances of survival than plants that originated vegetatively (Table 1).

The life expectancy of plants originating from seed varies between 0·2 years (just over 10 weeks) to 0·6 years (32 weeks) which compares with 1·2 to 2·1 years in vegetatively produced plants. The data clearly illustrate the view that seed production is a low-cost investment but with high risk and high returns. Vegetative propagules, on the other hand, are produced at a much higher price in a system that seems to regulate their production by the density of its own numbers, therefore avoiding a costly wastage of energy.

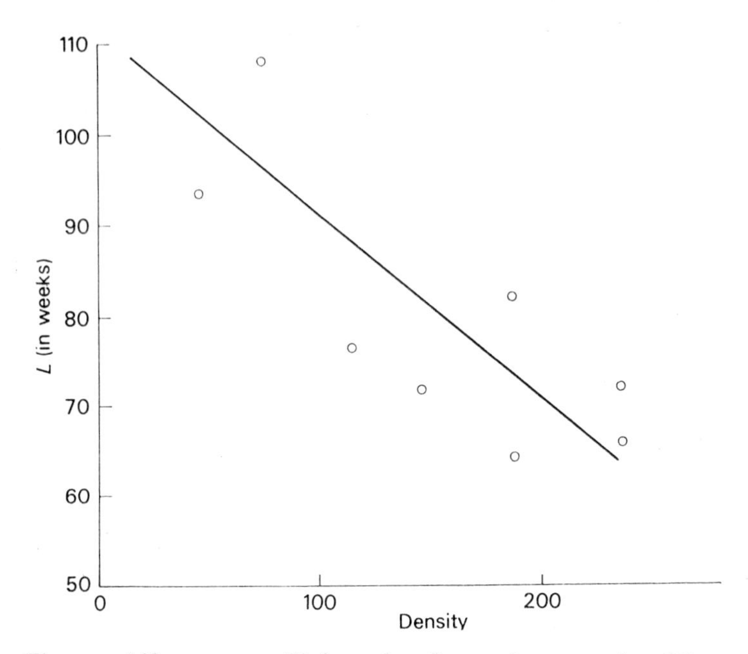

Figure 4. Life expectancy (L), in weeks, of vegetative propagules of *R. repens* as affected by its own species' density. Densities are average values for number of plants/m² observed at each site in April 1969, 1970 and 1971.

Table 1. Comparison of the life expectancies (in years) of plants of *R. repens* originating from seed or by vegetative reproduction in field populations.

Sites	A1	A2	A3	B1	B2	C1	C2	C3
Average densities (plants/m²)	271	219	163	128	50	185	222	73
seedlings	0·2	0·3	0·6	0·6	0·4	0·4	0·3	0·3
vegetative propagules	1·4	1·6	1·4	1·5	1·8	1·2	1·3	2·1

As the risk of mortality is density-regulated, the rate of increase of the population consequently also shows density-dependent regulation. Figure 5 shows that maximum rates of increase occurred in populations with a low density of *R. repens*, and that high densities were associated with low rates of increase. Clearly, the total number of plants per 1 m² is a rather crude measure of population density. Any measure of density needs in addition to be related to the capacity of the environment to maintain such numbers. If this 'carrying capacity' is high, relative to the actual density of a species,

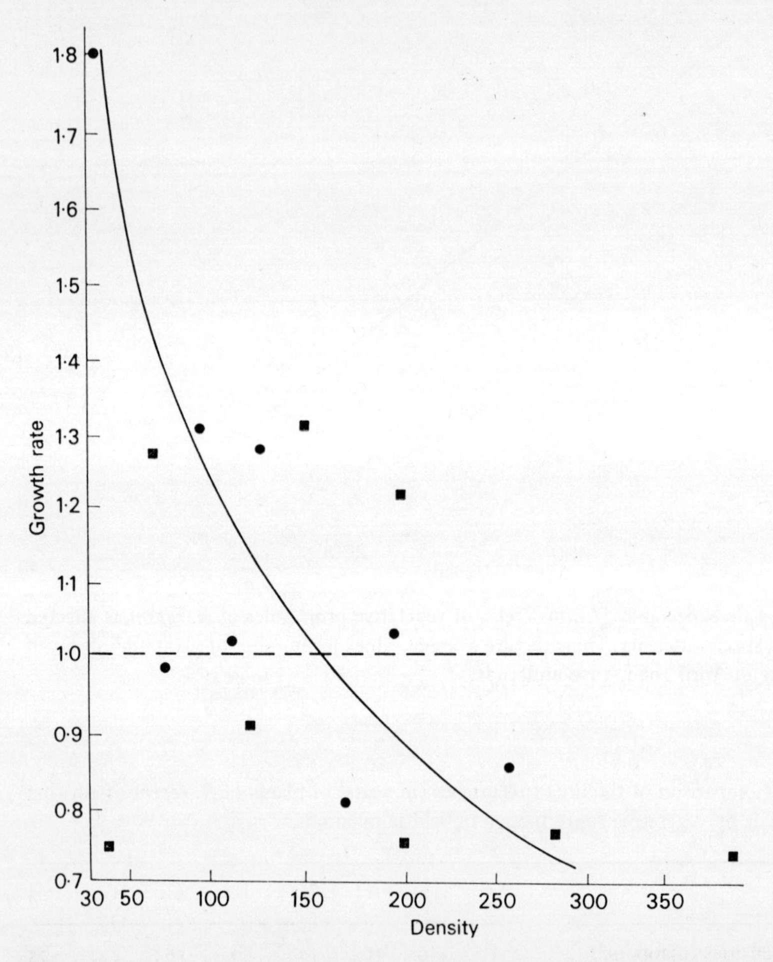

Figure 5. Rates of increase shown by populations of different densities of *R. repens*. Densities are the numbers of plants of the species per m², recorded in April of 1969 and 1970.

its population may be expected to show a high rate of increase; if, for the same actual density, the carrying capacity is low, the population may be expected to show a low rate of increase. The carrying capacity of the sites studied was unknown, but it was assumed that, as the field studied was a very stable pasture, the carrying capacity of a site did not change considerably from year to year. Then if the rate of increase of a population in a given year exceeded a value of 1 it might be expected that the population had increased beyond the carrying capacity of the site and that, consequently, the next year it would show a growth rate of less than 1, and *vice versa*. A negative correlation between the growth rates of the populations for two successive years in a site might be expected.

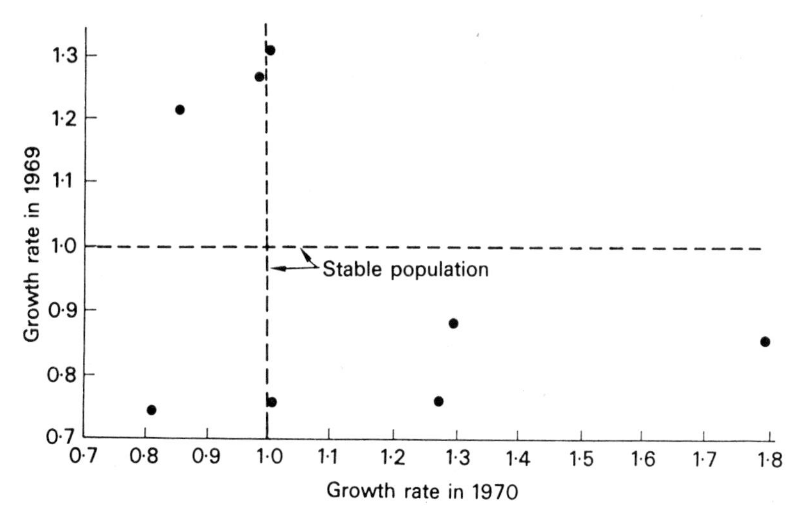

Figure 6. The relationships between the growth rates attained by populations of different densities of *R. repens*, during 1969 and 1970.

In Fig. 6, the population growth rates observed in 1969 have been plotted against those obtained in 1970, and show evidence of density-dependent regulation. Of the three species of buttercup, *R. repens*, with vegetative reproduction, showed the narrowest variation in population growth rates from year to year and from site to site (0·74 to 1·80), while *R. bulbosus*, reproducing exclusively by seed, showed variations from 0·9 to 8·17. Of the three buttercups, *R. repens* has the most stable populations. This stability is attained by a regulation of population size through alternately high and low rates of increase. It has been observed to attain near-stability in a successional stage within one year of the initiation of the successional process (Brook 1969). On the other hand, *R. bulbosus* is highly unstable and its populations behave in an explosive way, showing signs of inverse density-dependence.

Seed population dynamics

The diagram in Fig. 1 shows a plant population divided into two fractions: the growing and the dormant plants. For many species, especially annuals, the dormant fraction (seed bank in the soil) is numerically far greater than the actively growing population. Examples of this situation abound in plant literature (e.g. Robinson & Kust 1962, Roberts 1958, Brenchley & Warrington 1933). The great size of buried seed populations beneath different grasslands (e.g. Chippindale & Milton 1934, Milton 1936, Champness & Morris 1948, Kropàč 1966) pose two important questions: (a) how are these population levels achieved and maintained, and (b) how does it happen that the growing fraction of the population of a species is not more similar numerically to the dormant population. A demography of buried seed populations can indeed be envisaged as a viable research project in itself, and a mathematical model for the dynamics of seed populations in the soil has already been developed (Cohen 1966).

The mechanism which allows large populations of seed to accumulate and persist in the soil is dormancy. This has been slowly refined by natural selection to provide species with a vital buffering system by which their seed populations survive environmental catastrophes. It also allows a timing of germination in an environment that has periodicities—much in the way that diapause acts in insects. Harper (1957) distinguishes three kinds of dormancy—(a) innate, (b) induced, and (c) enforced dormancy. Innate dormancy is normally due to immaturity of the embryo; it can be overcome with a period of after-ripening. Induced dormancy develops in non-dormant seeds when an adverse factor acts upon the seed and produces a suspended animation that continues after the causal factor has ceased to act. Enforced dormancy is imposed by an exogenous factor (e.g. cold) and lasts only as long as the factor acts upon the seeds.

Many authors (e.g. Ødum 1965) have stressed the great longevity of seeds in the soil, but buried seeds are at risk and do not survive indefinitely. Roberts (1962) has shown that if fresh supplies of seed are prevented from entering an area of arable land, the populations of buried seeds decrease exponentially at a rate of 20 to 40% per annum, depending on the frequency of cultivations of the soil. At this rate it would take between 10 and 20 years to reduce the seed populations to 1% of the original size.

Despite the large numbers of seeds present in the soil and the limited duration of their dormant period, only a relatively small number of seedlings is observed to emerge. Wesson & Wareing (1967) and Black (1969) suggest that a complex relationship between the quantity and quality of light incident on the seed plays a very important role in breaking the dormancy of buried seeds.

The behaviour of seed populations of *R. repens* was studied in the same field and in plots adjacent to those where the observations on the demography of the growing plants were being made. Known amounts of seed were introduced into the permanent plots and recovered at intervals after the start of the experiment. Soil cores were obtained at each site on each sampling date and the seeds still remaining in the soil were extracted. The seeds were divided into the following categories: (a) *germinated:* the number of seedlings observed to emerge *in situ*; (b) *dormant-enforced:* the fraction of the seeds recovered from the soil cores which germinated when placed under appropriate conditions; (c) *dormant-induced:* those seeds that after being recovered from the soil cores failed to germinate under the appropriate conditions but were still alive (tetrazolium test); and (d) *dead:* all the empty and non-living seeds recovered from the soil cores. Therefore the total seed population S is

$$S = G + ED + ID + D$$

The changes throughout the year in each of these fractions is shown in Fig. 7. The seed bank of the 'average population' of *R. repens* in Fig. 1 received from the mature population about 0·5 seeds per plant present at the time of fruiting; this new seed represented about a quarter of the total living seed in the soil bank at the time. The number of seedlings observed to emerge represented only 6–8% of the living seeds present in the soil in May and June. The high mortality in the population of seed freshly shed between May and August was mainly due to predation, probably by voles. After August there was practically no decrease in the living seed population (see also Fig. 7). It is interesting to note that the fraction of seeds which were induced into dormancy increased progressively towards the winter, but declined towards the beginning of the summer, when seedlings of this species emerge in the field. The difference between the final and initial seed populations in the model is attributable to large differences in the seed output of the populations in the two years of the study.

Seeds of *R. repens* showed a much smaller germinating fraction than did those of the other two buttercups, and a much larger fraction of its seeds remained in enforced dormancy throughout the experiment. The information obtained in this experiment suggests that there is a considerable overlap of seed populations of this species produced in different years and stored in the soil. In contrast, the other two species, much more dependent on seedlings for the maintenance of population numbers, had their seed populations almost completely depleted by the time new seed was produced by the growing plants and was added to the seed bank.

These results support the theoretical findings (Cohen 1966, 1967) that species with little return on investment (i.e. a low probability that a seed will give rise to a successful reproducing adult) will have seeds that remain viable in the soil for long periods, and a low yearly germinating fraction. On the other hand, in a species of which the seeds have a high probability of producing a reproductive adult, the germinating fraction will be high and the ability to remain viable in the soil for long periods will be unimportant. In the first case selection will act to reduce the germinating fraction and to increase the dormant fraction.

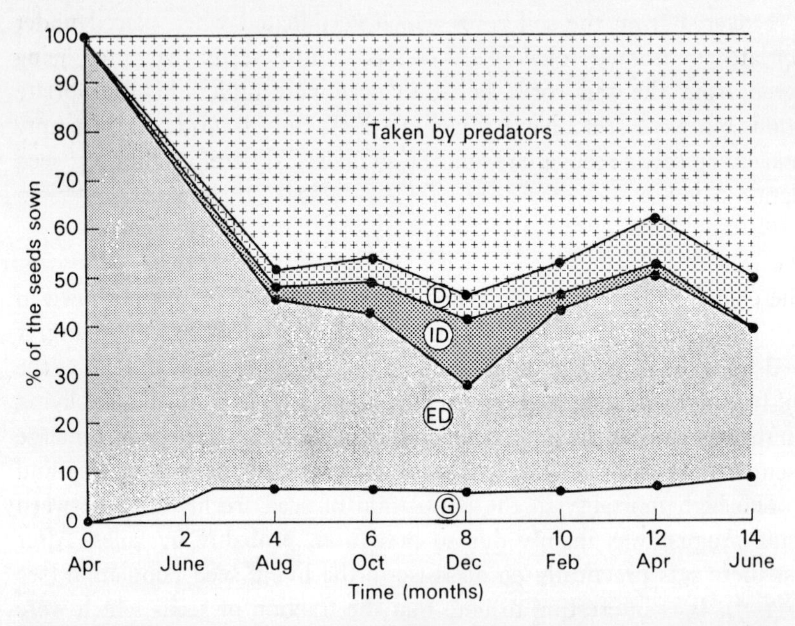

Figure 7. Variations in the size of the different fractions of the seed population of *R. repens* through time. G = seeds observed to germinate *in situ*; *ED*, seeds in enforced dormancy; *ID*, seeds in induced dormancy; *D*, fraction of the population that decays. The remainder is seed which has been removed from the area, mostly by predation. The curve for G is constructed with more frequent observations.

From the actuarial information obtained, it seems that populations of *R. repens* are regulated to a large extent by a density-dependent mortality which operates mainly at the initial phases of the yearly growth. Equally, these populations gave evidence of being stable and well buffered, being able to recover quickly after catastrophic events. Any attempt effectively to reduce the number of plants of this species in a grassland should take into consideration that reductions in the number of plants prior to or at the time of growth initiation will probably not effect any reduction in the density of the final population at the end of the growing season, but will probably increase it by releasing the density stress among the surviving plants. It also has to take into consideration that the seed bank provides an important and durable source of further recruitment to the population.

Numerous other agents not included in the population model play an important role in regulating the numbers of plants of *R. repens* and need to be studied further. The effects of grazing on the loss of foliage and the subtraction and redistribution of seed in the droppings of the grazers have not been quantified. Trampling greatly affects the chances of survival of newly recruited plants, and also the emergence and location of new seedlings. The effects of other predators on seed, seedling and mature plants have been largely ignored in the model, although it is clear that they may considerably affect the size and behaviour of the population.

This type of approach provides a means of extracting the kind of information which is basic and relevant for the understanding of the regulating mechanisms of plant populations in natural conditions. The populations are seen to be fluxes; the apparent stability observed in sporadic surveys is, in reality, the result of very dramatic changes in numbers of individuals. They also suggest that biotic elements play a more important role than the physical components of the environment in the regulation of populations. These studies, by identifying the main regulating phases in the chain of demographic events, disclose the weak links at which control of populations, either biologically or otherwise, is likely to be effective.

Acknowledgements

The present paper is part of a research project on plant demography carried out at the School of Plant Biology, U.C.N.W. The author wishes to thank sincerely Professor John L. Harper for suggesting the subject of research, his supervision on the work and his invaluable assistance and guidance. Dr Madhav D. Gadgil helped to develop a mathematical analysis of the data on which part of the information contained in this paper is based.

References

ALCOCK M.B., J.V. LOVETT & D. MACHIN (1968) Techniques used in the study of the influence of environment on primary pasture production in hill and lowland habitats. *The measurement of environmental factors in terrestrial ecology*, ed. R.M. Wadsworth, pp. 191–203.

ANSLOW R.C. & J.O. GREEN (1967) The seasonal growth of pasture grasses. *J. agric. Sci. Camb.* 68, 109–22.

ANTONOVICS J. (1966) *Evolution in adjacent populations.* Ph.D. Thesis, University of Wales.

ANTONOVICS J. (1972) Population dynamics of the grass *Anthoxanthum odoratum* on a zinc mine. *J. Ecol.* **60**, 351–65.

BLACK M. (1969) Light-controlled germination of seeds. *Dormancy and survival.* Symposia of the Society for Experimental Biology, XXIII, 193–217.

BRENCHLEY W.E. & K. WARRINGTON (1933) The weed seed population of arable soil. II. Influence of crop, soil, and methods of cultivation upon the relative abundance of viable seeds. *J. Ecol.* **21**, 103–72.

BROOK J.M. (1969) *Pattern and succession in weed populations.* Ph.D.Thesis, University of Wales.

CANFIELD R.H. (1957) Reproduction and life span of some perennial grasses of Southern Arizona. *J. Range Mgmt.* **10**, 199–203.

CHAMPNESS S.S. & K. MORRIS (1948) The population of buried viable seeds in relation to contrasting pasture and soil types. *J. Ecol.* **36**, 149–73.

CHIPPINDALE H.G. & W.E.J. MILTON (1934) On the viable seeds present in the soil beneath pastures. *J. Ecol.* **22**, 508–31.

COHEN D. (1966) Optimising reproduction in a randomly varying environment. *J. theor. Biol.* **12**, 119–29.

COHEN D. (1967). Optimising reproduction in a randomly varying environment when a correlation may exist between the conditions at the time a choice has to be made and the subsequent outcome. *J. theor. Biol.* **16**, 1–14.

HARPER J.L. (1957) The ecological significance of dormancy and its importance in weed control. *Proc. 4th int. Conf. Pl. Prot., Hamburg*, 415–20.

HARPER J.L. (1967) A Darwinian approach to plant ecology. *J. Ecol.* **55**, 247–70.

KROPÀČ Z. (1966) Estimation of weed seeds in arable soils. *Pedobiologica* (*Prague*), **6**, 105–28.

MILTON W.E.J. (1936) The buried viable seeds of enclosed and unenclosed hill land. *Welsh Plant Breeding Bull. 4*, Series H. 58–73.

ØDUM S. (1965) Germination of ancient seeds. *Dansk Bot. Ark.* **24** (2), 70 pp.

RABOTNOV T.A. (1956) The life cycle of *Heracleum sibiricum* L. *Byull. Mosk. Obshch. Ispyt. Prir.* (Biol.), **61**, 73–80.

RABOTNOV T.A. (1958) The life cycle of *Ranunculus acer* L. and *R. auricomus* L. *Byull. Mosk. Obshch. Ispyt. Prir.* (Biol.), **63**, 77–86.

ROBERTS H.A. (1958) Studies on the weeds of vegetable crops. I. Initial effects of cropping the weed seeds in the soil. *J. Ecol.* **46**, 759–68.

ROBERTS H.A. (1962) Studies on the weeds of vegetable crops. II. Effect of six years of cropping on the weed seeds in the soil. *J. Ecol.* **50**, 803–13.

ROBINSON E.L. & C.A. KUST (1962) Distribution of witchweed seeds in the soil. *Weeds*, **10**, 335.

SAGAR G.R. (1959) *The biology of some sympatric species of grassland.* D.Phil. Thesis, University of Oxford.

SAGAR G.R. (1970) Factors controlling the size of plant populations. *Proc. 10th Br. Weed Control Conf.* **3**, 965–79.

SAGAR G.R. (1974) On the ecology of weed control. Pp. 42–56 in *Biology in Pest and Disease Control*, eds. D. Price Jones and M.E. Solomon. Blackwell, Oxford.

SARUKHÁN J. (1971) *Studies on plant demography*. Ph.D. Thesis, University of Wales.

TAMM C.O. (1948) Observations on reproduction and survival of some perennial herbs. *Bot. Notiser*, **3**, 305–21.

TAMM C.O. (1956) Further observations on the flowering and survival of some perennial herbs. *Oikos*, **7**, 273–92.

VARLEY G.C. (1974) Population dynamics and pest control. Pp. 15–27 in *Biology in Pest and Disease Control*, eds. D. Price Jones and M.E. Solomon. Blackwell, Oxford.

WESSON G. & P.F. WAREING (1967) Light requirements of buried seeds. *Nature, Lond.* **213**, 600–1.

WILLIAMS O.B. (1970) Population dynamics of two perennial grasses in Australian semi-arid grassland. *J. Ecol.* 58, 869–75.

On the ecology of weed control

G. R. SAGAR *School of Plant Biology, University College of North Wales, Bangor, Caernarvonshire*

The purpose of this paper is to bring together in one place a number of the fragments of the fabric of weed control. Unfortunately some of the pieces are not yet properly fitted together, for the pattern is by no means completely designed and some of the critical threads still remain to be spun.

A major consequence of this situation is a recurring failure to reach integrating conclusions which is exposed in the disjunction of the paper. I have tried to outline the need for weed control and some of the ecological aspects of the methods used to achieve it. In a second part I have taken a simple population ecologist's approach to the regulation of weed populations and posed the questions of when and how such regulation occurs. The paper is completed by a brief outline of some of the methods which have been used in studies of population control in weeds.

Weed control

The crop systems in which weeds are found are always unnatural in the ecological sense. It is of course possible to argue that agriculture and horticulture represent a series of ecosystems specialized only by increased inputs and outputs and by deliberate adjustments of the balance between component members. This analogy may be used to emphasize that one of the adjustments is often simplification of the plant community. This is most evident in the monocultures of arable systems, but less clear in some grasslands (Elliott 1968). In a monoculture all weeds are aliens, and weeds which reduce crop yields reach population levels which are both commercially unacceptable and presumably below the levels at which biological regulation of those weeds occurs.

Since man first grew crops he has been involved in the removal of volunteer species which have become known as weeds. Huffaker (1957) listed the types of losses which weeds cause and this list emphasizes the advantages which the removal of weeds may give:

1 Improvement of yield and quality and hence value of the crop.
2 Reduction in the cost of cultivations.
3 Loss of toxic species.
4 Removal of alternate hosts for pests and pathogens (Moore & Thurston 1970, Franklin 1970, Heathcote 1970, van Emden 1970).

Two further consequences of weed control have also been suggested:

1 Reduction in species richness and its consequences (Fryer & Chancellor 1970 a & b).
2 The instability of monoculture.

Any argument for a reduction in the intensity of weed control must be an argument for a reduction in the intensity of agriculture. Weed control is essential, but two aspects need discussion—the degree of control required and the manner of achieving it.

The degree of control

The simplest proposition is that all unsown or non-planted species which grow on an area of agricultural land are weeds and should be totally removed. This proposition is suspect on at least three counts.

1 It is known that some crops can tolerate the presence of some other species without there being economic losses, at least in the short term (Evans 1968, 1969). Such species are therefore only weeds by a definition which depends on their being unsown.
2 It is tacitly assumed that weeds never give advantage to crops even over the long term. Such an assumption is unwarranted, for there is no experience over a sufficiently long period of time to permit judgement. This dilemma can only be, and indeed is being resolved by current studies of crop production in weed-free environments (Woodford 1963, Robinson 1966). There is little evidence that the loss of weeds has detrimental effects, although some methods of removal may cause mechanical damage of the crop plants.
3 Removal of weeds only from cropped land leaves a re-infestation potential from surrounding areas. The extension of weed control practices into uncropped land poses delicate problems (Way 1968).

It is not possible to reach a conclusion about the degree of control required, partly because much of the argument is economic, and partly, because we require longer experience of the weed-free environment. We can however, be sure that, if weed control is not complete, weed populations will expand every time control is relaxed (e.g. Selman 1970) and that the costs of control measures will continue *ad infinitum*. On the other hand, if a policy of total weed control is adopted there may be a time (Elliott 1968, but see Fryer & Chancellor 1970b) when there are no weeds.

One further comment is necessary, and it has relevance to other parts of this paper. When some major change of agricultural practice comes into operation it does so initially on a small scale, thereafter expanding at a rate determined by the degree of success of the pilot trials. It therefore follows that dangerous ecological consequences are most likely to be revealed on a small scale and sufficiently early to allow avoidance of large-scale catastrophe.

The manner of control

Control methods may be divided into three groups, biological, cultural and chemical. The ecological consequences of each must be considered.

Biological control

Before certain species became weeds their populations were regulated, for where the eco-geographical origins of weeds can be identified there is little sign of the expression of geometric increase. Transoceanic escape of species has given us examples (Harper 1957) of rates of spread (and increase in population size) of much the same type as occur locally, for a short time at least, if weed control is relaxed in an agricultural situation. The precise nature of the controls which have been relaxed may be open to dispute, for there is uncertainty about the agents which control or regulate plant populations. Taking biological control in a strict sense as control of a weed population by predators or parasites (other than man and his domestic animals), we must consider three points.

1 The agent chosen must be utterly incapable of transferring its attentions to any crop plant either at present in existence or likely to be brought into existence by importation or breeding (Zwölfer & Harris 1971). The screening procedures prior to importation of a new organism are very rigorous, but clearly they cannot be totally predictive of either the evolution of the organism or the engineering of the plant breeder. Despite this fear there do not appear to have been any examples of deliberately imported insects becoming pests (Dodd 1954).

2 According to Wapshere (1970b) there are at least three prerequisites before the introduction of insects is likely to succeed in effecting weed control, viz. a high density of the weed species, intraspecific competition of the weed

population and an increased habitat range of the weed in the country of introduction. These possibilities seem unlikely to apply in a crop-weed situation in Western Europe, unless its agriculture suffers extreme recession, or there is invasion by some new weed species.

3 By the laws of the game, a weed-free environment could not be an aim of the biological method; the weed-agent relationship must be to some extent oscillatory in time or space or both. The aim would have to be containment on average at an (economically) acceptable population level (Zwölfer 1968).

There are several wide-ranging accounts of the ecology of biological control and many of these may be located through DeBach (1964).

Cultural control

There are three separate types of cultural methods. The first is rotation, which was introduced to solve several problems, one of which was weed control, and another the maintenance of nitrogen fertility. Much of the second problem is now soluble through the use of artificial fertilizers, and the contribution towards weed control has been steadily diminishing for 20–25 years. The ecology of rotation is fearful, being replete with unecological activity. Above all, rotation coupled with the plough increases the weed-seed bank of the soil in terms of both numbers of species and numbers of seeds (Harper 1957). As a very long-term ecological experiment, rotation seems to have proved that a continuing sequence of unnatural and very severe edaphic and biotic upheavals has only very limited ecological consequences. It is both pertinent and proper to ask how a symposium of the present sort, held 25 years after rotation was introduced, but otherwise in the present state of public awareness, would have regarded that agricultural advance.

Cultivation of arable land remains a major agricultural operation. Originally very important in weed control, and indeed still absolutely essential if herbicides are not used, its ecological consequences are not difficult to find. Removal of weeds by cultivation results in the loss of food plants for animals (and plants) of the next trophic level, and some degree of food web interference must result, and, if cultivation is sufficiently severe and frequent, the composition of the weed flora is slowly changed, although the seed bank dampens the rate of change.

A third aspect of cultural control is management, which in grassland for example, through the use of fertilizers and the control of grazing or cutting, may have enormous effects on weed control, effects which can be expressed and interpreted in ecological terms (Jones 1933, Charles 1968, Harper 1971).

If rotation is ecologically acceptable then cultivation and management must also be accepted. The problem with rotation is its inhibition of the evolution of efficiency and specialization, both aspects of economics.

Chemical control

Control of weeds by herbicides is a very young science. Although history (Fryer & Evans 1968a) shows some interest early in this century, we have only 25 or so years of experience, and this is altogether too short a period to prove ecological consequences. Nevertheless, there have been a number of attempts. The loss of food plants and consequential food web interference (Potts 1970) is a threat, as phytoxicity may be in unintended places through drift, mishandling or accident (Macfadyen 1957, Way 1968). The evolution of resistance (Harper 1956, Hammerton 1968), long-term changes in the flora, and possibly reduced species richness (Moore 1970) through the greater efficiency of weed control, have all given cause for concern (see Fryer & Chancellor 1970a). One ecologically fascinating consequence of the selective properties of herbicides is that in many situations crop and weed are becoming more closely related.

Rather strangely, it has not been the directly predictable consequences that have yet provided the real problems (Madel 1970), but rather exploitation of the freedom from the need to rotate. Monoculture of arable crops has long been practised, but continuous monoculture of the same annual plant has been exceptional in Britain. Is there something peculiar about the ecology of annuals beyond the fact that as members of a succession they are replaced by longer-lived species, or are the current fears of disaster through build-up of pathogens, parasites and crop-predators ill-founded?

Real evidence of a general need to reduce reliance on the use of chemicals for weed control is lacking. There have been accidents with herbicides, there has been misuse, there is fear of the unproven and some resistance to progress; but above all, herbicides seem to have been enveloped into the pesticide cloud, and have been assumed to have the worst properties of both DDT and the organo-phosphates.

It is infinitely more difficult to prove no effect than effect. Accidents are reported; non-accidents are not, nor is it an asset to have DNOC as a skeleton in the cupboard.

More generally it seems to me impertinent to single out any of the three approaches to weed control as any more or less likely to be bad ecologically. All can be shown theoretically to be undesirable if a strict and narrow ecological view is taken. In practice, weed control comes from a combination of methods. Against a background of biological control, chemical and cultural methods are often deliberately combined (Charles 1968, Allen 1968).

Population regulation of weeds

In his classical paper on the biological control of weeds, Huffaker (1957) wrote of 'a pressing need for some common meeting ground in the consideration of measurements of abundance of plants and animals'. At the same time he recognized the dilemma that plant ecologists have through plasticity. However, at risk of grave oversimplification, I intend now to think of plant populations in terms of numbers of individuals. Two questions need to be posed.

1 When are weed populations regulated?
2 How are weed populations regulated?

When are weed populations regulated?

The skeleton of Fig. 1 is a diagram of life cycles, and can potentially at least be described numerically (see Sarukhán 1974, this volume). At the beginning of the year, all the individuals of the population of an annual weed of an arable crop are present as seeds distributed through the top 17–23 cm of the soil. The seeds are in various forms of dormancy and the population is slowly decreasing (Roberts 1970a). A proportion of these seeds germinate, the proportion depending on the species, the environmental conditions or the season (Fryer & Evans 1968a), but in general being low (10–20%). After germination, some of the seedlings die before emergence, and yet more succumb before establishment. Losses of seed from the soil may for example be 22–36% per annum, of which only 0·3–9% may be accounted for as emerged seedlings (Roberts & Dawkins 1967). Pre-emergence use of herbicides may also reduce these values. Following establishment, and unless post-emergence herbicides are used, plant numbers may remain reasonably constant (although this is the phase where numbers are most meaningless); but with the onset of reproduction, numbers begin to increase, with seed output often, if not always, being regulated by interference (Harper 1960, Bate *et al.* 1970). Chancellor & Peters (1970) have shown that for wild oat (*Avena fatua*) seed production per 23·9 plants/m^2 ranged between 393 and 4784 depending on the companion crop. The number of seeds produced which finally return to the seed bank, to join the seeds which have remained dormant throughout the growing season, may be reduced by losses through export when the crop is harvested (see Wilson 1970), through predation by animals, or through pathogens (e.g. Huffaker 1964).

Thus the life cycle of an annual weed of arable land has a long phase when the size of the population is continuously and naturally declining, followed by a short explosive phase of multiplication against a background reserve of dormant but viable members of the population; presumably this latter fraction often constitutes the majority of the members of the population.

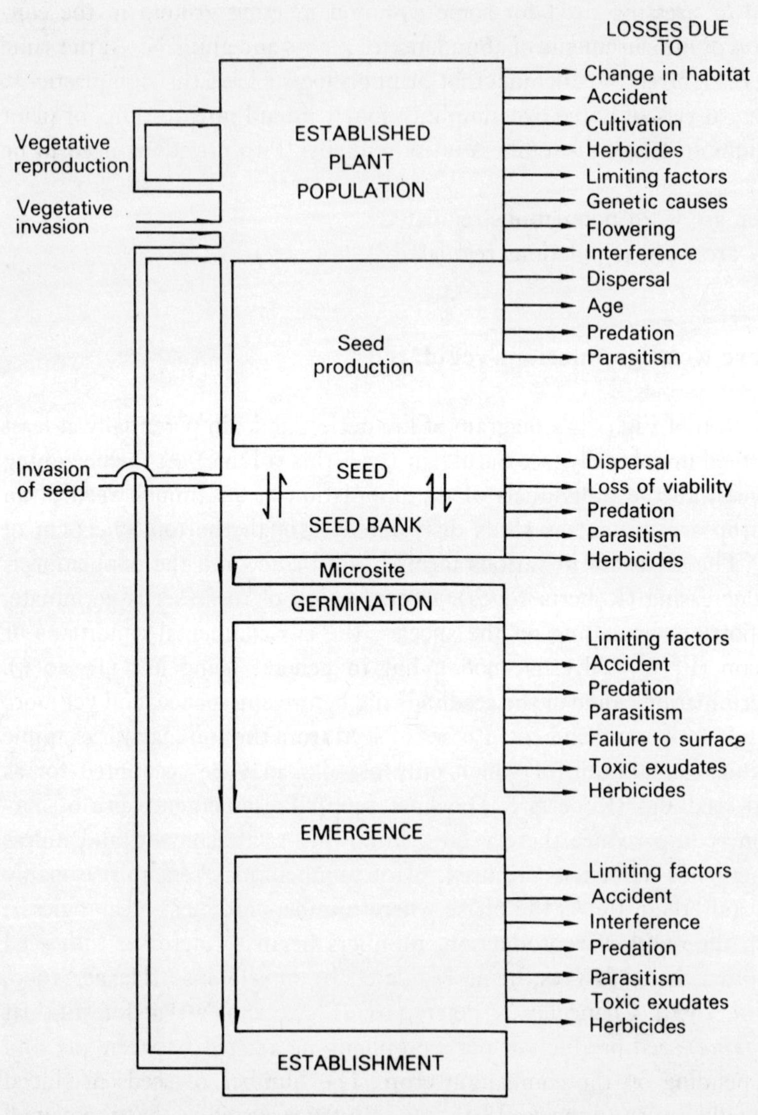

Figure 1. Some of the factors which may control the size of plant populations (modified from Sagar 1970).

A perennial weed which reproduces only vegetatively is likely to be at a minimum population size immediately after the last cultivation before a crop is sown. Thereafter vegetative reproduction begins, the rate of multiplication being regulated more or less continuously (Cussans 1968, Thurston & Williams 1968).

The only conclusion which can reasonably be drawn is that control of numbers of a weed population appears to occur at all stages of the life cycle, and in the absence of life-tables we can only guess about the stages at which most mortality occurs. We cannot even begin to guess at which are the most significant density-related phases.

How are weed populations regulated?

In theory, crop and weed share the same environmental resources, although in practice co-habitation may occur without crop yields being economically reduced. For weeds of economic importance, control is of two sorts, direct and indirect. In direct control the weed plant is removed physically or chemically and the resources of the environment are then more wholly, and almost instantly, available to the crop. The indirect approach is Darwinian, to give the crop an advantage, be it ever so small, so that in the succeeding struggle it wins (Sagar 1968b). At the level of the individual plant, biological, chemical and cultural methods of control may show both direct and indirect elements. In the short term the chemical approach is the most direct, and management of grassland, for example, almost always indirect. At population level all methods of control are indirect, save when total sterilization is achieved by steam or chemicals as in some high-value horticultural systems.

In practice, weed populations are controlled by a combination of agencies, and it is a truism that in agriculture control occurs at several stages of the life cycle (Sagar 1970). *Avena fatua* populations may be reduced by natural mortality of the grains on or in the soil, by pre-emergence herbicides, by cultivation before the crop is sown, by herbicides acting or applied post-emergence, and by roguing the crop chemically or by hand during the phase of seed production (Fryer & Evans 1968b). Vigorous interference by the crop is always a prerequisite of the population control of this weed (Thurston 1962). It is worth recording that roguing the crop of wild oats is carried out in part to reduce the level of subsequent infestations of the weed.

It is relevant to those interested in biological control to note that a farmer growing arable crops is unlikely to accept a control method which fails to operate before the weed has interfered with the growth and development of the crop. It follows that the most acceptable regulating agents would be effective on the seed in the soil or on the young seedlings.

Qualitatively, we understand the population regulation of wild oat; certain aspects are also on record quantitatively; but the identification of the critical density-dependent regulators does not appear to have been attempted, partly because of a failure to recognize the concept, partly because it seems very doubtful whether it is constant from field to field or year to year (e.g. Thurston 1959), partly because populations of this weed in arable land may never be allowed to reach the size at which any natural regulation occurs, and partly because of the view that the identification of such regulators would not necessarily lead to improved control (Sagar 1968a).

Studies of population regulation of weeds

Harper (1960) presented the results of studies of the regulation of plant numbers of *Agrostemma githago* and *Papaver* spp. The weeds were sown as seeds at different densities and with different companion species. Density-related responses were isolated (*cf.* Palmblad 1968). The fact that companion species also affected the number of individuals emphasizes the truism that population control is through a complex of limiting and regulating factors, the latter being density related.

My own studies of *Plantago lanceolata* (Sagar 1959, 1970) showed some control of population size at virtually all stages of the life cycle, with major losses through failure to germinate and in the mortality of seedlings, but they yielded remarkably little information on the causes of post-emergence seedling mortality. Similarly the work of Putwain (1970) on *Rumex acetosa* and *R. acetosella*, notable for the difference exposed between the species, was again not really designed to identify the regulating or controlling factors. Foster (1964), studying *Bellis perennis*, confirmed failure to germinate and seedling mortality as contributing to the failure to achieve significant increases in population growth, but, like Putwain (1970) and Sagar (1959), showed the significance of vegetative reproduction as a contributor to population dynamics. In one of Foster's experiments the addition of fungicides and insecticides improved the chances of survival of seedlings (Sagar 1970). More recently Mortimer (unpublished) has confirmed this improvement with other species in grassland.

All of these studies, and of course that of Cameron (1935) on *Senecio jacobaea*, exposed the control which other components of a community had over the species studied. In the experimental approaches, removal of some or all of these other components led to increases in population size of the species remaining (Sagar & Harper 1961, Harper 1968). This type of response can lead to the suggestion that although other plant species may play a very significant part in controlling the size of a population, actual regulation takes place intraspecifically. Foster (1964) set up a series of different management treatments in established grassland containing *Bellis perennis* and showed, not only that the size of *Bellis* populations could be increased by selectively suppressing companion species, but also that the *Bellis* populations stabilized at a new level after only a short time. In none of these studies were the regulating factors isolated. Using hindsight, it seems that the observations taken and the experiments made were inadequate for this purpose.

Approaches to the study of population control in plants

When organized biological control is contemplated, entomologists search for organisms which can be seen to damage the target species, on the sound theoretical assumption that 'all organisms that in some way reduce the fitness (i.e. reduce reserves, reduce competitiveness, reduce seeding ability) have the potential to act as biological control agents' (Wapshere 1970a). Knowing that the agent has geometric potential for increase, and assuming that its population size is regulated from below by its food and not from above, then clearly the plant population can be regulated. The successes of the biological control methods confirm this (DeBach 1964). It would be fascinating to have sufficiently detailed information to allow the compilation of life-tables for at least some of these weeds, if only to identify the break-even point (*cf.* Lashley 1969, who reported that a 98% loss of seed was required to prevent increases in the population size of *Carduus nutans*). It seems that from these examples may come a simple yet complete account of the population dynamics of a plant.

In what are undoubtedly more complex situations there are at least two types of approach.

Data accumulation

Quantitative information about the size of plant populations is gathered at frequent intervals. Ideally this should be (and with plants in the vegetative phase can be) related to individuals, and detailed life-tables can be constructed, thus allowing the identification of the phases when control is effected (Harper 1967, Sagar 1970, Sarukhán 1970, 1974, this volume). By itself the approach will usually fail to identify the mechanisms by which populations are controlled, and it is to this end that some future work must be undertaken (Harper 1970). Census studies are notoriously tedious and demanding but there do not appear to be any short-cuts.

One of the most complete studies which falls into this category is the series of Roberts and his co-workers who have actuarial data for the loss of seeds of many species from horticultural soils (Roberts 1970a). Over a wide range of densities, for a given species, losses are of a fairly constant percentage per year. The data show not only the persistence of populations in the seed bank, but also that the losses are often non-regulatory. The causes of the losses need further attention.

Experimental approaches

Manipulation of populations or their environments by treatments imposed by the experimenter has been a much more popular approach than the actuarial. By various means, it is possible to go far in identifying the phases in the life cycle when populations are being controlled and in some cases the nature of the agents concerned. Progress along these lines may be achieved, for example, by adding seed to a habitat (Sagar & Harper 1960, Cavers & Harper 1967), by increasing the established plant population by transplants, by removing other species (Harper 1968), by changing the edaphic, biotic or climatic factors of an environment (e.g. Jones 1933 a–e) or indeed by using pesticides and by other management practices.

Conclusion

It must be acknowledged that the practice of weed control has outstripped understanding of its achievement. The ecologist has been a wondering bystander, and his studies of weeds have been primarily in the quest for understanding and knowledge. Even the exponents of biological control have proceeded in an *ad hoc* fashion, for there is little evidence that they have sought, or indeed needed to seek, for organisms which were regulatory in the native sites.

Weed control is essential; in developed agricultures it must be achieved before significant damage is caused to the crop. It is at present done by a combination of cultural and chemical means against a background of biological control. There does not seem to be any ecological case for a reduction in the use of any of the existing methods of control. Economic considerations are likely to dominate the decisions.

The regulation of weed populations, although usually achieved, is not well understood in the sorts of terms which animal ecologists use in studies of animal populations. What part phytophagous insects play in the regulation of weed populations in Britain and Western Europe remains to be seen (Huffaker 1957, citing Brues 1946). Approaches to the identification of the biological agents of control may have to be like those of Chater (1931) who accounted for the fate of no less than 891 *Ulex* seedlings in 1929.

Finally it may be pertinent to ask whether the entomological and mycological sections of the British Ecological Society's Biological Flora accounts are not of more value than they may at first sight appear.

Acknowledgements

I wish to thank the members of the Ecological Discussion Group at Bangor for their contribution to the formulation of this paper.

References

ALLEN G.P. (1968) The potential role of selective herbicides in grassland improvement. *Proc. 9th Br. Weed Control Conf.* 1231–7.

BATE P.G., J.G. ELLIOTT & B.J. WILSON (1970) The effect of barley population and row width on the growth of *Avena fatua*, Wild Oat. *Proc. 10th Br. Weed Control Conf.* 826–30.

BRUES C.T. (1946) *Insect Dietary*. Harvard University Press, Cambridge, Mass. (From Huffaker 1957.)

CAMERON E. (1935) A study of the natural control of ragwort (*Senecio jacobaea* L.). *J. Ecol.* **23**, 265–322.

CAVERS P.B. & J.L. HARPER (1967) Studies in the dynamics of plant populations. I. The fate of seed and transplants introduced into various habitats. *J. Ecol.* **55**, 59–71.

CHANCELLOR R.J. & N.C.B. PETERS (1970) Seed production by *Avena fatua* populations in various crops. *Proc. 10th Br. Weed Control Conf.* 7–11.

CHARLES A.H. (1968) Control of weed grasses by the selective effects of fertiliser application and management. *Proc. 9th Br. Weed Control Conf.* 1223–30.

CHATER E.H. (1931) A contribution to the study of the natural control of gorse. *Bull. ent. Res.* **22**, 225–35.

CUSSANS G.W. (1968) The growth and development of *Agropyron repens* (L.) Beauv. in competition with cereals, field beans and oilseed rape. *Proc. 9th Br. Weed Control Conf.* 131–6.

DEBACH P. (Ed.) (1964) *Biological Control of Insect Pests and Weeds*. Chapman & Hall, London.

DODD A.P. (1954) Biological control of weeds. In Weed Control Conf., Roseworthy Agric. Coll., Session, 5 Aug. 1954. Pp. 121–3. (From Huffaker 1964.)

ELLIOTT J.G. (1968) Herbicides in modern crop production (b) Extensively grown crops. *Proc. 9th Br. Weed Control Conf.* 1053–68.

VAN EMDEN H.F. (1970) Insects, weeds and plant health. *Proc. 10th Br. Weed Control Conf.* 942–52.

EVANS S.A. (1968) Project NAE 30. Results and Conclusions. *Proc. 9th Br. Weed Control Conf.* 1165–6.

EVANS S.A. (1969) Spraying of cereals for the control of weeds. *Expl Husb.* **18**, 103–9.

FOSTER J. (1964) *Studies on the population dynamics of the daisy* (Bellis perennis L.). Ph.D. Thesis, University of Wales.

FRANKLIN M.T. (1970) Interrelations of nematodes, weeds, herbicides and crops. *Proc. 10th Br. Weed Control Conf.* 927–33.

FRYER J.D. & R.J. CHANCELLOR (1970a) Herbicides and our changing weeds. In *The Flora of a Changing Britain*, ed. F. Perring, 105–18. Classey, Hampton.

FRYER J.D. & R.J. CHANCELLOR (1970b) Evidence of changing weed populations in arable land. *Proc. 10th Br. Weed Control Conf.* 958–64.

FRYER J.D. & S.A. EVANS Eds. (1968a) *Weed Control Handbook: Vol.* 1, *Principles.* Blackwell, Oxford.

FRYER J.D. & S.A. EVANS Eds. (1968b) *Weed Control Handbook:* Vol. 2, *Recommendations.* Blackwell, Oxford.

HAMMERTON J.L. (1968) Past and future changes in weed species and weed floras. *Proc. 9th Br. Weed Control Conf.* 1136–46.

HARPER J.L. (1956) The evolution of weeds in relation to resistance to herbicides. *Proc. 3rd Br. Weed Control Conf.* 179–88.

HARPER J.L. (1957) Ecological aspects of weed control. *Outlook Agric.* 1, 197–205.

HARPER J.L. (1960) Factors controlling plant numbers. In *Biology of Weeds*, ed. J. L. Harper, 119–32. Blackwell, Oxford.

HARPER J.L. (1967) A Darwinian approach to plant ecology. *J. Ecol.* 55, 247–70.

HARPER J.L. (1968) The regulation of numbers and mass in plant populations. In *Population Biology and Evolution*, ed. R.C. Lewontin, 139–58. Syracuse University Press.

HARPER J.L. (1970) The population biology of plants. In *Population Control*, ed. A. Allison 32–44. Penguin Books, Harmondsworth.

HARPER J.L. (1971). Grazing, fertilizers and pesticides in the management of grasslands. In *The Scientific Management of Animal and Plant Communities for Conservation*, eds. E. Duffey & A.S. Watt, *Symp. Brit. Ecol. Soc.* 11, 15–31. Blackwell, Oxford.

HEATHCOTE G.D. (1970) Weeds, herbicides and plant virus diseases. *Proc. 10th Br. Weed Control Conf.* 934–41.

HUFFAKER C.B. (1957) Fundamentals of biological control of weeds. *Hilgardia* 27, 101–57.

HUFFAKER C.B. (1964) Fundamentals of biological weed control. In *Biological Control of Insect Pests and Weeds*, ed. P. DeBach. Chapman & Hall, London.

JONES M.G. (1933) Grassland management and its influence on the sward (see below).

JONES M.G. (1933a) I. Factors influencing the growth of pasture plants. *Emp. J. exp. Agric.* 1, 43–57.

JONES M.G. (1933b) II. The management of a clovery sward and its effects. *ibid.* 1, 122–8.

JONES M.G. (1933c) III. The management of a grassy sward and its effects. *ibid.* 1, 223–34.

JONES M.G. (1933d) IV. The management of poor pastures. *ibid.* 1, 361–6.

JONES M.G. (1933e) V. Edaphic and biotic influences on pastures. *ibid.* 1, 366–7.

LASHLEY R. (1969) Musk thistle control. *Proc. 24th N. cent. Weed Control Conf.* 99–100. (From Roberts 1970b.)

MACFAYDEN A. (1957) *Animal Ecology. Aims and Methods.* Pitman, London.

MADEL W. (1970) Herbicides and conservation in the Federal Republic of Germany. *Proc. 10th Br. Weed Control Conf.* 1079–88.

MOORE F.J. & J.M. THURSTON (1970) Interrelations of fungi, weeds, crops and herbicides. *Proc. 10th Br. Weed Control Conf.* 920–6.

MOORE N.W. (1970) Pesticides and conservation. *Proc. 10th Br. Weed Control Conf.* 1032–5.

PALMBLAD I.G. (1968) Competition in experimental populations of weeds with emphasis on the regulation of population size. *Ecology* 49, 26–34.

POTTS G.R. (1970) The effects of the use of herbicides in cereals on the feeding ecology of partridges. *Proc. 10th Br. Weed Control Conf.* 299–302.

PUTWAIN P.D. (1970) The population dynamics of *Rumex acetosa* L. and *R. acetosella* L. *Proc. 10th Br. Weed Control Conf.* 12–19.

ROBERTS H.A. (1970a) Viable weed seeds in cultivated soils. *Rep. natn. Veg. Res. Sta. 1969*, 25–38.

ROBERTS H.A. (1970b) Weed Science in the service of vegetable production. *Proc. 10th Br. Weed Control Conf.* 901–8.

ROBERTS H.A. & P.A. DAWKINS (1967) Effect of cultivation on the numbers of viable weed seeds in the soil. *Weed Res.* 7, 290–301.

ROBINSON D.W. (1966) A comparison of chemical and cultural methods of weed control in gooseberries. *Proc. 8th Br. Weed Control Conf.* 92–102.

SAGAR G.R. (1959) *The biology of some sympatric species of grassland.* D.Phil. Thesis, University of Oxford.

SAGAR G.R. (1968a) Weed Biology—a future. *Neth. J. agric. Sci.* 16, 155–64.

SAGAR G.R. (1968b) Factors affecting the outcome of competition between crops and weeds. *Proc. 9th Br. Weed Control Conf.* 1157–62.

SAGAR G.R. (1970) Factors controlling the size of plant populations. *Proc. 10th Br. Weed Control Conf.* 965–79.

SAGAR G.R. & J.L. HARPER (1960) Factors affecting the germination and early establishment of three plantains (*Plantago lanceolata*, *P. media* and *P. major*). In *The Biology of Weeds*, ed. J.L. Harper, 235–45. Blackwell, Oxford.

SAGAR G.R. & J.L. HARPER (1961) Controlled interference with natural populations of *Plantago lanceolata*, *P. major* and *P. media*. *Weed Res.* 1, 163–76.

SARUKHÁN J. (1970) A study of the population dynamics of three *Ranunculus* species. *Proc. 10th Br. Weed Control Conf.* 20–25.

SARUKHÁN J. (1974) Demographic studies on grassland weed species. Pp. 28–41. In *Biology in Pest and Disease Control*, eds. D. Price Jones and M.E. Solomon. Blackwell, Oxford.

SELMAN M. (1970) The population dynamics of *Avena fatua* (wild oats) in continuous spring barley. *Proc. 10th Br. Weed Control Conf.* 1176–88.

THURSTON J.M. (1959) A comparative study of the growth of wild oats (*Avena fatua* L. and *A. ludoviciana* Dur.) and of cultivated cereals with varied nitrogen supply. *Ann. appl. Biol.* 47, 716–39.

THURSTON J.M. (1962) The effect of competition from cereal crops on the germination and growth of *Avena fatua* in a naturally infested field. *Weed Res.* 2, 192–207.

THURSTON J.M. & E.D. WILLIAMS (1968) Growth of perennial grass weeds in relation to the cereal crop. *Proc. 9th Br. Weed Control Conf.* 1115–23.

WAPSHERE A.J. (1970a) The assessment of biological control potential of the organisms attacking *Chondrilla juncea* L. *Proc. 1st Int. Symp. Biol. Contr. Weeds*, 81–9.

WAPSHERE A.J. (1970b) In discussion. *Proc. 1st Int. Symp. Biol. Contr. Weeds*, p. 95.

WAY M.J. (1968) Vegetation control and wildlife. *Proc. 9th Br. Weed Control Conf.* 989–94.

WILSON B.J. (1970) Studies of the shedding of seed of *Avena fatua* in various cereal crops and the presence of this seed in the harvested material. *Proc. 10th Br. Weed Control Conf.* 831–6.

WOODFORD E.K. (Ed.) (1963) *Crop Production in a Weed-free Environment.* Symp. 2 Br. Weed Control Council. Blackwell, Oxford.

ZWÖLFER H. (1968) Some aspects of biological weed control in Europe and North America, *Proc. 9th Br. Weed Control Conf.* 1147–56.
ZWÖLFER H. & P. HARRIS (1971) Host specificity determination of insects for biological control of weeds. *Ann. rev. Ent.* 16, 159–78.

Section 2
Approaches to control

The use of biological methods in pest control

F. WILSON *Formerly, Australian Commonwealth Scientific and Industrial Research Organisation, Sirex Unit, Imperial College Field Station, Silwood Park, Sunninghill, Ascot, Berkshire, now at Department of Zoology, Imperial College, London*

The origins of this Symposium are rather complex. It was organized by the British Ecological Society, together with other learned societies, on the proposal of the Biological Control Subcommittee of the British National Committee for Biology. The Subcommittee had been formed in 1970 to provide continuity for U.K. developments relative to the International Organization for Biological Control, a Commission of the International Union of Biological Sciences. Changes under way in IOBC led to it being re-created as a global body in 1971, the motivation for this important change being the conviction of interested biologists that biological control and other biological forms of pest control could be exploited to a much greater extent, and that this was essential to the emergence of satisfactory long-term solutions to pest problems. During the Subcommittee's work, it became evident that there was wide interest amongst U.K. scientists both in biological forms of control, especially integrated control, and in the work of the restructured IOBC. The Symposium was planned so as to allow a broad presentation of U.K. research in these areas, and wide-ranging discussion of future research possibilities.

Solutions to pest problems have to be found that will accord with long-term public interests, for the potential hazards are too great to allow pest control to become purely a function of pesticides (Wilson 1971). In the words of the IBP Working Group on Biological Control (Worthington 1971), there is 'increasing realization that the recent era of nearly sole reliance on pesticides for insect control has been a major failure—due to the attendant problems of pesticide resistance, inducement of secondary pest species, the ever-increasing need for more pesticides (pesticide addiction) and mounting costs, yet poor insect control and severe harm to non-target species and the environment. In short, a more ecological approach has become a recognized necessity.'

No one imagines that pesticides will be discarded rapidly. On the contrary, their use in the world as a whole may well increase for the foreseeable future. What is required now is a much accelerated exploration of the alternatives, a more rational use of pesticides, including increased selectivity (Jones 1974; this volume, pp. 178–95), and the harmonious integration of chemical and other forms of control. These last are most often likely to be forms of control that are biology-based.

Biological control

Basis of biological control

The term 'biological control' was first used by H. S. Smith (1919) to denote insect pest control by the use of natural enemies. The scope of the term was extended later to cover pests of any kind. More recently, the term has been employed to include various biology-based forms of control that have no connection with natural enemies. I propose to use the term in Smith's sense but to include pests of any kind, and to employ the phrase 'biological forms of control' to cover all methods of control that are biology-based.

Methods of pest control develop from the interactions of scientific knowledge and social and economic needs. Biological control emerged in the second half of the nineteenth century when a better appreciation of the interactions of organisms (particularly a growing knowledge of the natural enemies of insects) coincided with an urgent need to find solutions for severe infestation of crops by immigrant insect species. These pests were mainly species spread by commerce to areas colonized by Europeans. For example, in North America 44% of the important pests were of foreign origin.

The cause of such severe infestations was attributed by some to the insects having escaped from a 'balance of nature' existing in native habitats, but others doubted the importance of biotic regulation. Forbes (1880) wrote: 'Hence as a general thing, the real limits of a species are not set by its organic environment, but by the inorganic; and the removal of the organic checks upon a species would not finally diminish its average numbers.' However, ten years later the introduction of the predator *Rodolia cardinalis* into California permanently and completely solved the serious problem of cottony cushion scale (*Icerya purchasi*) on citrus, and biological control was established as a valid method.

The original conception of the basis of biological control was the re-establishment in an invaded habitat of a balance of nature existing in the native habitat. However, no attempt was made to investigate such a balance (which indeed is becoming technically possible only today), and no one attempted, or ever will attempt, to introduce the complexity of a native balance of nature into an invaded habitat. Nevertheless, the idea persists and finds expression in some form even with leading authorities.

Doutt & DeBach (1964) have written: 'Applied biological control attempts to restore the natural balance by duplicating the conditions in the pest's native home through the importation of natural checks, for by introducing effective natural enemies, the pest population may often be reduced to the low status that it formerly held in its endemic area. This successful restoration of the natural balance exemplifies the ecological basis for biological control.'

While, of course, these authors are fully aware of the facts and scope of biological control, and one can give different meanings to particular phrases in the quotation, in my view this passage tends to perpetuate an incorrect view of biological control and to lead to important misconceptions as to its potentialities. Let us examine some well-known examples.

In the control of cottony cushion scale, already cited, or in that of the green vegetable bug (*Nezara viridula*) in Australia by the egg-parasite, *Asolcus basalis*, we have illustrations of biological control near to Doutt and DeBach's description—immigrant pests controlled by natural enemies from the native habitat. We cannot, however, say that the natural balance of the native habitat has been reconstructed in the new habitat, because control has been provided by only one member of the biota participating in the natural balance of the native habitat.

In the example of the biological control of prickly pears (*Opuntia inermis* and *O. stricta*) in Australia by *Cactoblastis cactorum*, immigrant pests from North America were controlled by a moth from South America whose natural hosts were other species of *Opuntia*, and the controlling agent had no connection with the natural balance of the native habitat.

In another instance, the coconut moth (*Levuana iridescens*) was controlled in Fiji by a tachinid parasite, *Ptychomyia remota*. The pest was a native Fijian species, the parasite was imported from Malaya, where its host was a moth of another genus, and the parasite had no connection with the initial natural balance of the native habitat.

There was a different situation again with an immigrant pest, the oriental fruit moth (*Grapholitha molesta*), in the eastern United States and Canada. Here a native American parasite, *Macrocentrus ancylivorus*, adopted the immigrant as a host and, aided by periodic mass colonization of laboratory-bred stocks of the parasite, provided a substantial degree of control.

An adequate definition of biological control, therefore, must cover the following:

(a) The pest controlled may be an immigrant species or a native species.
(b) The controlling natural enemy may originate in the native habitat of the pest, the area invaded by the pest, or elsewhere.
(c) The natural host of the controlling natural enemy may be the pest, other species of the same genus as the pest, or species of other genera.

The definition should also include the notions that control may, and usually will, be provided by more than one species of natural enemy, and that human manipulation of one kind or another may improve the level of control provided. Such a view of biological control is very different from that of the restoration of the natural balance by duplication of the conditions of the pest's native home. Rather it involves the creation of a new natural balance by the establishment and/or manipulation of a very restricted number of natural enemy species whose intrinsic characteristics allow them to serve a regulatory role. The balance achieved is peculiar to the new habitat, where the complex of biotic and abiotic factors is very different from that in the native habitat.

This broader view has the advantage that the possibilities of biological control are seen to extend well beyond the natural balances of native habitats. Attention can be directed to new host-natural enemy combinations that may be brought about to achieve desired ends. For example, research on the biological control of the sugarcane moth borer (*Diatraea saccharalis*) in Barbados proceeded for decades without success, but the recent establishment of *Apanteles flavipes*, an Asiatic parasite of graminaceous borers of other genera, has, quite unexpectedly, provided a level of control of considerable economic importance (Alam *et al.* 1971).

Scope of biological control

Knowledge of the ecological foundations of biological control has grown with, and substantially as, a direct or indirect result of biological control research. Limitations in ecological knowledge have always made it difficult to foresee extensions in the scope of the method. Its range of possible usefulness has often been assumed to be more or less coextensive with its area of proven effectiveness, and its scope has grown by research exploration into novel applications.

For example, from the earliest years biological control was pursued whole-heartedly and effectively in certain Pacific Islands and in California, and it was deduced that the method was appropriate for insular areas and 'ecological islands' but not for continents (Imms 1931). This view has long been discarded, and a review has shown that biological control successes have been recorded from 31 islands and 34 continental countries (DeBach 1964b).

Much of the early work on biological control was done in tropical or subtropical areas, and it has often been considered that the method is appropriate only for such climates; on the contrary, many successful results have been obtained in temperate or cold areas in Canada, eastern USA, New Zealand and Tasmania. Similarly, the numerous successes obtained with orchard and forest pests suggested that biological control was unsuited for pests of annual crops, but, to take some examples from Australia, complete or useful biological control has been obtained there with the green vegetable bug, the greenhouse white-fly (*Trialeurodes vaporariorum*), the cabbage moth (*Plutella maculipennis*) and the cabbage butterfly (*Pieris rapae*).

Taylor (1955) took a gloomy view of the future of biological control. He concluded that, while the best of methods when it worked, it had seldom done so, there was little future for it in continental areas, most pest susceptible to the method had already been controlled, successes would become increasingly rare, and the method was incompatible with chemical control. Since that forecast, there have been many important successes and exciting developments. One need mention, as examples, only the control of winter moth (*Operophthera brumata*) in Canada, and of Crofton weed (*Eupatorium adenophorum*) in Hawaii and elsewhere, the commercial production of viral and bacterial insecticides, and the work on the integrated control of cotton pests in Peru, and of the spotted alfalfa aphid (*Therioaphis maculata*) in California, to see how little has this pessimism been justified by events.

More recent views are that in every area where biological control has been seriously tried there have been major successes (Munroe 1971), and that there have been few complete failures in the biological control of insects where intensive, well-supported programmes have been followed (Bartlett & van den Bosch 1964). DeBach (1964b) commented that complete biological control has occurred in all types of plant environments, and that over a period of time successes will be proportional to the amount of research and natural enemy importation. DeBach adds that there are no mystical or peculiar features that make Hawaii or Fiji or California outstandingly favourable for biological control. It so happens that most work has been done by relatively few countries.

Comparatively few attempts have been made to control native pests by introduced natural enemies, and there has been consequent doubt about the prospects of successful results. However, Pimentel points out that such pessimism ignores both the possibility of managing native natural enemies and the potential of exotic natural enemies. He comments (Pimentel 1966) that introduced natural enemies from allied hosts have often provided control of pests, although relatively few attempts have been made to use such enemies, and he attributes their effectiveness to the absence of ecological homoeostasis between host and natural enemy. The biological control of native pests by introduced natural enemies is an area that would probably well repay thorough exploration.

Manipulation of natural enemies and the habitat

Increasing attention is being given to methods of enhancing the effectiveness of natural enemies, both introduced and native. These methods may involve compensating for defects in adaptation of natural enemies to the habitat, or providing requisites of the natural enemies that are not met adequately by the habitat. Effective control may be facilitated by the periodic release of laboratory-reared or field-collected natural enemies. Such releases are intended either to swamp pest numbers promptly (inundative release), or to obtain control over a period of several generations (inoculative release). The controlling agents used may be predacious insects such as coccinellids, parasites such as *Trichogramma*, or pathogens. The performance of natural enemies may also be improved by other means, such as modification of cultural practices or of the environment, or, in theory at least, by the laboratory production of better adapted strains of natural enemies. This general approach to biological control is of particular importance for the development of integrated forms of control.

Biological control of weeds

Cussans (1974; this volume, pp. 97–105) deals with the biological control of weeds; it is necessary now only to refer to a few generalities.

Deficiencies in ecological knowledge of the effects of insects on plant abundance led Thompson (1930) and others to conclude that biological control could play only a very limited role in weed control. For many years, research was confined to a few countries, such as Australia and New Zealand, and a few islands, such as Hawaii and Fiji, but a good many countries are now actively involved, and many strikingly successful results have been obtained. There is growing acceptance of the view that most plants, particularly higher plants, have associated phytophagous insect species that are capable of greatly reducing the abundance of their hosts in countries where the plants are immigrants (Wilson 1964), and that with many weeds such insects can be used without risk to crops. The general view of biological control workers is that biological control is at least as effective against weeds as against insects. There is much research in progress on the biological control of aquatic weeds, and some of it is very promising, but an important question for determination is the extent to which aquatic plants possess oligophagous insect enemies.

There is also evidence that plant pathogens can be effective agents for the control of weeds, and that pathogens already present in an area can be exploited for weed control (Wilson 1969). For example, oak trees and persimmon, which are weeds in parts of U.S.A, are being controlled effectively by fungal pathogens. Viruses are being used to control blooms of green-blue algae in sewers in U.S.A, and in hydroelectric dams in U.S.S.R. Fungi are being used to control dodder in U.S.S.R. and skeleton weed (*Chondrilla juncea*) in Australia.

There is a large and promising field for research in the use of insects, pathogens, fish, and other organisms for the control of immigrant weeds, and in the exploitation of alien natural enemies of related species to control native weeds.

Range of pests

The pests available for consideration as subjects for biological control are extremely diverse. Though most research has been directed towards insects and mites, and to weeds from algae to trees, the range includes plant pathogens, nematodes, snails, ticks, birds, and mammals. This volume contains other papers on some of these groups; here it is desired to mention only a few points.

The extension of biological control research in these fields is highly desirable. The great economic importance of many of the pests concerned means that successful research brings valuable economic gains. For example, the control of the rabbit (*Oryctolagus cuniculatus*) in Australia by myxoma virus produced an enormous saving for agriculture that would easily pay for all the biological control research that Australia could conceivably undertake for a generation. The myxoma virus was, of course, native to America, where its hosts were of another genus, *Silvilagus*. This emphasizes the great potential of pathogens for the control of new hosts (whether immigrant or native) in new areas, because of the absence of evolved resistance to or tolerance of the pathogen in the new host. One wonders what prospects there may be for similar pathogenic control of other economically important vertebrate pests, such as rats and mice.

In the important area of the biological control of plant pathogens, the scientific difficulties are evidently very great, but there seem to be promising features. For example, observations that plant nematode populations increase after the application of pesticides has suggested that important natural enemies of these nematodes exist. We have too the qualified optimism of Garrett (1965) concerning the control of plant pathogens; he said: '. . . I now believe that eventual prospects for biological control are brighter than ever before, partly because we now comprehend the magnitude of the problems to be solved.'

Range of natural enemies

The natural enemies used for biological control have so far been predominantly oligophagous species of entomophagous or phytophagous insects, with increasing attention being given to viral and bacterial pathogens, which is discussed by Tinsley & Entwistle (1974; this volume, pp. 115–29). In addition, some use has been made of fungi, nematodes, snails, mites, fish, amphibia, birds and mammals. In general, it is unlikely that polyphagous natural enemies will often be introduced into new countries, as they tend to have undesirable side-effects. The main exception, perhaps, is in the use of fish to control weeds and mosquitoes. Greater use is also likely to be made in the future of fungi for weed control and of nematodes for insects. An important recent development has been the liberation in Australia of the nematode, *Deladenus siricidicola*, to sterilize females of *Sirex noctilio*. An area of particular importance, of course, is the use of bacterial and viral pathogens to control insects and perhaps some vertebrate pests. Vertebrate predators are unlikely to be imported into new regions to any significant extent for pest control, because of their general polyphagy. There is, however, growing appreciation of their importance in native habitats (Buckner 1966, Munroe 1971), and increasing attention is likely to be given to their conservation and encouragement in native habitats.

Advantages and successes in biological control

Biological control has many advantages. Generally, such control is automatic and permanent, and its cost is non-recurrent, restricted to the research undertaken, and relatively very low. Biological control is virtually free of the disadvantages of chemical control, and the risks are small when the work is undertaken, as it always should be, by responsible institutions. The development of host resistance to introduced parasites and predators is virtually unknown, though attenuated strains of viruses may appear. Under some circumstances, as when periodic releases or microbial sprays are employed, the position may be somewhat different.

Taking into account only biological control successes against insects up to the year 1961, there are some 220 examples of success, ranging from partial to complete control, and affecting 110 pest species in over 60 countries (DeBach 1964b). The very great majority of these successes have been with immigrant species, and a good many have been with scale insects in orchards, these problems having attracted considerable research effort. There has been a wide range in the level of control provided, but, of course, even partial effectiveness reduces dependence on pesticides.

The economic benefit from biological control has been considerable, making the overall cost of the research trifling by comparison. For example, in California, the reported saving to agriculture through the biological control of the spotted alfalfa aphid was, in 1958 alone, greater than the total Californian budget for biological control research for the years 1923 to 1959 (DeBach 1964a). Bearing such things in mind, and also the urgency of the need to find alternatives to the chemical control of pests, it is quite out of proportion that the expenditure on biological control research may be at the level of perhaps one-thousandth of the cost of pesticide research and use (adequate figures are lacking).

Other biological forms of control

It is probably fair to say that biological control has so far proved to be the most important and effective application of ecological knowledge to pest problems, but some other biological forms of control have been long in existence and have made major contributions to agriculture. In recent years there has been a stream of ideas leading to suggestions for new forms of control—sterile male release, genetical control, behavioural control, hormonal control, and so on. Other contributors discuss many of these.

Two of the longer established methods, the breeding of resistant varieties of crops and cultural control, have had extremely valuable practical results, but they remain far from adequately developed or exploited. This is the more to be regretted as they are ideal forms of control in that they aim at evading pest attack. No less important potentially is the sterile male technique, which has added the new dimension of eradication to the control of insects, and it seems very probable that other methods, such as genetical control, which have not yet proved themselves in practice, are likely to attain great importance later.

Two of the contributions that will attract very wide interest are those dealing with integrated control. Whether this is a new form of pest control or, as some suggest, an old form in modern clothing, no one is likely to dispute the importance of achieving the comprehensive control of the pests and potential pests of crops, while maximizing reliance on the biological elements in control and minimizing and rationalizing pesticide use. Kennedy (1968) raised the question whether such ecologically complicating proposals could survive the economic pressure to simplify agricultural production, and it is to be hoped that the Symposium will go some way towards clarifying this.

Biological control under modern agricultural conditions

Biological control research has as its main practical aim to increase the geographical range and/or abundance of natural enemies in order to control particular pests in particular places. It is the intrinsic properties of natural enemies that are important, rather than merely what they do or achieve in native habitats (Wilson 1964). Preadaptation of a natural enemy to a new environment, and conversely, suitability of the new environment for the natural enemy, are essential, except insofar as the environment and/or the natural enemy may be conveniently managed to compensate for inadequacies.

The fitness of a natural enemy to serve a regulatory role in respect of its host depends on properties evolved in the original natural habitat, which, often enough, has vanished or become greatly modified. The present native habitat of a natural enemy may comprise agroecosystems or even city suburbs that are vastly different from the original habitat. The natural enemy may then be established in other parts of the world, where flora, fauna and land use may be very different again. Yet in all these diverse habitats the natural enemy may prove able to regulate the numbers of its host.

The needs of a natural enemy have little to do with the complexities of the biota of the original natural habitat, but the natural enemy has a limited number of specific needs which must be met in the new habitat if it is to be established and provide effective control. These needs include a suitable physical environment, supply of hosts, freedom from efficient natural enemies, and a few other requirements, such as supplementary adult foods. In principle, natural enemies of pests cannot face much greater obstacles to establishment than can immigrant pests. It is, therefore, a mistake to regard the varied conditions of the original natural habitat as being necessary for the effective operation of the natural enemy, or to think that simplified agricultural systems necessarily make successful biological control impracticable. Natural enemies have proved effective over a very wide range of agricultural, horticultural and forestry circumstances. To this extent, one can agree with Chant (1964), who wrote: 'Perhaps we should abandon common concepts of communities, ecosystems, natural balance, etc. and recognize that what we are trying to do is completely unnatural and artificial.'

However, to suggest that the needs of natural enemies are probably few is not to imply that they are simple and obvious. One has only to consider the complexities of climate and weather in relation to synchronization of host and natural enemy. Entomologists have had great difficulty in defining natural enemy requirements and in forecasting the likelihood of establishment and effective control. Introduction is a ready test of adaptation to the environment and of capacity to control. It does nothing, however, to explain why perhaps three-quarters of natural enemies do not become established. What is needed is the development of a system of laboratory analysis of the environmental needs and tolerances of natural enemies, so that suitable habitats can be defined and the deficiencies of unsuitable habitats recognized.

Though at present we seem to be far from that level of technique, there is no reason why biological control research should not accept the challenge presented by the trend towards simplicity in agriculture. The main problem will probably be to ensure the continuity of host and plant-habitat necessary for the maintenance of the natural enemy population. But for biological control, as for integrated control, as Way (1966) has commented, required elements of diversity can be reintroduced or added when we know what is needed.

It should be stressed, however, that there should be a general cultural flexibility, and not an expectation that at every point the pest control method should conform to the exigencies of mass production techniques. These techniques and profitability should not be the only criteria; the need to avoid the adverse side-effects of pesticides and to devise control methods that will have long-term effectiveness must also be given its proper weight in assessing the overall situation.

In devising systems of pest control, their effect on conservation problems can no longer be ignored. Mellanby (1974; this volume, pp. 349–53) deals with the topic later; here it is desired to draw attention only to two points. First, concern about the environment is a social force that greatly strengthens moves to create ecologically harmless forms of pest control; and, secondly, modern modes of land use and agricultural techniques, especially pesticide use, involve hazards for the conservation of innumerable species of insects and other organisms that serve various important biological roles, not least important being the great diversity of natural enemies of pests. In my opinion, these natural enemies constitute a natural resource of the greatest importance for the present and future control of pest species (Wilson 1971).

International aspects

Pests and pesticide pollution recognize no frontiers. We can no longer afford to look at pest problems purely from a local or national point of view. The overlap in national interests is so great that rationalization of research at the international level becomes increasingly necessary. The advantages of such co-operation have been seen by those participating in the biological control programme of the International Biological Programme, and by those who have profited from the wide-ranging field research of the Commonwealth Institute of Biological Control.

To the growing world desire for the development of better forms of pest control, all countries, particularly scientifically and economically powerful countries, are under obligation to respond in the common interest. Signs of this response have been the restructuring of the International Organization for Biological Control (a Section of the International Union of Biological Sciences), and the rapid growth in support for its activities which is occurring all over the world. The creation of regional sections of IOBC in different parts of the world, which is now well advanced, should make an important contribution to the development of solutions for regional pest problems. Western European countries are particularly fortunate that, in terms of working and study groups, and other scientific activities, the West Palaearctic Regional Section is so advanced and well organized. The opportunity there exists for Britain to make a major scientific contribution to the European effort. The interests of IOBC extend to all biological forms of control of pests of any kind, and it is only a matter of time before the present predominance of entomological interest is substantially reduced.

Conclusion

We can have confidence that the high levels of biological knowledge now being approached and the social need for better long-term forms of pest control will in time lead us out of present difficulties, though these may well become exacerbated during the next decade or two. For this confidence to be justified, several things are necessary.

Firstly, there must be concern about the ecological consequences that may result from the continuation of old pest control policies. Secondly, there must be an awareness of the vast amount of research on biological forms of control that must now be done, partly because of past sins of omission. Thirdly, it must be appreciated that biological forms of pest control, though highly profitable to the community, are seldom suitable for commercial exploitation; and, consequently, that nearly all such research must be undertaken or sponsored by governments or international agencies. Fourthly, a sense of urgency needs to be instilled into decision-making, so that a much higher rate of progress can be achieved in the near future. Finally, scientists must accept the responsibilities of formulating new and better approaches and solutions to pest problems, and of convincing administrators of the importance of supporting the research involved.

For everyone concerned, the comment of C. L. Wilson (1969) is appropriate: 'At this stage new and imaginative thinking should be encouraged.' In re-thinking research policy on pest control, it should be appreciated that we are only at the beginning of the exploration of the potential of biological forms of pest control. Supposed limits to the applicability of biological control have been steadily pushed back, and there seems to be no area in which further progress is clearly barred. Great agricultural gains could follow the extended application of biological control to native pests, plant pathogens and vertebrate pests. Considerable benefits could also follow from the further development of mass production methods for microbial insecticides, and from the extended application of the many other biological forms of pest control. It is my conviction that we are at the beginning of a revolution in pest control techniques. Britain, with her great scientific and technical skills, and her wide-ranging world connections and commitments, would seem well placed to play a leading role in such a development.

References

ALAM M.M., F.D. BENNETT & K.P. CARL (1971) Biological control of *Diatraea saccharalis* (F) in Barbados by *Apanteles flavipes* Cam. and *Lixophaga diatraeae* T.T. *Entomophaga* 16 (2), 151–8.

BARTLETT B.R. & R. VAN DEN BOSCH (1964) Foreign exploration for beneficial organisms. In *Biological Control of Insect Pests and Weeds*, ed. P. DeBach. Chapman & Hall, London, 283–304.

BUCKNER C.H. (1966) The role of vertebrate predators in the biological control of forest insects. *Ann. Rev. Ent.* 11, 449–70.

CHANT D. (1964) Strategy and tactics of insect control. *Canad. Ent.* 96, 182–201.

CUSSANS G.W. (1974) The biological contribution to weed control. In *Biology in Pest and Disease Control*, eds. D. Price Jones and M.E. Solomon. Blackwell, Oxford, 97–105.

DEBACH P. (1964a) The scope of biological control. In *Biological Control of Insect Pests and Weeds*, ed. P. DeBach. Chapman & Hall, London, 3–20.

DEBACH P. (1964b) Successes, trends and future possibilities. In *Biological Control of Insect Pests and Weeds*, ed. P. DeBach. Chapman & Hall, London, 673–713.

DOUTT R.L. & P. DEBACH (1964) Some biological concepts and questions. In *Biological Control of Insect Pests and Weeds.* Chapman & Hall, London, 118–42.

FORBES S.A. (1880) On some interactions of organisms. *Bull. Illinois Nat. Hist.* 1 (3), 3–17.

GARRETT S.D. (1965) Towards biological control of soil-borne plant pathogens. In *Ecology of Soil-borne Plant Pathogens. Prelude to Biological Control*, eds. K.F. Baker & W.C. Snyder. John Murray, London, 4–17.

IMMS A.D. (1931) *Recent Advances in Entomology*. J. & A. Churchill, London

JONES D. PRICE (1974) Selectivity in pesticides: significance and enhancement. In *Biology in Pest and Disease Control*, eds. D. Price Jones and M.E. Solomon. Blackwell, Oxford, 178–95.

KENNEDY J.S. (1968) The motivation of integrated control. In *Integrated Pest Control*, ed. D.W. Empson. *J. appl. Ecol.* 5, 489–516.

MELLANBY K. (1974) The conservationists' viewpoint. In *Biology in Pest and Disease Control*, eds. D. Price Jones and M.E. Solomon. Blackwell, Oxford, 349–53.

MUNROE E.G. (1971) Status and potential of biological control in Canada. In *Biological programmes against insects and weeds in Canada. Tech. Comm.* 4 (CIBC).

PIMENTEL D. (1966) Population ecology of insect invaders of the Maritime Provinces. *Canad. Entom.* 98, 887–94.

SMITH H.S. (1919) On some phases of insect control by the biological method, *J. Econ. Ent.* 12, 288–92.

TAYLOR T.H.C. (1955) Biological control of insect pests. *Ann. Appl. Biol.* 42, 190–6.

THOMPSON W.R. (1930) The principles of biological control. *Ann. Appl. Biol.* 17, 306–38.

TINSLEY T.W. & P.F. ENTWISTLE (1974) The use of pathogens in the control of insect pests. In *Biology in Pest and Disease Control*, eds. D. Price Jones and M.E. Solomon. Blackwell, Oxford, 115–29.

WAY M.J. (1966) The natural environment and integrated methods of pest control. *J. Appl. Ecol.* 3 (Suppl.), 29–32.

WILSON C.L. (1969) Use of plant pathogens in weed control. *Ann. Rev. Phytopath.* 7, 411–34.

WILSON F. (1964) The biological control of weeds. *Ann. Rev. Ent.* 9, 225–44.

WILSON F. (1971) *Biotic agents of pest control as an important natural resource. 4th Gooding Memorial Lecture*. Central Association of Bee-Keepers (Ilford, Essex), 12 pp.

WORTHINGTON E.B. (1971) Insecticides. *Nature, Lond.* 234, No. 5323, 55.

Pest resistance in fruit breeding

THE LATE R. L. KNIGHT* & F. H. ALSTON
East Malling Research Station, East Malling, Maidstone, Kent

Breeding for resistance to pests has not, until relatively recently, excited much interest in biologists. This is probably mainly because of the assumption that such resistance would almost inevitably break down; moreover, many entomologists have been content to rely on man-made insecticides, rather than on any hereditary capacity of the plant to resist attack. Rachel Carson's 'Silent Spring', though often criticized, at least provided the jolt that was needed to create a rather livelier interest in resistance breeding, and the continuing recognition of unsuspected dangers from some insecticides, coupled with the build-up of populations of pests resistant to them, has stimulated interest still further.

Good variety and species collections are an almost indispensable part of any soundly-based programme on breeding for resistance. It is much better to have a thorough survey and choose a *number* of sources of strong resistance—working with all of them in parallel, initially, and casting aside those that show least promise as the project progresses. Plant breeding is full of disappointments, and it is foolhardy to pin all one's faith on the first good source of resistance that is found to a particular pest.

The main reason for working on several sources of resistance in parallel is that it is usually only after one or two generations of crossing that some idea can be obtained as to the complexity of inheritance of the resistance. The case for major *versus* minor gene control has been a bone of contention for years—both have their advantages and disadvantages. Where it is already present in a commercially acceptable type, then polygenic resistance is reasonably easy to handle. Where species crossing is involved, especially between widely different species, a major 'universal' gene, if available, is much easier to manage.

So often the only sources of resistance are commercially inferior plants, so that several backcrosses to more valuable parents are required to produce

* The Editors record with regret that Dr. R. L. Knight died on 15 February 1972.

a commercially valuable type. In such circumstances the *more complex* the control, the *less likely* is final success. This is particularly true with clonally propagated crops such as our temperate fruits, most of which are highly heterozygous.

Relative permanence of pest resistance

Much of the despondent outlook towards breeding for pest resistance probably stems from the experience of cereal breeders who have had the misfortune to work with the rusts and smuts, groups widely known for their capacity to produce an almost infinite variety of biologic races. Fortunately very few pests appear to have a comparable capacity for variation. Thus the *Phylloxera* resistance, first recorded in vines by Riley in 1875, is still widely used. According to Boubals (1966a) the resistance of *Vitis rupestris* and of *V. riparia* to *Phylloxera* depends on the formation of cork to isolate attacked tissue; this supports the earlier findings of Becker (1959). *Vitis rupestris* was the parent of the St George rootstock which, according to Winkler (1962), was first recorded as resistant about 1885. The *riparia* × *rupestris* stocks, the earliest resistant hybrids to be released, have been available for over 50 years and are still as resistant as they were initially. At least eight races of *Phylloxera* have been claimed to exist (Börner & Schilder 1934) as judged by the response of cross-bred *riparia-vinifera* hybrids, and the inheritance of resistance in these two species is considered to be polygenic (Boubals 1966b, Breider 1939).

The resistance of the apple variety Winter Majetin to the woolly aphid (*Eriosoma lanigerum*), first noted by Lindley in 1831, is still as valid today as it was 140 years ago. This form of resistance has been shown at East Malling to be inherited as a polygenic character with only a low proportion of seedlings in progenies derived from Majetin showing the full expression. Unpublished work by A.B. Beakbane at East Malling suggests that resistance depends, at least in part, on the thickness of collenchyma cell walls, width of the collenchyma sheath, and the width and structure of the phellem.

The Northern Spy variety of apple is about 170 years old. It was for many years widely grown in the U.S.A. both as a scion variety and as a rootstock. It is markedly resistant, though not immune, to the woolly aphid, and resistant rootstocks derived from it have been distributed from East Malling throughout the world as the MM series. By contrast with Winter Majetin, the resistance of Northern Spy does not appear to be anatomical or structural. This variety was originally thought to show complex inheritance by Crane *et al.* (1936), largely because they were considering only the so-called 'immune' plants in interpreting segregation data. Such plants have since been shown to be *resistant* but not *immune*, and an interpretation by Knight *et al.* (1962) both of Crane's data and of more recent information showed that resistance is controlled by a single dominant gene. This assessment was based on a broader classification—'resistant' *versus* 'susceptible' (the two groups being quite distinct in response)—instead of on a count of so-called 'immunes' *versus* 'non-immunes'. Various sources of resistance exist other than Northern Spy, and there is evidence of strain differentiation of the aphid in relation to some of the sources; nevertheless the resistance of Northern Spy itself, and of the derived MM rootstocks, has remained of value throughout the apple growing regions of the world.

The raspberry variety Lloyd George, discovered in 1918 and imported into America in 1929, was soon found to be resistant to the American form of the rubus aphid, first called *Amphorophora rubi* but now given specific status as *Amphorophora agathonica* (Kennedy *et al.* 1962). Schwartze & Huber (1939) considered that this resistance was controlled by two or more genes. More recently Daubeny (1966) has shown that a single dominant gene is involved. This resistance is still as effective in N. America as it was when first discovered.

Jassid resistance, in both Old World diploid and New World allotetraploid cottons, evolved largely as a result of natural selection in peninsular India. Its full expression depends on the presence of hairs of adequate length and density on the underside of the leaf lamina. This character was bred into African varieties of upland cotton by Parnell and his co-workers in the thirties (Parnell *et al.* 1949), and remains just as effective today. This form of resistance depends on the presence of H_1, a dominant A-genome key gene conferring leaf hairiness (as opposed to the more or less glabrous condition). This gene by itself confers little jassid resistance, so that for full expression of this character its action has to be supported by major modifying genes and minor genes giving adequate length and density to the hairs (Knight 1952, 1954, 1955, Knight & Sadd 1953, 1954, Saunders 1961, 1965a, b). A second source of jassid resistance, the nature of which is not fully understood, occurs in the wild glabrous-leaved species, *Gossypium armourianum*.

Examples of mite resistance, known to have existed for many years, are rare in the literature. However, the gooseberry (a congener introduced into Britain in the thirteenth century) has been grown alongside black currants at least since the eighteenth century; nevertheless, it still remains immune to the gall mite, *Cecidophyopsis ribis*.

Breeding for nematode resistance seems to have proved more akin to the rust story with cereals, though in potato the digenic resistance of *Solanum vernei* appears promising (Howard 1969), and in peach the Okinawa rootstocks imported into America in 1953 (Sharpe 1957) still appear to retain satisfactory resistance to the root-knot nematodes *Meloidogyne incognita* and *M. javanica* (Sharpe *et al.* 1968).

Perhaps the question of bird resistance is rather outside the scope of this paper, but it is worth noting that in *Sorghum*, according to Doggett (1957) and others, a considerable degree of avoidance of weaver bird (*Quelea*) damage can be obtained by incorporating the gene for 'goose neck'. This form of avoidance is far from absolute in its effect but it has the merit that it is unlikely to be broken down.

Recent work on pest resistance in temperate fruit crops

All the examples we have so far given of pest resistance in relation to plant breeding were chosen to illustrate the degree of permanence that has been achieved over a range of pests. Proof of permanence requires a certain amount of delving into the past. We should now like to turn to present-day work and to limit the subject to temperate fruit crops, since detailed surveys on all aspects of the genetics and breeding of most of these are available (Knight & Keep 1958, Knight 1963, 1966, 1969, Knight *et al.* 1972).

Raspberry

Amphorophora rubi

A considerable part of the raspberry breeding programme at East Malling is concerned with virus avoidance through resistance to the main aphid vector, *Amphorophora rubi*. In the course of this work 14 genes (Table 1) have been delineated (Keep & Knight 1967, Keep *et al.* 1970, Knight, Keep & Briggs 1959, Knight *et al.* 1960) some of which have provided the means of classifying the aphid into four biotypes (Briggs 1959, 1965a, b). Wide and repeated search, coupled with a programme of interbreeding between these strains of the aphid, has not resulted in the discovery of any additional ones. It would appear that such variability as exists is controlled by segregation at two gene loci in the aphid (Briggs 1965a).

Initial surveys disclosed a number of sources of *Amphorophora* resistance in *Rubus* spp. and raspberry cultivars. These sources were used in a breeding programme and almost all of them had an oligogenic basis for their resistance. No polygenic type of resistance was found of a strength approaching the near-immunity that is the minimum requirement to prevent virus spread in a clonally propagated perennial crop. The location and identification of resistance genes in the raspberry enabled a classification to be made of biologic races of the vector, the whole process leading to the discovery of four genes, each of which confers near immunity to all British races of the aphid under field conditions. Although four aphid races are recognized,

Table 1. Relationship between *Rubus* resistance genes and strains of *A. rubi* (From Keep) *et al.* 1970

Genes	Origin	Strain of *A. rubi* 1	2	3	4
A_1	Baumforth A (*R. idaeus*)	R	S	R	S
A_2	Chief (*R. strigosus* phenotype)	S	R	S	S
A_1A_3	Chief (*R. strigosus* phenotype)	R	R	R	S
A_3A_4	Chief (*R. strigosus* phenotype)	S	R	S	S
A_5	Chief (*R. strigosus* phenotype)	R	S	S	S
A_6	Chief (*R. strigosus* phenotype)	R	S	S	S
A_7	Chief (*R. strigosus* phenotype)	R	S	S	S
A_8	*R. strigosus* L 518	R	R	R	R
A_9	*R. strigosus* L 518	R	R	R	R
A_{10}	Cumberland (*R. occidentalis*)	R	R	R	R
A_{K4a}	*R. idaeus*	R	R	R	R
A_{L503}	*R. occidentalis*	R	R	R	R
$A_{cor \cdot 1}$	*R. coreanus*	?	R	R	R
$A_{cor \cdot 2}$	*R. coreanus*	?	R	S	?

only two of these (strains 1 and 3) are common. Strain 2 occurs sporadically in crops carrying the gene A_1 (conferring resistance to strains 1 and 3) but winter survival of strain 2 appears extremely poor and it never reaches the abundance of the common races. Strain 4 has also shown poor winter survival in most years at East Malling. Nevertheless, although initial products of the raspberry breeding programme carrying A_1 only (e.g. Malling Orion) are being released, many of the subsequent cultivars now going on trial carry resistance to all known races.

Aphis idaei

In Britain it is rare for *Aphis idaei* to become established as a major pest of raspberry, so that, although it is a vector of certain viruses, little work on breeding for resistance has been done. Baumeister (1961) found strong resistance in the German cultivars Klon 4a and Klon 72a, the North American variety Eaton and the old English variety Magnum Bonum, while Hill (1953) and Rautapää (1967) recorded *Rubus fruticosus* as resistant. These sources should be of considerable interest to workers in Yugoslavia where this aphid is a serious pest of raspberry.

Other pests

Resistance has been recorded in the raspberry, blackberry, and allied species to the raspberry beetle (*Byturus tomentosus*), the cane maggot (*Pegomyia rubivora*), the raspberry sawfly (*Empira tridens*), the red necked cane borer (*Agrilus ruficollis*), the cicada (*Cicadetta montana*), the raspberry cane midge (*Thomasiniana theobaldi*), the dryberry mite or raspberry leaf and bud mite (*Aceria gracilis*), the blackberry mite (*A. essigi*), the spider mite (*Tetranychus althaeae*) and to the eelworm (*Longidorus elongatus*) (Knight *et al.* 1971). Although none of these sources of resistance has been widely utilized in breeding, they represent a valuable potential for future work.

Blackcurrant and gooseberry

Gall mite, Cecidophyopsis ribis

By far the most important pest of black currant is the gall mite *C. ribis*, vector of reversion virus. Indeed Legowski & Gould (1965) suggested that even with only 6% of buds galled at least 30% and probably nearer 80% of bushes are likely to show reversion symptons within one to two seasons. Considerable work on breeding for resistance to this pest as a means of avoiding the virus has been done in Russia, mainly using *Ribes nigrum sibiricum* as donor (Knight *et al.* 1972). In Scotland, Anderson & Adams (1970) and Anderson (1971) have identified a dominant gene in this species governing *resistance* to this pest, and at East Malling a level of resistance akin to *immunity* is being sought from the gooseberry. As F_1s of blackcurrant × gooseberry are sterile, fertile colchiploids of BBGG composition were made and back-crossed to diploid blackcurrant, giving resistant triploids. By crossing these again with blackcurrant, resistant diploids were obtained which have remained immune although grown in an infection plot, with their branches interlacing with those of susceptible controls, for seven or more years. Fifty-four derivatives selected for resistance in the next backcross have so far maintained their resistance in an infection plot for three years. These have now been propagated for selection for commercial qualities and further backcrosses have been made (Knight *et al.* 1971; Knight *et al.* 1972).

Chrysanthemum leaf and bud eelworm, Aphelenchoides ritzemabosi

Boys (1972) has reported what appears to be immunity to *A. ritzemabosi* in a third backcross of gooseberry to blackcurrant.

Other pests

Work by Keep & Briggs (1971) has delineated sources of resistance to the aphids *Hyperomyzus lactucae*, *H. pallidus*, *Nasonovia ribisnigri*, *Aphis grossulariae*, *Cryptomyzus ribis* and *C. korschelti*. Work on utilizing a number of these sources in breeding blackcurrants and gooseberries is in progress. Sources of resistance are also known to the currant clearwing moth (*Synanthedon tipuliformis*), the common gooseberry sawfly (*Nematus ribesii*), the blackcurrant leaf midge (*Dasyneura tetensi*), and the red spider mite (*Tetranychus urticae*) (Knight *et al.* 1971).

Apple

Surveys of pest resistance in the genus *Malus* and in apple cultivars based on published literature and on field observations have been made at East Malling (Briggs 1967, Briggs & Alston 1967, 1969, Knight 1963, 1966). These surveys indicate sources of resistance to, or avoidance of, the principal apple pests in this country. Work on breeding for resistance or avoidance to these pests has recently been outlined (Knight & Alston 1969, Alston 1971).

The apple is unique amongst temperate commercial crops in the extent of its dependence on a complex spray programme to control pests and diseases. This influences the whole outlook on the economics of plant breeding for resistance to any one pest. Thus the gene Sm_h from *Malus robusta* (Alston & Briggs 1970) offers a promising source of resistance to the rosy apple aphid (*Dysaphis plantaginea*) but the inclusion of such resistance in breeding a new cultivar would not affect the spray programme unless it could be accompanied by adequate resistance to sawfly (*Hoplocampa testudinea*) from *M. zumi* (Briggs 1967), to the rosy leaf-curling aphid (*Dysaphis devecta*), using the gene Sd_1 from Cox's Orange Pippin or Sd_3 from *M. robusta* (Alston & Briggs 1968, Alston 1971), and to woolly aphid using the gene Er from Northern Spy.

It is fortunate that major genes exist governing the control of so many of these pests. Any attempt to combine polygenic resistances to each, and still to include all the important commercial attributes, could only fail.

A different, yet complementary, approach lies in the use of late-leafing and -flowering cultivars, the breeding of which was initiated by Tubbs and Tydeman (Tydeman 1958). Such varieties almost invariably avoid damage by apple-grass aphid (*Rhopalosiphum insertum*), apple sucker (*Psylla mali*), and rosy-apple aphid (Briggs & Alston 1967). Damage from other spring pests is also avoided (Franca 1933, Wildbolz & Staub 1963).

Table 2. Pest resistance/avoidance in *Malus* and its relevance to a routine spray programme. (After Alston 1971)

Pests	No. of sprays	Resistance	Avoidance
Rhopalosiphum insertum			
Psylla mali			
'Caterpillars'	1		late flowering
Dysaphis plantaginea		Sm_h	
Dysaphis devecta		Sd_1, Sd_2, Sd_3	
Eriosoma lanigerum	1	Er	
Hoplocampa testudinea		ex *M. zumi*	
Panonychus ulmi	2	*Malus* spp.	predators
Laspeyresia pomonella		Cultivars	

Thus it will be seen from Table 2 that the provision of late-leafing/-flowering cultivars would save one spray application, whilst still controlling spring pests. Moreover, the omission of one insecticidal spray would affect the predator position in the orchard and thus help to control red spider mite.

The need for incorporating resistance to several pests complicates the plant breeding approach. Nevertheless, we have already bred at East Malling selections which combine major genes for resistance to woolly aphid, rosy leaf-curling aphid, and rosy-apple aphid, as well as to apple mildew (*Podosphaera leucotricha*) and collar rot (*Phytophthora cactorum*). These have not yet fruited. Other selections carry late flowering, combined with major gene resistance to rosy leaf curling aphid and scab (*Venturia inaequalis*); several of these have attractive fruit.

Pear

Work on pest resistance in pears at East Malling was initiated considerably later than that on apples, but sources of resistance to pear leaf midge

(*Dasyneura pyri*) and pear bedstraw aphid (*Dysaphis pyri*) (Briggs & Alston 1971) and pear sucker (*Psylla pyricola*) (Westigard *et al.* 1970) have been located and are being investigated. Sources of resistance to pear leaf blister mite (*Eriophyes pyri*) (Suša 1967) as well as to woolly aphid (*Eriosoma pyricola*), pear thrips (*Taeniothrips inconsequens*), and the nematode *Meloidogyne* sp. are also known (Knight 1963). In addition apparent resistance to damage from bullfinches has been noted in the cultivar Doyenné du Comice by Becker & Gilbert (1958), and attributed by Child (1968) mainly to avoidance through late bud burst.

Prunus

Despite the numerous sources of pest resistance available in this genus (Table 3), breeding work seems to have been confined to root-knot nematode resistance in peaches (Sharpe *et al.* 1968). The resistance to *Meloidogyne incognita* and *M. javanica* shown by the Okinawa peach rootstocks has been found to be under independent oligogenic control (Sharpe *et al.* 1969). Transference of these genes to almond is in progress (Kester *et al.* 1970).

Table 3. Groups of *Prunus* spp. and cultivars in which resistance (R) to various pests has been reported (Data from Knight 1969)

	Apricots	Cherries	Peaches and Nectarines	Plums, Damsons, etc.
Aphids				
Brachycaudus helichrysi				R
Hyalopterus amygdali			R	
H. pruni	R		R	R
Myzus cerasi		R		
Phorodon humuli				R
Pterochlorus persicae	R			
Rhopalosiphum nymphaeae			R	R
Beetles				
Capnodis tenebrionis	R			R
Galerucella cavicollis		R		
Popillia japonica				R
Xyleborus saxeseni				R
'Plum curculio'				R
Fruit flies				
Ceratitis capitata			R	
Rhagoletis cerasi			R	
R. cingulata			R	
R. fausta			R	

Table 3 (*contd.*)

	Apricots	Cherries	Peaches and Nectarines	Plums, Damsons, etc.
Moths				
Euzophera semifuneralis				R
Gelechia demissae			R	
Grapholitha molesta			R	R
Laspeyresia funebrana	R		R	R
Sanninoidea exitiosa			R	
S. opalescens				R
Sawflies				
Hoplocampa flava				R
H. minuta				R
Scales				
Aspidiotus perniciosus		R	R	R
Pseudaulacaspis pentagona			R	
Mites				
Bryobia pratensis	R			
Panonychus ulmi	R	R		R
Nematodes				
Meloidogyne incognita			R	
var. *acrita*			R	R
M. javanica			R	R
Meloidogyne sp.	R	R	R	R
Pratylenchus vulnus	R	R		R
*P. penetrans**			R	
Radopholus similis			R	

*Mountain & Patrick 1959.

Malo (1967) attributes the resistance of Okinawa and Nemaguard rootstocks to occlusion of the invasion points whilst Kochba (1969) considers the resistance of Nemaguard to depend on low cytokinin activity.

Conclusion

In this paper we have dealt mainly with clonally propagated perennial crops, several of which are highly heterozygous; moreover they are characterized by having a long juvenile phase before cropping. Such factors place a prerequisite on attempts to search for and use strong oligogenic types of resistance. Whilst in annual crops the incorporation of low levels of polygenic resistance may facilitate all-round biological control (van Emden 1966), such a course could drastically handicap breeding progress in fruit crops. Where virus avoidance is the objective, it could be positively harmful.

The value of wild species as sources of pest resistance is often disregarded, despite ample evidence of its successful utilization, and seldom is enough weight given to the rapidity with which the genome of the donor species can be eliminated in a properly planned breeding programme (Knight & Alston 1971).

Resistance breeding has often been channelled into an investigation of the *nature* of the resistance. Can we make a plea that at least an equal if not greater effort should be put into expanding the *plant breeding* approach?

References

ALSTON F.H. & J.B. BRIGGS (1968) Resistance to *Sappaphis devecta* (Wlk.) in apple. *Euphytica* **17**, 468–72.

ALSTON F.H. & J.B. BRIGGS (1970) Inheritance of hypersensitivity to rosy apple aphid *Dysaphis plantaginea* in apple. *Can. J. Genet. Cytol.* **12**, 257–58.

ALSTON F.H. (1971) Integration of major characters in breeding commercial apples. *Proc. Angers Fruit Breed. Symp., 1970*, pp. 231–48.

ANDERSON M.M. & T.G. ADAMS (1970) Black currant. *Rep. Scott. hort. Res. Inst. for 1969*, p. 38.

ANDERSON M.M. (1971) Resistance to gall mite (*Phytoptus ribis* Nal.) in the Eucoreosma section of *Ribes*. *Euphytica* **20**, 422–6.

BAUMEISTER G. (1961) (Investigations on the resistance of different raspberry varieties to the virus vectors *Amphorophora rubi* (Kalt.) and *Aphis idaei* (v.d. Goot)). *Züchter* **31**, 351–7.

BECKER H. (1959) (Investigations on the attack of rootstock vines by *Phylloxera*). *Proc. 4th int. Congr. Crop Prot., Hamburg* **1**, 783–5.

BECKER P. & E.G. GILBERT (1958) Selective bird damage on pears in winter. *Jl R. hort. Soc.* 83, 509–15.

BÖRNER C. & F.A. SCHILDER (1934) (On the breeding of phylloxera and mildew resistant vines. II. The behaviour of phylloxera towards the vines of the Naumberg collection). *Mitt. biol. Reichsanst. Ld- u. Forstw.* **49**, 17–84.

BOUBALS D. (1966a) (A study of the distribution and causes of resistance to radicicolous phylloxera in the Vitaceae). *Annls Amél. Pl.* **16**, 145–84.

BOUBALS D. (1966b) (Inheritance of resistance to radicicolous phylloxera in the vine). *Annls Amél. Pl.* **16**, 327–47.

BOYS P.A. (1972) *Rep. E. Malling Res. Stn for 1971*, pp. 108–9.

BREIDER H. (1939) (The inheritance of resistance of the vine to *Phylloxera vastatrix* Planch. I. The behaviour of F_3 generations obtained from selfings of resistant and susceptible F_2 species hybrids). *Z. PflZücht.* **23**, 145–68.

BRIGGS J.B. (1959) Three new strains of *Amphorophora rubi* (Kalt.) on cultivated raspberries in England. *Bull. ent. Res.* **50**, 81–7.

BRIGGS J.B. (1965a) The distribution, abundance, and genetic relationships of four strains of the rubus aphid (*Amphorophora rubi* (Kalt.)) in relation to raspberry breeding. *J. hort. Sci.* 40, 109–17.

Briggs J.B. (1965b) The importance of strains of *Amphorophora rubi* (Kalt.), the rubus aphid, in the problem of breeding for resistance in the raspberry. *Proc. 12th int. Congr. Ent., London*, p. 532.

Briggs J.B. (1967) Sources of pest resistance in the genus *Malus. Rep. E. Malling Res. Stn for 1966*, pp. 166–9.

Briggs J.B. & F.H. Alston (1967) Pest avoidance by late-flowering apple varieties. *Rep. E. Malling Res. Stn for 1966*, pp. 170–1.

Briggs J.B. & F.H. Alston (1969) Sources of pest resistance in apple cultivars. *Rep. E. Malling Res. Stn for 1968*, pp. 159–62.

Briggs J.B. & F.H. Alston (1971) *Rep. E. Malling Res. Stn for 1970*, p. 127.

Child R.D. (1968) Factors influencing bullfinch damage to pear bud fruits. *Rep. Long Ashton Res. Stn for 1967*, pp. 110–15.

Crane M.B., R.M. Greenslade, A.M. Massee & H.M. Tydeman (1936) Studies on the resistance and immunity of apples to woolly aphis, *Eriosoma lanigerum* (Hausm.). *J. Pomol.* 14, 137–63.

Daubeny H.A. (1966) Inheritance of immunity in the red raspberry to the North American strain of the aphid, *Amphorophora rubi* Kltb. *Proc. Am. Soc. hort. Sci.* 88, 346–51.

Doggett H. (1957) Bird resistance in *Sorghum* and the *Quelea* problem. *Fld Crop Abstr.* 10, 153–6.

van Emden H.F. (1966) Plant insect relationships and pest control. *Wld Rev. Pest Control* 5, 115–23.

Franca A. (1933) (Regarding a variety of apple escaping severe infestation by the apple moth). *Note Fruttic.* 11 (3), 50–2.

Hill A.R. (1953) Aphids associated with *Rubus* species in Scotland. *Entomologist's mon. Mag.* 49, 298–303.

Howard H.W. (1969) Breeding potatoes resistant to cyst-nematode. *Proc. 5th Br. Insectic. Fungic. Conf., Brighton*, pp. 159–63.

Keep E. & R.L. Knight (1967) A new gene from *Rubus occidentalis* L. for resistance to strains 1, 2 and 3 of the rubus aphid, *Amphorophora rubi* Kalt. *Euphytica* 16, 209–14.

Keep E., R.L. Knight & J.H. Parker (1970) Further data on resistance to the rubus aphid *Amphorophora rubi* (Kltb.). *Rep. E. Malling Res. Stn for 1969*, pp. 129–31.

Keep E. & J.B. Briggs (1971) A survey of *Ribes* species for aphid resistance. *Ann. appl. Biol.* 68, 23–30.

Kennedy J.S., M.F. Day & V.F. Eastop (1962) A conspectus of aphids as vectors of plant viruses. [*Publ.*] *Commonw. Int. Ent., Lond.* 114 pp.

Kester D.E., C.J. Hansen & B.F. Lownsberry (1970) Selection of F_1 hybrids of peach and almond resistant and immune to root-knot nematodes. *HortScience* 5, 349.

Knight R.L. (1952) The genetics of jassid resistance in cotton. I. The genes H_1 and H_2. *J. Genet.* 51, 47–66.

Knight R.L. & J. Sadd (1953) The genetics of jassid resistance in cotton. II. Pubescent T 611. *J. Genet.* 51, 582–5.

Knight R.L. (1954) The genetics of jassid resistance in cotton. IV. Transference of hairiness from *Gossypium herbaceum* to *G. barbadense. J. Genet.* 52, 199–207.

Knight R.L. & J. Sadd (1954) The genetics of jassid resistance in cotton. III. The Kapas Purao, Kawanda *punctatum* and Philippines Ferguson group. *J. Genet.* 52, 186–98.

Knight R.L. (1955) The genetics of jassid resistance in cotton. V. Transference of hairiness from *Gossypium arboreum* to *G. barbadense. J. Genet.* 53, 150–3.

KNIGHT R.L. & E. KEEP (1958) Abstract bibliography of fruit breeding and genetics to 1955. *Rubus* and *Ribes*—A survey. *Tech. Commun. Commonw. Bur. Hort. Plantn Crops* **25**, 254 pp.

KNIGHT R.L., E. KEEP & J.B. BRIGGS (1959) Genetics of resistance to *Amphorophora rubi* (Kalt.) in the raspberry. I. The gene A_1 from Baumforth A. *J. Genet.* **56**, 261–80.

KNIGHT R.L., J.B. BRIGGS & E. KEEP (1960) Genetics of resistance to *Amphorophora rubi* (Kalt.) in the raspberry. II. The genes A_2–A_7 from the American variety, Chief. *Genet. Res.* **1**, 319–31.

KNIGHT R.L., J.B. BRIGGS, A.M. MASSEE & H.M. TYDEMAN (1962) The inheritance of resistance to woolly aphid, *Eriosoma lanigerum* (Hausm.), in the apple. *J. hort. Sci.* **37**, 207–18.

KNIGHT R.L. (1963) Abstract bibliography of fruit breeding and genetics to 1960. *Malus* and *Pyrus*. *Tech. Commun. Commonw. Bur. Hort. Plantn Crops* **29**, 535 pp.

KNIGHT R.L. (1966) Progress in breeding apples and pears. *Proc. Balsgård Fruit Breed. Symp., Fjalkestad 1964*, pp. 129–46.

KNIGHT R.L. (1969) Abstract bibliography of fruit breeding and genetics to 1965. *Prunus*. *Tech. Commun. Commonw. Bur. Hort. Plantn Crops* **31**, 649 pp.

KNIGHT R.L. & F.H. ALSTON (1969) Developments in apple breeding. *Rep. E. Malling Res. Stn for 1968*, pp. 125–32.

KNIGHT R.L., E. KEEP & J.H. PARKER (1971) *Rep. E. Malling Res. Stn for 1970*, pp. 100–1.

KNIGHT R.L., J.H. PARKER & E. KEEP (1971) Abstract bibliography of fruit breeding and genetics 1956–1969. *Rubus* and *Ribes*. *Tech. Commun. Commonw. Bur. Hort. Plantn Crops* **32**, 449 pp.

KNIGHT R.L. & F.H. ALSTON (1971) [Progress at East Malling in breeding for resistance to apple mildew, *Podosphaera leucotricha*]. *Jugosl. Vocarstvo*, pp. 17–18, 317–24.

KNIGHT R.L., E. KEEP, J.B. BRIGGS & J.H. PARKER (1972) *Rep. E. Malling Res. Stn for 1971*, p. 108.

KOCHBA Y. (1969) Effects of endogenous factors in roots of stone fruit trees on their resistance to nematodes. *Res. Rep. Hebrew Univ. Jerusalem Sci. Agric.* **1**, p. 588 (Abstr.).

LEGOWSKI T.J. & H.J. GOULD (1965) Relationship between gall mite, reversion and yield on commercial plantations of black currants. *Pl. Path.* **14**, 31–4.

LINDLEY G. (1831) *Guide to the orchard and kitchen garden*. Longman, Rees, Orme, Brown and Green, London, 111 pp.

MALO S.E. (1967) Nature of resistance of Okinawa and Nemaguard peach to the root-knot nematode *Meloidogyne javanica*. *Proc. Am. Soc. hort. Sci.* **90**, 39–46.

MOUNTAIN W.B. & Z.A. PATRICK (1959) The peach replant problem in Ontario. VII. The pathogenicity of *Pratylenchus penetrans* (Cobb, 1917) Filip. and Stek. 1941. *Can. J. Bot.* **37**, 459–70.

PARNELL F.R., H.E. KING & D.F. RUSTON (1949) Jassid resistance and hairiness of the cotton plant. *Bull. ent. Res.* **39**, 539–75.

RAUTAPÄÄ J. (1967) Studies on the host plant relationships of *Aphis idaei* v.d. Goot and *Amphorophora rubi* (Kalt) (Hom., Alphididae). *Ann. Agric. Fenn.* **6**, 174–90.

RILEY C.V. (1875) (Results of the introduction of *Phylloxera* resistant vines into France). *Bull. Soc. ent. Fr.* 141–2.

SAUNDERS J.H. (1961) The mechanism of hairiness in *Gossypium*. I. *Gossypium hirsutum*. *Heredity, Lond.* **16**, 331–48.

SAUNDERS J.H. (1965a) The mechanism of hairiness in *Gossypium*. 3. *Gossypium barbadense* —the inheritance of upper leaf lamina hair. *Emp. Cott. Grow. Rev.* **42**, 15–25.

SAUNDERS J.H. (1965b) The mechanism of hairiness in *Gossypium*. 4. The inheritance of plant hair length. *Emp. Cott. Grow. Rev.* **42**, 27–32.

SCHWARTZE C.D. & G.A. HUBER (1939) Further data on breeding mosaic-escaping raspberries. *Phytopathology* **29**, 647–8.

SHARPE R.H. (1957) Okinawa peach shows promising resistance to root-knot nematodes. *Proc. Fla St. hort. Soc.* **70**, 320–2.

SHARPE R.H., C.O. HESSE, B.F. LOWNSBERRY & C.J. HANSEN (1968) Breeding peaches for root-knot nematode resistance. *HortScience*, **3**, 92.

SHARPE R.H., C.O. HESSE, B.F. LOWNSBERRY, V.G. PERRY & C.J. HANSEN (1969) Breeding peaches for root-knot nematode resistance. *J. Am. Soc. hort. Sci.* **94**, 209–12.

SUŠA V.I. (1967) (The resistance of pears to pear leaf blister mite). *Zborn. nauk. Prac'. umans'k. sil's'kogosp. Inst.* **15**, 268–72.

TYDEMAN H.M. (1958) The breeding of late flowering apple varieties. *Rep. E. Malling Res. Stn for 1957*, pp. 68–73.

WESTIGARD P.H., M.N. WESTWOOD & P.B. LOMBARD (1970) Host preference and resistance of *Pyrus* species to the pear psylla, *Psylla pyricola* Foerster. *J. Am. Soc. hort. Sci.* **95**, 34–6.

WILDBOLZ T. & A. STAUB (1963) (On differences in aphid infestation determined by the stage of growth of apple buds). *Schweiz. Z. Obst- u. Weinb.* **72**, 184–8.

WINKLER A.G. (1962) *General viticulture*. University of California Press, Berkeley, 633 pp.

Plant breeding for disease resistance

F. G. H. LUPTON *Plant Breeding Institute, Cambridge*

Biological control is normally concerned with circumstances where the action of a pest organism is limited in its effects on its host by the intervention of one or more further species. Put in another way, it is concerned with interceptor organisms operating between the attacking pest and its target. The plant breeder, on the other hand, prevents or limits the effects of the pest by changing the genotype of the target organism. The control so effected is, however, entirely biological and is in many ways the most effective method for the control of plant diseases because, by adapting the plant genotype to a resistant condition, the breeder provides a form of resistance which will remain effective unless a similar adaptation occurs in the genotype of the pathogen. It is also the best form of control from the grower's viewpoint, because someone else has paid for the work on which the resistance of his crop depends. Although this may be to some extent reflected in the price of the seed, it is unlikely to be more than a fraction of the cost of control by sprays or seed dressings.

If it is to be acceptable to the grower, it is essential that a disease resistant variety should yield at least as well as other varieties when grown in disease free conditions. This may be a limiting factor—it takes from ten to fifteen years to produce a new cereal variety, measuring from the time the initial cross is made until seed is produced on a sufficient scale to be available on the general market, and that the time taken to produce new varieties of other crops may be very much longer. A resistant variety must therefore be able to compete with other varieties which have become available during this time. It must also be appreciated that, when looking for sources of disease resistance, the plant breeder is limited to the known genetic variability of his crop, although this may be extended by collections from its centres of origin, or by exposing it to mutagenic radiation. The limited extent of the available genetic variability is particularly important in breeding for resistance to diseases in which the pathogen is itself readily adaptable to changes in the genotype of the host. This will be referred to in more detail later.

Successful applications

There have, however, been cases where the plant breeder has been able to provide what appears to be a permanent solution to disease problems. One of the most frequently mentioned concerns the resistance of potatoes to wart disease (*Synchytrium endobioticum*). This disease was extremely serious in parts of Britain in the early years of this century, until it was discovered that certain varieties were not attacked. At first, control was achieved by legal restrictions on growing susceptible varieties; it has since been shown that resistance is determined by a simply inherited genetic factor (Black 1935), which has been used successfully in the breeding of all subsequent varieties. This example is of particular interest in that it concerns resistance to a soil-borne organism, which it would have been extremely difficult to control by other means. The same consideration applies in the control of eyespot of wheat (*Cercosporella herpotrichoides*) where the resistance of Cappelle Desprez and its derivatives has been maintained for nearly twenty years, though the mechanism of resistance in these two cases is different. The resistance to *Synchytrium* is due to a hypersensitive reaction, in which the pathogen kills a localized zone of host cells and is unable to survive in the dead cells so formed (Glynne 1926), while that to *Cercosporella* is due to a slower growth of the pathogen through the leaf sheaths of resistant varieties (Macer 1966).

Many other cases in which long-term resistance has been introduced by plant breeders are concerned with virus diseases, where for example, resistance to viruses A, B, C and X of potatoes, determined in each case by single dominant genetic factors, conferring hypersensitive or top necrotic reactions, has been used extensively in plant breeding (Howard 1960). Mention may also be made of the work on breeding onions for resistance to smudge (*Colletotrichum circinans*), where it was shown by Link & Walker (1933) that the resistance occurred in varieties with pigmented outer scales containing catechol and protocatechuic acid.

Appearance of new physiological races

In many cases, however, the introduction of single major genes for disease resistance has been followed by the appearance of strains of the pathogen capable of attacking previously resistant varieties. One of the best-known examples concerns the breeding for resistance to the cereal rusts in Britain and North America. Biffen (1906) reported that resistance to yellow rust of wheat (*Puccinia striiformis*) was inherited as a single recessive character, thus demonstrating for the first time that resistance to a plant disease was determined by the Mendelian laws of inheritance. In 1910 he released the variety Little Joss, demonstrating the application of these laws in practical plant breeding. The existence of physiologic races in yellow rust was first demonstrated by Gassner & Straib (1932), but their economic importance was not clearly appreciated until 1952, when the previously resistant variety, Nord Desprez, was severely attacked by an apparently new race of the pathogen, race 2B. Since this date, the history of wheat breeding in Western Europe has been marked by the appearance of a succession of new races of *P. striiformis*, each virulent over a wider spectrum of major genes for resistance. The most noteworthy of these are race 8B attacking Heine VII (1955), race 60 attacking Rothwell Perdix (1966), race 3/55 attacking Opal (1967), race 58C attacking Maris Envoy (1968), race 3/55D attacking Maris Beacon (1969), and race 1B attacking Maris Ranger (1970).

Similar cases of the breakdown of resistance due to the appearance or spread of physiologic races of the pathogen are well known in other crops. The problem was first appreciated by wheat breeders in North America, where the previously resistant variety Ceres was devastated by race 56 of stem rust (*Puccinia graminis*). The subsequent history of wheat breeding in that continent has until recent years been largely concerned with the introduction, by backcrossing, of new genes for resistance, as new races capable of attacking previously resistant varieties have developed.

Similar patterns are seen in breeding for resistance to powdery mildew (*Erysiphe graminis*) in cereals, and to late blight (*Phytophthora infestans*) in potatoes. In the latter case, much of the breeding work was, until recently, based on genes for resistance derived from the wild species, *Solanum demissum*. Black *et al.* (1953) showed that resistance in this species was determined by four major genes and pointed out that these genes, occurring singly or in groups, should be able to determine resistance to 2^4, that is to 16 races of blight. All these races have since been identified, and a further two genes found. The speed with which these races have appeared has shown the impracticability of exploiting major gene resistance and has led breeders to develop other approaches.

It has been observed that the appearance and spread of new races of pathogenic fungi has commonly followed the widespread cultivation of a previously resistant variety. This suggests that natural selection for new virulent forms of the pathogens has taken place. These may have existed as rare components of the previous pathogen population, or may have arisen *de novo* by mutation or by sexual or parasexual recombination. The development of such a new race has been well demonstrated in the case of tobacco mosaic virus (Pelham *et al.* 1970). Tomato varieties with monogenic resistance to this virus were first grown commercially in Britain in 1966, but during the next three years a new strain of virus (strain 1), capable of attacking these varieties, became widespread. Pelham and his colleagues collected samples of virus infected leaves from tomato growers in 1967 and 1968, recording in each case whether they had grown resistant varieties in any of the previous three years, and compared the virus isolates obtained from these samples with isolates from samples taken before 1966 (Table 1). All the pre-1966 samples were of strain 0, to which the new varieties were resistant. Samples collected from growers using resistant varieties comprised an increasing proportion of the resistance-breaking strain 1, but samples from growers who had used resistant varieties but had reverted to previously susceptible varieties were mostly of strain 0, suggesting that strain 1 competes poorly with strain 0, when a variety susceptible to strain 0 is grown.

Table 1. Frequency of detection of tobacco mosaic virus strains (after Pelham *et al.* 1970)

Varieties grown			Strains detected, 1968		
1966	1967	1968	No virus	Strain O	Strain 1
R	R	R	4	1	13
R	R	S	0	9	8
R	S	S	0	5	0
S	S	S	3	21	2

A new approach to breeding for disease resistance was therefore necessary. Vallega (1959) suggested that wheat breeders should introduce genetic factors for resistance to each of the races of rust separately into a single acceptable variety, by backcrossing. Only one of the isogenic lines so formed would be released for cultivation at any one time, and each line would be replaced whenever there were signs of the development of a strain of rust capable of attacking it. Another possibility, suggested by Borlaug (1958), was that pure line varieties should be replaced by mixtures of such isogenic lines. Both methods have the disadvantages of reliance on major gene resistance and, at the same time, of creating an environment in which the pathogen is likely to develop strains virulent over all known sources of resistance. Indeed, Borlaug's mixture of isogenic lines provides conditions in which the pathogen can successively accumulate genes to form races virulent over an ever-wider range of resistance genes.

Breeding for non-race-specific resistance

The failure of major gene resistance to provide a satisfactory method for control of many plant pathogenic fungi has led breeders to seek other sources of resistance, that are less likely to succumb to new forms of the pathogen.

Although major genes are more convenient to handle, particularly in a back-crossing programme, it has been noticed that more satisfactory results have often been obtained where breeding has been based on selection for a more generalized field resistance. Van der Plank (1963) drew attention to the distinction between such generalized resistance which he termed 'horizontal' in contrast to the simply-inherited 'vertical' resistance so widely used by plant breeders. He pointed out that before the introduction of vertical or race-specific resistance, many varieties had possessed a reasonable level of field resistance, but that this had been lost in the course of back crossing programmes designed to introduce major genes. To illustrate this point, he contrasted the situation regarding the wheat rusts, with that concerning *P. sorghi*, the cause of maize rust in North America. Although widespread, maize rust has never caused serious loss, and Van der Plank maintained that this was because selection had always been based on field resistance rather than on race-specific-resistance. This viewpoint is supported by Hooker & le Roux (1957), who tested the reactions of 85 varieties of maize to 15 isolates of *P. sorghi*. Only one of the maize varieties showed race-specific-resistance (in this case to only two of the isolates) but 16 varieties showed a high level of non-race-specific resistance in the field.

Because of the difficulties of using race-specific resistance, plant breeders are now increasingly turning their energies to breeding for non-race-specific resistance. In many cases they are discarding selections showing major gene resistance, because of the danger that such resistance may break down leaving a crop completely unprotected. This situation can be illustrated by consideration of recent work in breeding for resistance to yellow rust (*P. striiformis*) in wheat, where, as has been mentioned previously, numerous new physiologic races have arisen in recent years, following the introduction of varieties with major gene resistance. It has, however, been noted that a number of the older varieties of wheat, such as Browick, though susceptible as seedlings to all races to which they have been exposed, never appear to be seriously attacked in the field as mature plants (Table 2). Furthermore, this resistance is heritable and, although it may be polygenically determined, it can readily be used in a plant breeding programme (Lupton & Johnson 1970). Fortunately some of the more advanced hybrids, such as the high-baking-quality winter wheat Maris Widgeon, show a high level of non-race-specific resistance, so that it is not necessary for the breeder to resort to very poorly adapted varieties for his breeding programme. It has also been found that contrasting levels of field resistance may be shown by sister varieties with the same major gene complement. This may be illustrated by comparison of the winter wheat varieties Maris Beacon, Maris Huntsman and Maris Nimrod. These varieties were all obtained from the cross ((CI 12633 × Cappelle Desprez^5) × (Hybrid 46 × Cappelle Desprez)) × Professeur Marchal^2, and were selected for

Table 2. Infection of wheat varieties with yellow rust in the field (after Lupton & Johnson 1970) (% leaf area)

	Races 3/55 and 60 prevalent		Races 3/55, 58C and 60 prevalent
	1967	1968	1969
Browick	0·1	0	0·1
Cappelle Desprez	25	8	8
Little Joss	0–1	1–2	0·1
Maris Envoy	0	0	50
Maris Ranger	0	0	0
Maris Settler	0	0	25
Maris Widgeon	2–3	1–2	0·1
Rothwell Perdix	90	50	60

major gene resistance inherited from Hybrid 46 and Professeur Marchal. When Maris Beacon was first grown on a field scale, a multiplication field was found to contain a number of heavily-attacked foci of yellow rust. These were caused by a previously unknown race of *P. striiformis*, which was designated race 3/55D. This race was also capable of attacking seedlings of the two sister varieties, but adult plants of Maris Nimrod, and in particular of Maris Huntsman, showed an acceptable level of field resistance, as shown in Table 3 (Lupton & Bingham 1971). This was further demonstrated in an experiment in which artificial foci of infection were placed in the middle of large plots of these varieties. It was then found that the rust spread much

Table 3. Yellow rust development on winter wheat varieties (after Lupton & Bingham 1971) (% leaf area attacked)

	Cappelle Desprez	Maris Beacon	Maris Nimrod	Maris Huntsman
15 June	4	15	4	1
25 June	4	20	2	1
6 July	7	34	2	< 1
Grain yield (LSD, P = 0·05, 12·4)				
	100	87	117	115

more slowly in the plots of Maris Huntsman and Maris Nimrod than it did in those of Maris Beacon (Johnson *et al.* 1971).

Because of the dangers of new races of *P. striiformis* appearing on new varieties with major gene resistance, it is now the practice at the Cambridge Plant Breeding Institute to test all F_3 families for seedling rust reaction, and to retain only those which are susceptible to at least one physiologic race. These are then tested for rust reaction as adult plants, and those showing a reasonable level of field resistance are retained for selection on the basis of other characters. It is hoped that the resistance of varieties bred in this way will prove more stable than that of varieties with race-specific resistance. Some doubt was however thrown on this subject when severe attacks were observed in 1971 on the variety Joss Cambier, which had previously shown an acceptable level of field resistance although it was susceptible to several races at the seedling stage (Macer 1972).

Epidemiology of non-specific resistance

The widespread use of non-race-specific resistance has caused breeders to study more carefully the epidemiology of the diseases with which they are concerned, in order to find the most effective way of selecting for resistance. In some cases, a point of weakness may readily be found in the life cycle of the pathogen. When breeding for resistance to loose smut (*Ustilago nuda*) in barley, for example, it was found that some varieties are normally pollinated apogamously, while in others the lodicules expand, causing the flowers to open at the time of anthesis. As infection of loose smut is caused by spores which land on the stigmas at flowering time, normally apogamous varieties are rarely infected (Pedersen 1960).

A more detailed analysis is often necessary. Thus Russell (1972), working on resistance to downy mildew (*Peronospora farinosa*) of sugar beet, found that although a simply-inherited hypersensitive form of resistance to this disease is known, selection for resistant genotypes rapidly led to the appearance of a virulent physiologic race of the pathogen, so that the use of this form of resistance in a breeding programme would be undesirable. He therefore analysed the successive stages at which the life cycle of the fungus might be broken, investigating the possibility of selecting for resistance at each stage. He distinguished five stages at which this may be done.

First, he found that germination of conidia could be stimulated by an exudation from the leaves of the host and that the amount of exudation varied between genotypes. Secondly, there may be differences in resistance of the host to germ tube penetration, though such differences are more often due to differences in stage of crop ontogeny (the first-formed leaves being more susceptible than those formed later in the growth of the crop) than to differences between genotypes.

The third stage concerns rate of spread of the parasite within the host, which may be determined by varietal differences in the concentration of nutrients or of fungitoxic substances within the plant. Fourthly, varieties may show differences in tolerance of the pathogen, that is, differences in the extent of stunting or growth distortion, implying, in the case of resistant varieties, a less marked imbalance of growth substances in the host plant. Finally, some genotypes may show resistance to sporulation, this probably being associated with a lower concentration of certain sugars in the leaves.

Genotypes of sugar beet may show resistance at one or more of these five stages, so that the breeder has a wide range of characters amongst which to select. If an adequate background of resistance is assured, he may risk including the simply inherited hypersensitive resistance in the genotype of his new varieties, knowing that, should this fail, an adequate level of field resistance would remain.

It has been suggested that recent developments in the production of highly specific fungicides, often systemic in action, has materially changed the necessity for disease control by plant breeding (Brian 1972), and that the breeder might more usefully concentrate his efforts in other directions. There have, however, been numerous reports in the literature of the breakdown of control by such chemicals, because of the evolution of new physiologic races of the pathogens concerned (Wolfe 1971). Indeed, Wolfe suggests that resistance conferred by systemic fungicides, where the pathogen is exposed to decreasing concentrations of chemical for long periods of time, is more likely to be broken down by new races of pathogens than resistance to a non-systemic fungicide, where the pathogen is controlled by a brief exposure to a high concentration of chemical.

Integrated control

There remains, however, the possibility of integrated control, in which the effects of disease on a moderately resistant variety are ameliorated by applications of a suitable fungicide. Johnson *et al.* (1971) reported the use of this technique in the control of powdery mildew in a spring barley trial comprising varieties with contrasting levels of mildew resistance. They found that the effect of fungicide on mildew development was more pronounced in the more resistant than in the highly susceptible varieties, though the effects on yield were masked by the development of an epidemic of brown rust (*Puccinia hordei*). This developed most severely on the sprayed plots, possibly because the food reserves of the unsprayed plots had been depleted as a result of mildew attack. Similar results were found by Clifford *et al.* (1971) who compared the effects of increasing doses of the fungicide ethirimol on a range of barley varieties with different levels of mildew resistance. They found that, although the most resistant variety was least responsive to fungicide, it required less chemical to give virtually complete control. Few other cases of the combination of genetic resistance with chemical control of disease have been reported, but with the widespread exploitation of non-race-specific resistance, further examples of this type of integrated control may be anticipated.

Conclusion

The plant breeder has undoubtedly made a most valuable contribution to the control of plant disease, both in this country and elsewhere. But, as has been pointed out, some of the early concepts of the ease with which this could be done have proved illusory and in many cases it has been necessary to turn from simply inherited race-specific to more complex non-race-specific resistance. In all cases, however, when new diseases arise, sources of resistance must be sought either amongst world collections of cultivated varieties, or by sending expeditions to areas where related wild species may be found, or possibly by the induction of resistant mutations in susceptible varieties.

As modern agricultural techniques spread to ever-increasing areas of the world, and high-yielding varieties replace the indigenous forms previously cultivated, the conservation of the natural variability available in cultivated crops becomes a matter of increasing urgency. I would like, therefore, to conclude with a plea for greater attention to the conservation, in this country and abroad, of varietal collections of all the principal agricultural crops, and where possible of their related wild progenitors.

References

BIFFEN R.H. (1906) Mendel's laws of inheritance and wheat breeding. *J. agric. Sci. Camb.* 1, 4–48.

BLACK W. (1935) Studies on the inheritance of resistance to wart disease (*Synchytrium endobioticum*). *J. genet.* 30, 127–46.

BLACK W., C. MASTENBROEK, W.R. MILLS & L.C. PETERSEN (1953) A proposal for an international nomenclature of races of *Phytophthora infestans* and of genes controlling immunity in *Solanum demissum* derivatives. *Euphytica* 2, 173–79.

BORLAUG N.E. (1958) The use of multilineal or composite varieties to control airborne epidemic diseases of self-pollinated crop plants. *Proc. 1st Int. Wheat Genet. Symp., Winnipeg, Canada*, pp. 12–27.

BRIAN P.W. (1972) The metabolic background of disease resistance. *Proc. 6th Eucarpia Congress, Cambridge, 1971* (in press).

CLIFFORD B.C., I.T. JONES & J.D. HAYES (1971) Interactions between systemic fungicides and barley genotypes: their implications in the control of mildew. *Proc. 6th Br. Insectic. Fungic. Conf.* 287–94.

GASSNER G. & W. STRAIB (1932) Die Bestimmung der biologischen Rassen des Weizengelbrostes (*Puccinia glumarum* f. sp. *tritici* (Schmidt) Erikks. u. Henn.). *Arb. biol. Abt. (Anst.-Reichsanst.) Berl.* **20**, 141–163.

GLYNNE M.D. (1926) Wart disease of potatoes: the development of *Synchytrium endobioticum* (Schilb.) Perc. in immune varieties. *Ann. appl. Biol.* **13**, 358–59.

HOOKER A.L. & P.M. LE ROUX (1957) Sources of protoplasmic resistance to *Puccinia sorghi* in corn. *Phytopathology* **47**, 187–191.

HOWARD H.W. (1960) Potato cytology and genetics, 1952–59. *Bibliogr. Genet.* **19**, 87–210.

JOHNSON R., P.R. SCOTT & M.S. WOLFE (1971) Pathology. *Rep. Pl. Breed. Inst. 1970*, 100–110.

LINK K.P. & J.C. WALKER (1933) The isolation of catechol from pigmented onion scales and its significance in relation to disease resistance in onions. *J. biol. Chem.* **100**, 379–84.

LUPTON F.G.H. & R. JOHNSON (1970) Breeding for mature-plant resistance to yellow rust in wheat. *Ann. appl. Biol.* **66**, 137–43.

LUPTON F.G.H. & J. BINGHAM (1971) Winter wheat. *Rep. Pl. Breed. Inst. 1970*, 34–8.

MACER R.C.F. (1966) Resistance to eyespot disease (*Cercosporella herpotrichoides* Fron) determined by a seedling test in some forms of *Triticum*, *Aegilops*, *Secale* and *Hordeum*. *J. agric. Sci. Camb.* **67**, 389–96.

MACER R.C.F. (1972) The resistance of cereals to yellow rust. *Proc. roy. Soc. B.* (in press).

PEDERSEN P.N. (1960) Methods of testing the pseudo-resistance of barley to infection by loose smut, *Ustilago nuda* (Jens.) Rostr. *Acta agric. Scand.* **10**, 312–32.

PELHAM J., J.T. FLETCHER & J.H. HAWKINS (1970) The establishment of new strains of tobacco mosaic virus resulting from the use of resistant varieties of tomato. *Ann. appl. Biol.* **65**, 293–97.

RISSELL G.E. (1972) Components of resistance to disease in sugar beet. *Proc. 6th Eucarpia Congress, Cambridge, 1971* (in press).

VALLEGA J. (1959) Comportamiento de las variedades en cultivo frent e las variaciones de las pobliaciones parasites. *Robigo*, *8*, 7–9.

VAN DER PLANK J.E. (1963) *Plant Diseases: epidemics and control.* Academic Press, New York & London, 349 pp.

WOLFE M.S. (1971) Fungicides and the fungus population problem. *Proc. 6th Br. Insectic. Fungic. Conf.* (in press).

The biological contribution to weed control

G. W. CUSSANS *Agricultural Research Council, Weed Research Organization, Begbroke Hill, Oxford*

Reference to past work on this subject, notably the review by DeBach (1964), leaves no doubt that there is a potential for biological control of weeds and that some spectacular successes have already been achieved. The control of prickly pear cacti (*Opuntia* spp.) by the moth *Cactoblastis cactorum* and by cochineal insects (*Dactylopius* spp.) is probably the best known, followed by the control of St. John's Wort (*Hypericum perforatum*), mainly by the beetle *Chrysolina quadrigemina*.

However, there is a wide gulf between recognizing a potentiality for biological control and achieving the practical objective. As the proceedings of a recent international symposium clearly indicate (Simmonds 1969), a great deal of research is necessary to establish safety and suitability before introducing a new phytophage. The potential hazards of introducing such an agent may be greater than the potential hazard from a herbicide. Furthermore, there is, at present, very little dissatisfaction with the use of herbicides. The problem of resistant strains within susceptible species has not occurred with weeds although there has been an analogous problem of resistant species attaining dominance when populations of species more susceptible to herbicides have been reduced. Similarly, the use of herbicides has not generally been beset by many problems of toxicity to animals, persistence in food cycles, etc. Problems have occurred, however, and the very proper insistence on additional safety testing will, inevitably, increase the cost of developing new chemicals and may reduce the number of compounds entering commercial use.

It is necessary, therefore, to define rather carefully the situations in which biological control may supplement, or replace, chemical or mechanical methods of control.

Pre-requisites of success

To date, most biological control projects have been concerned with weeds of uncultivated land, often rangeland, although there has been an extensive programme of work on *Chondrilla juncea*, a weed of the wheat-growing areas of Australia.

Furthermore, the most successful agents of biological control have been those used to control introduced weed species. It has even been suggested by a number of authors that biological weed control is only appropriate when and where an introduced weed species achieves dominant status in the vegetation. However, as Huffaker (1964) points out, there is no biological reason why native weeds should not be equally or even more susceptible to introduced parasites or predators.

While previous workers may have taken too limited a view of the potentialities of biological control, it is suggested that any substantial increase in the biological contribution is dependent upon the following essential pre-requisites of success.

Economic dominance of a single species

In most cultivated land the annual flush of seedlings, in sufficient numbers to interfere with economic cropping, is produced by between 5 and 20 weed species. Fewer species occur on uncultivated land but, even here, one is more often concerned with the control of weed associations than with a single species. Furthermore, herbicide technology has responded to this problem quite adequately; 'broad spectrum' chemicals have been developed to deal with many weed associations. Conversely, the biological control of complete weed associations would be very difficult to achieve except in a few instances where non-selective suppression of herbage would suffice. Examples of the latter might include the use of sheep to graze between orchard trees and the use of grass carp (*Ctenopharyngodon idella*) for the control of water weeds.

There are circumstances, however, when a single weed species may attain dominance. This may be the result of selection pressures by herbicides or by management techniques; it can be due to a population explosion of an introduced species, or to a combination of any or all of these factors.

It is at this point that economics must be considered. A weed species which becomes dominant in this way may prove easy and cheap to control with a new compound or herbicide mixture. Alternatively, a weed may be difficult to control, and its numbers increase without becoming a threat to economic cropping. The build up of the field pansy (*Viola arvensis*) and the speedwells (*Veronica* spp.) in cereal crops in the U.K. are examples of this.

Only species which are both difficult to control and of serious economic importance can really be considered economically 'dominant' and therefore suitable subjects for the attempted application of biological control methods.

Economic considerations are vital but, whereas biological facts are, or should be, true for all time, economic facts are susceptible to change both with time and from country to country. The one general conclusion to be drawn is that the maximum scope for biological control exists in situations where application of herbicides is physically difficult and where the financial returns are likely to be low in relation to costs. These could include control of aquatic weeds and weed control in remote areas and in topographically rugged country.

Crops tolerant of continuing, low weed infestation

Although the level of control achieved biologically has often been remarkably complete, generally it is unusual for biological agents to achieve the high degree of short-term control expected from herbicides. More often the weed is suppressed or reduced over a long term. However, a level of weed suppression which may be quite adequate in highly competitive crops such as cereals and grassland may not be so in less competitive crops such as beans and sugar beet following in rotation. Ideally, but not essentially, a biological control agent should be introduced into a monoculture which provides a comparatively stable ecological background in which it can attain an equilibrium, for example, the rangeland in which so many successful introductions have been made. The biological reduction of the population of an economically significant weed to a low level may not be acceptable if the species is, at that level, toxic to livestock, e.g. ragwort (*Senecio jacobaea*) or causes taint of produce, e.g. wild onion (*Allium vineale*).

Weed status maintained in all situations

It is self-evident that, whereas chemical and cultural control methods are applied only to those areas where control is necessary, biological control agents must usually be expected to establish themselves wherever their food occurs. Studies of the safety of potential biological control agents should not, therefore, be confined to conventional crop species. Useful attributes of the target weed and of closely related wild species should also be considered. This can be a difficult and complex aspect of biological control. In Hawaii, for example, the introduction of *Cactoblastis cactorum* destroyed some ornamental cacti and the tree cactus (*Opuntia megacantha*) which was valuable as a source of cattle feed during drought. This caused considerable economic hardship.

The objections may not only be economic ones. It is suggested below that bracken (*Pteridium aquilinum*) could be a suitable subject for biological control. This plant may, however, be considered valuable or aesthetically desirable in some situations. Certainly the biological control of bracken would be deprecated by conservationists if there were also a risk of destruction of other ferns.

All the successful examples of biological weed control in cropped land reported in the literature do meet these three suggested pre-requisites. Other species should be selected on the same basis to ensure success. In the United Kingdom some examples would be, in grassland, bracken (*Pteridium aquilinum*) and, in arable land, wild oats (*Avena fatua* and *A. ludoviciana*) and the couch grasses (*Agropyron repens* and *A. gigantea*).

Agencies of biological weed control

The effects of other plants

Weed populations may be influenced by other plants directly by competition, or indirectly by the production of toxic exudates. The latter has not been considered very widely as a means of weed control but, in a recent symposium, Holm (1971) reviewed a number of cases of biochemical interactions between plants on agricultural land, and Hanawalt (1971) described in detail the inhibition of germination of annual plants by exudates from two species of manzanita shrub (*Arctostaphylos glauca* and *A. glandulosa*). Although there appears to be no immediate prospect of exploiting such biochemical phenomena, they may prove to be the basis of the equilibria obtaining in some existing weed associations.

The effect of competition by crop plants on weeds has been studied far more extensively. Salisbury (1942) showed that competition reduced seed production by *Ranunculus bulbosus*, and Harper & Gajic (1961) studied the plasticity of *Agrostemma githago* in response to crop competition. A similar effect was noted by Thurston (1962) on *Avena fatua*.

Work at the Weed Research Organization has shown that crop competition has an equally dramatic effect on rhizome production by *Agropyron repens* (Cussans 1968, 1970). In this case, not only is the total growth of rhizome affected; so, too, is the distribution of this growth during the season.

Further work has confirmed Thurston's results on *Avena fatua* and, with both these species, has demonstrated that the most important factor within the farmer's control is crop population. In a series of experiments, doubling the seed rate of barley from 80 to 160 lb/ac approximately halved the number of *Avena fatua* seed/unit area and the amount of *Agropyron repens* rhizome/unit area (Cussans & Wilson, paper in preparation). This was despite the fact that, in the absence of weeds, increasing the crop population had very little effect on crop yield.

Wells (1969) also found a marked inverse relationship between crop population and weed survival when he studied competition between lucerne and *Chondrilla juncea*. When lucerne density was adequate, this weed was very effectively controlled by breaking out of an alternate wheat/fallow system into two or three years of lucerne.

There must be many other instances in which plants, which are used or, at least, managed for the good of agriculture, have the effect of suppressing other plants which would otherwise become problem weeds. The use of shade plants and cover plants in tropical plantation crops would provide many such examples.

This subject has been reviewed at some length because the significance of these phenomena has possibly not been adequately appreciated in the context of biological control. However, some of the difficulties inherent in their exploitation are considerable and, in some ways, illustrative of the difficulties of biological control in general. For example, increasing the population of cereal crops may lead to increased severity of diseases, notably mildew (*Erysiphe graminis*). Yet again, the effectiveness of barley in suppressing weeds must be, in some part, due to the rapid production of a very high tiller population. Many of these tillers make no contribution to the final yield but die after having attained considerable size, the number dying being greatest at high plant populations. For this reason, Bingham (1970) believed there was a good prospect for increasing potential yield in cereals by breeding for dwarf habit and for greater economy of tiller production. Such development could reduce the effectiveness of the barley crop as a weed suppressor; alternatively, manipulating the crop to achieve more effective weed control might conflict with attaining the maximum yield potential.

The effects of other biological agencies

Insects have been the organisms most widely exploited for biological weed control although other arthropods have also been used, e.g. the spider mite *Tetranychus desertorum*. Fungi and viruses also have a great potential for use and one rust, *Puccinia chondrillina*, has been introduced into Australia for the control of *Chondrilla juncea*.

At one time there was vigorous debate on the type of predator most likely to be a successful control agent. Stem-boring and gall-forming insects were, for example, considered superior to leaf-eating types. Huffaker (1964) considered it unwise to prejudge the merits of potential control agents in this way and the comparable success of predators of widely differing feeding behaviour supports this. Different agents of control may, of course, act together. However, the control of weeds by animal organisms and plant pathogens is, broadly, effected in three ways.

Attack upon the vegetative organs

This has been the most widely studied category and all the examples given earlier fall into it. Organisms may have a direct effect on the plant or, by retarding growth, may render the host weed more vulnerable to crop competition. There may be a further interaction in that the physical damage caused by animal organisms may lead to invasion by secondary parasites such as bacterial soft rots. Dodd (1940) considered such secondary parasitism to be an important aid to control of *Opuntia* spp. by *Cactoblastis*.

To survive at all, an introduced organism must be reasonably adapted to the weed's environment but, for optimum weed control, the life history of the organism must also be synchronized with the growth phases of the plant. Close synchronization is believed to have been the reason for the success of *Chrysolina quadrigemina* compared to *C. hyperici* in controlling *Hypericum perforatum* in North America (Holloway 1964).

Clearly this need for synchronization is even greater with biological agents in the second category.

Attack upon flowers and developing seeds

These organisms have been exploited to a lesser extent but they have been studied extensively (Zwölfer 1970) and do provide a source of extremely specific predators. One disadvantage is that weed competition has occurred before the predator intervenes. Another is that perennial plants are seldom very dependent on spread by seed. However, any reduction or prevention of seed spread is valuable, even in perennial species. It is suggested that this type of predator may be ideally suited for use in association with partially effective chemical herbicides. Potential seed production, already reduced by the herbicide, could be further reduced by the phytophage.

Attack upon shed seed and separated propagules

Sagar (1970), reviewing the factors controlling the size of plant populations, pointed to the loss of population potential which continually occurs. Many weed plants are capable of producing thousands of seeds. *Agropyron repens* can, in a season, multiply its rhizome weight at least 300-fold in the absence of competition and some 5- to 12-fold in the presence of a crop. These high potential increases are, however, matched by considerable natural loss of propagules, so that the actual population increases are much lower.

A study of these natural sources of population loss should be fruitful in leading to an increased biological contribution to control. For example, in our studies, losses of seed of *Avena fatua* in excess of 95% have been recorded and, of this loss, a very high proportion occurred in the first 3–4 months after shedding from the parent plant. It is believed that a number of factors, including biological agents, is involved here and studies are proposed. proposed.

The rhizome of *Agropyron repens* is quite resistant to microbial attack while undisturbed but, after damage by cultivation, rotting does occur and may make an important contribution to ultimate control. Whether the responsible organisms could usefully be stimulated by some means other than cultivation is not known.

Interaction of agencies

Reference has already been made to some interactions between primary and secondary biological agents, and between an agent and a competing crop. These must always be considered carefully if only because, in practice, they may already be occurring. Goeden *et al.* (1967) instance the control of *Opuntia littoralis* and *O. oricola*, two native prickly pears on the island of Santa Cruz, by the introduction of the cochineal insect, *Dactylopius opuntiae*. These had become serious pests as a result of wild sheep overgrazing and denuding the natural range vegetation of the normally competitive grass components. The introduction of *Dactylopius* was also accompanied by destruction of the wild sheep so that the restored grass swards not only allowed of more profitable beef production but curbed the recovery of the prickly pear.

It is not unreasonable to assume that, in almost every weed-crop situation, there must be some interaction between the effects of the weed control agents, the natural population regulation mechanisms, and crop and weed competition. The problem is to recognize these interactions, in order that they can usefully be exploited.

Conclusions

Although a complete review of this vast subject has not been attempted, it is suggested that certain conclusions, not necessarily in agreement with those of previous writers, may justifiably be drawn.

First, any substantial increases in the biological contribution to weed control will be dependent upon three essential pre-requisites of success: 1, the economic dominance of a single species; 2, crops tolerant of a continuing, low weed infestation; and 3, weed status maintained in all situations.

Second, the comparative success of equally effective biological, chemical and cultural means of weed control will be largely determined by their cost and logistic problems.

Third, it follows that the number of situations when biological agents can make a technically and economically successful contribution to weed control will be limited.

Fourth, when a weed is selected for biological control, the necessary preliminary research must include the biology of the weed and *all* the natural constraints on population increase. The construction of a population dynamics model is likely to be helpful.

References

BINGHAM J. (1971) Plant breeding: arable crops. In *Potential Crop Production*, eds. P.F. Waring and J.P. Cooper, 273–294. Heinemann, London, 387 pp.

CUSSANS G.W. (1968) The growth and development of *Agropyron repens* (L.) Beauv. in competition with cereals, field beans and oilseed rape. *Proc. 9th Br. Weed Control Conf.* 131–136.

CUSSANS G.W. (1970) A study of the competition between *Agropyron repens* (L.) Beauv. and spring sown barley, wheat and field beans. *Proc. 10th Br. Weed Control Conf.* 337–43.

DEBACH P., ed. (1964) *Biological Control of Insect Pests and Weeds.* Chapman & Hall, London, 844 pp.

DODD A.P. (1940) *The Biological Campaign against Prickly Pear.* Commonwealth Prickly Pear Board, Brisbane, 117 pp.

GOEDEN R.D., C.A. FLESCHNER & D.W. RICKER (1967) Biological control of prickly pear cacti on Santa Cruz Island, California. *Hilgardia* 38, 579–606.

HANAWALT R.B. (1971) Inhibition of annual plants by *Arctostaphylos. Proc. Nat. Acad. Sci. Symp. on Biochemical Interactions Among Plants.* Washington D.C. 1971, 33–8.

HARPER J.L. & D. GAJIC (1961) Experimental studies of the mortality and plasticity of a weed. *Weed Res.* 1, 91–104.

HOLLOWAY J.K. (1964) Projects in biological control of weeds. In *Biological Control of Insect Pests and Weeds*, ed. P. DeBach. Chapman & Hall, London, 650–70.

HOLM L. (1971) Chemical interactions between plants on agricultural lands. *Proc. Nat. Acad. Sci. Symp. Biochemical Interactions Among Plants.* Washington D.C. 1971, 95–101.

HUFFAKER C.B. (1964) Fundamentals of biological weed control. In *Biological Control of Insect Pests and Weeds*, ed. P. DeBach. Chapman & Hall, London, 631–48.

SAGAR G.R. (1970) Factors controlling the size of plant populations. *Proc. 10th Br. Weed Control Conf.* 965–79.
SALISBURY E.J. (1942) *The Reproductive Capacity of Plants.* Bell & Sons, London, 244 pp.
SIMMONDS F.J., ed. (1969) *Proc. 1st Internat. Symp. Biological Control of Weeds.*
THURSTON J.M. (1962) The effect of competition from cereal crops on the germination and growth of *Avena fatua* L. in a naturally-infested field. *Weed Res.* **2**, 192–207.
WELLS G.J. (1969) Skeleton weed (*Chondrilla juncea*) in the Victorian Mallee 1. Competition with legumes. *Aust. J. exp. Agric. Anim. Husb.* **9**, 521–7.
ZWÖLFER H. (1970) Current investigations on phytophagous insects associated with thistle and knapweeds. *Proc. 1st Internat. Symp. Biol. Control of Weeds*, 63–7.

Diversification of crop ecosystems as a means of controlling pests

J. P. DEMPSTER *The Nature Conservancy, Monks Wood Experimental Station, Huntingdon*
T. H. COAKER *Department of Applied Biology, University of Cambridge*

Many ecologists have noted that there appears to be greater stability in populations of animals in complex ecosystems than in more simple ones (e.g. Voûte 1946, MacArthur 1955, Elton 1958). The evidence for this is not experimental but is based on four broad generalizations: (1) theoretical and laboratory studies of the interaction between an animal population and that of a single enemy species show great instability; (2) simple communities on islands appear to be colonized more easily by invading species than complex communities in continental areas; (3) in the extremely diverse tropical forests, animal numbers fluctuate far less violently than in the simpler ecosystems of coniferous forests, and (4) in natural ecosystems outbreaks of species are far less common than in more simple agro-ecosystems. However, the importance of diversity has not gone unchallenged, which is not surprising, since the concepts of diversity and of stability can be interpreted in a number of ways (Turnbull & Chant 1961, Zwölfer 1963, Watt 1964, 1965). Southwood & Way (1970) suggest that only 'trophic diversity', i.e. a large number of species at a higher trophic level of a food web, is likely to promote stability at the lower trophic level. In practice it is not stability that we require in pest populations so much as their maintenance below the level at which economic damage occurs. In this, diversity may play an important role, since the greater the number of subtractive factors affecting a population, the smaller its size is likely to remain. Any inadequacies of one factor may be made good by the action of others.

Besides this numerical effect of diversity, there is another potential advantage of it in agro-ecosystems. Smallman (1965) pointed out that a diversity of selection pressures acting on a species will reduce the chance of a pest getting out of control by developing resistance to any one control agent, such as an insecticide. This may well prove to be a far stronger argument for diversity in crops than that of stabilizing pest numbers.

Modern agricultural practices reduce the diversity within the crop ecosystem by tending towards monocultures of single plant species from which weed species are eliminated. Crop plants also tend to be of uniform age and character (genotype) which leads towards their synchronized development. The crop ecosystem is further simplified by the use of pesticides and, whether or not the dependence of stability on diversity is accepted, there have been many examples of pest outbreaks caused by the elimination of natural enemies by pesticides. Probably the best-known example of this is the increased pest status of the fruit tree red spider mite (*Panonychus ulmi*) which has resulted from orchard spray programmes (Collyer 1953). This type of effect is not confined to insecticides; for example, the recent upsurge of certain facultative pests of sugar beet probably owes its origin to the efficient elimination by herbicides of weeds which serve as alternative food plants for these pests (Dunning 1971). The need to procure maximum seedling survival from lower-density precision sowing of the crop, than was obtained before monogerm seed was available, also adds to the difficulties arising from the uniformity of modern agricultural practices. There are invariably sound practical and economic reasons for this trend towards decreased diversity within crops and the demand for increased productivity will undoubtedly ensure its continuance. On the other hand, changes in agricultural practice have marked effects on insect numbers and small increases in diversity might be used within an integrated approach to pest control. This diversity can be sought at two levels: first, by diversifying the habitats surrounding the crop; and secondly, by diversifying the crop habitat itself.

Attempts to show whether the presence of non-crop habitats adjacent to crops is beneficial or detrimental to pest control have been inconclusive (Lewis 1965, van Emden 1965). Wild habitats may act as reservoirs for both pests and their enemies, while sheltering effects of windbreaks, such as hedges, may concentrate airborne insects onto a crop. Pollard (1971) has shown that the influence of wild habitats on the syrphid predators of the cabbage aphid (*Brevicoryne brassicae*) on neighbouring crops is extremely small, probably because the syrphids are such active fliers that they disperse over far greater distances than between the immediate surrounding vegetation and the adjacent crop. Insects vary considerably in their powers of dispersal, and the distance travelled will determine the impact that the proximity of wild habitat will have on the numbers of pest, or beneficial species, in the crop. Most species are probably too mobile for pest control to be achieved by the management of habitats surrounding the crop. For some pest species there may, therefore, be a better chance of controlling their numbers, and the damage that they cause, by increasing diversity within the crop itself. Any modification within the crop must be compatible with current agricultural practice, but quite small changes in the method of growing the crop can sometimes greatly alter its attractiveness to a pest and its enemies.

Experiments with brassica crops

During the past two years, the possibility of modifying the way in which brassica crops are grown to make them less suitable for insect pests has been studied. This work arose from earlier findings that the presence of weed within a crop greatly reduced the incidence of cabbage aphid (Smith 1969) and the cabbage caterpillar (*Pieris rapae*) (Dempster 1969). In addition, Pimentel (1961) had shown that pest infestation levels per plant were lower on dense plant populations than on sparse ones. Demptser (loc. cit.) found, however, that any advantage in terms of pest numbers in the presence of weed was outweighed by competition effects which reduced yield, and suggested that undersowing the crop with clover might provide the advantages of weed cover without competition with the crop. In 1970 an experiment was made at Monks Wood Experimental Station to test this suggestion. An area of 3,600 m^2 was ploughed, rotovated and divided into four equal plots (A–D). Two of these (A and C) were sown with Kersey White Clover at a rate of 22·5 kg/ha (20 lb/ac) on 14 May. The centre of each plot was planted with 100 (10 × 10) Brussels sprout plants, var. Ashwells Dark Green, on 23 May, before the clover had germinated. The sprout plants were spaced at 1 m intervals on plots A and D and at 0·56 m (21 inch) intervals on plots B and C. Plots B and D were kept free from weed by hoeing between the sprout plants, and the surrounds of the plots by spraying with a total herbicide. The clover failed to keep down perennial weeds such as creeping thistle and these were cut to about 5 cm twice during the summer.

Throughout the summer, counts were made twice a week on the number of eggs and larvae of *Pieris rapae* on 20 plants per plot. Ground-living predators were studied by means of 25 pitfall traps containing 2% formalin arranged in a 5 × 5 grid over each plot.

Table 1. *Pieris rapae* larvae on Brussels sprouts under different treatments in 1970

	A Clover 1 m spacing	D Hoed 1 m spacing	C Clover 0·56 m spacing	B Hoed 0·56 m spacing
Eggs/plant (Mean ± SE)	4·55 ± 0·78	2·65 ± 0·56	7·65 ± 2·01	3·20 ± 0·73
Survival to instar III (%)	27·2	42·6	22·9	43·3

The distribution of eggs laid by *Pieris* was patchy, but there was no significant difference between the numbers laid on the different plots (Table 1). The larger number laid on plot C was the result of eggs being laid particularly heavily on two plants (36 and 28 eggs, respectively). As the older caterpillars of *Pieris* do most damage to the crop, survival beyond the third instar is important in determining pest status. It can be seen from Table 1 that survival of larvae to the third instar was lower on those plots planted with clover (A and C). Mortality of the young caterpillars of *Pieris* is mainly due to arthropod predators (Dempster 1967, 1968). By far the most important of these are a ground beetle, *Harpalus rufipes*, and a harvest spider, *Phalangium opilio*. Both of these are nocturnal, spending the day in the soil and litter under the plants, and both are more abundant in weedy crops (Dempster 1969). The numbers of these were significantly greater in pitfall trap catches from the clover plots (A and C) than from the hoed plots (B and D) (Table 2).

Table 2. Ground-living predators in Brussels sprouts plots in 1970 (mean nos in 25 pitfall traps per day ± SE)

	A Clover 1 m spacing	D Hoed 1 m spacing	C Clover 0·56 m spacing	B Hoed 0·56 m spacing
Phalangium opilio	6·00 ± 1·1	2·6 ± 0·5	4·1 ± 0·6	0·6 ± 0·2
Harpalus rufipes	26·5 ± 2·6	9·9 ± 1·1	22·0 ± 1·7	15·8 ± 1·2
Bembidion spp.	2·2 ± 0·5	6·3 ± 0·6	2·4 ± 0·4	9·8 ± 1·0
Feronia melanaria	4·4 ± 0·6	1·9 ± 0·4	5·3 ± 1·0	1·3 ± 0·2
Staphylinidae	17·8 ± 1·2	3·1 ± 0·8	13·4 ± 1·5	7·5 ± 1·1

Pitfall traps do not give an accurate measure of population size since their catch depends upon the activity of the animals and this is affected by the type of plant cover on an area. The presence of clover is likely to have impeded movement and in consequence reduced the numbers trapped on those plots. Activity is also dependent upon microclimate and this will also have been affected by plant cover. Although pitfalls give a poor measure of population size, they probably give a reasonable measure of predatory activity, since this is also dependent upon both numbers and mobility of the predators. There is little doubt that the differences in survival of *Pieris* larvae, recorded in Table 1, reflect differences in the activity of these predators, especially *Harpalus*.

Two counts of the cabbage aphid (*Brevicoryne brassicae*) were made during the summer. Numbers were extremely small on all plots and there were no significant differences between treatments. The numbers of the immature stages of cabbage root fly (*Erioischia brassicae*) were not studied, but examination of the roots of ten plants per plot on 23 December gave a somewhat higher root-damage index (Rolfe 1969) on the hoed plots (Table 3). This index would include damage mostly from the second and third generations of the fly, as earlier damage is often difficult to recognize as late in the year as this.

Yield was assessed in December by cutting 10 plants at ground level from one diagonal on each plot and comparing their fresh weights (Table 3).

Table 3. Cabbage root fly damage and crop yields in Brussels sprouts under different treatments in 1970

	A Clover 1 m spacing	D Hoed 1 m spacing	C Clover 0·56 m spacing	B Hoed 0·56 m spacing
Mean cabbage root fly damage index	35·7	49·2	32·3	43·5
Mean fresh wt/plant (kg)	2·19 ± 0·20	1·26 ± 0·20	1·81 ± 0·20	1·18 ± 0·15

The plants from the clover plots were significantly heavier than those from the hoed plots. There was little difference in the quality of the sprouts from the different treatments, except that those from the clover plots were larger. It is difficult to explain the greatly improved yield from the clover plots, unless it was due to differences in cabbage root fly damage during the earlier stage of growth of the plants.

In 1971 the experiment was repeated at the University of Cambridge Farm. As before, an area of 3,200 m² was ploughed, cultivated and divided into four equal-sized plots. Two of these (A and C) were sown with Broad-leaf Red Clover (22·5 kg/ha) on 16 April. Red clover having a more rapid growth than Kersey White was used in an attempt to get early cover of the soil, in time for the first generation of cabbage root fly. In the centre of each plot 110 (10 × 11) sprout plants, var. Cambridge No. 5, were planted at 1 m spacing and 110 (10 × 11) cauliflower plants, var. All Year Round, were planted at 1 m × 0·5 m spacing. In each plot three rows (27 plants) of both brassicas were treated with a standard root drench of chlorfenvinphos against cabbage root fly. The clover had fully emerged at the time of planting the brassicas, but it did not provide full cover until three weeks later. Plots B and D were kept free from weeds by hoeing and by spraying the surrounds with herbicide.

No *Pieris rapae* occurred on any of the sampled plants by the end of July, when searches were discontinued. Immigrant alatae of *Brevicoryne brassicae* were trapped in seven yellow water-traps per plot between 18 June and 15 July. Very many more were caught on the hoed plots than on those covered with clover (Table 4). A single count of aphids on 10 plants/plot on 30 July showed significantly more on the hoed plots.

Cabbage root fly eggs were counted around five plants on each plot once a week (Hughes & Salter 1959). Significantly more eggs per plant were found on the bare plots than on the clover ones (Table 5) during the sampling period, which covered the first and part of the second generation oviposition periods. The different levels of cabbage root fly attack up to the time of harvesting the cauliflower crop are shown by the root damage indices in Table 5. These were lower on the clover plots (A and C) than on the hoed ones. The insecticide treatment gave a far better protection on the clover plots than on the bare plots, but why this should be so is unknown. It may have been due to chlorfenvinphos having greater biological effectiveness in the moister soil under the clover, or loss of insecticide by volatilization may have been lower on these plots.

Table 4. *Brevicoryne brassicae* on Brussels sprouts under different treatments in 1971

	C Clover	A Clover	D Hoed	B Hoed
No. alatae/7 traps, 18 June–15 July	91	84	349	344
Mean no./plant, 30 July (derived from log (n + 1) aphids/plant)	0·7 ± 0·9	0·2 ± 0·2	9·4 ± 5·9	12·0 ± 7·3

Table 5. Cabbage root fly on Brussels sprouts and cauliflowers under different treatments in 1971

	C Clover	A Clover	D Hoed	B Hoed
Mean no. eggs/plant, 5 May–14 July	169·6 ± 32·1	161·6 ± 33·5	208·2 ± 33·2	257·0 ± 23·5
Root damage index on 28 June at harvest of cauliflowers without insecticide	65·7	68·2	94·0	91·2
With insecticide	9·5	10·0	36·9	36·6

Ground-living predators were studied between 5 May and 14 July by means of twenty pitfall traps placed 1 m apart in a double row across the centre of each plot, 15 cm from a brassica plant. The numbers of the commonest species caught are shown in Table 6. *Bembidion* was trapped in larger numbers on the hoed plots, but there was no difference in the numbers of the staphylinids *Aleochara* spp. Trapping stopped on 14 July, before *Harpalus* and *Phalangium* occurred in any number on the plots.

Red clover grows very much more vigorously than white clover, and by

Table 6. Ground-living predators in Brussels sprouts and cauliflower plots under different treatments in 1971 (mean nos in 25 pitfall traps per day ± SE)

	C Clover	A Clover	D Hoed	B Hoed
Bembidion spp.	12·1 ± 2·8	13·4 ± 3·1	21·1 ± 4·7	16·2 ± 4·8
Trechus spp.	1·9 ± 0·2	2·5 ± 0·5	1·6 ± 0·4	0·8 ± 0·2
Feronia spp.	3·5 ± 0·7	5·4 ± 0·5	0·9 ± 0·2	1·4 ± 0·7
Agonum dorsale	0·6 ± 0·2	1·2 ± 0·3	0·6 ± 0·3	0·3 ± 0·2
Aleochara spp.	9·3 ± 1·2	10·9 ± 2·7	9·4 ± 0·9	8·2 ± 1·2

Table 7. Yields from cauliflowers and Brussels sprouts under different treatments in 1971

	C Clover	A Clover	D Hoed	B Hoed
Cauliflowers harvested 14–30 June (curd diameter × no)				
without insecticide	53	76	126	136
with insecticide	79	107	131	147
Brussels sprouts harvested 8 August (mean fresh wt, kg/plant)				
without insecticide	0·57 ± 0·14	0·58 ± 0·14	2·39 ± 0·54	2·38 ± 0·62
with insecticide	1·74 ± 0·26	0·96 ± 0·11	2·42 ± 0·60	2·59 ± 0·54

mid-summer it had swamped the brassica plants. The latter showed signs of stress from competition with the clover by mid-June, and this is reflected in the very poor crop yields from these plots (Table 7). The yield from both the sprouts and cauliflowers was greatly reduced. After the harvesting of the cauliflowers in late June, the plants were replaced with sprouts on 22 June, after the clover had been cut to ground level. These also failed after being swamped by regrowth of the clover.

Discussion

Although different clovers were used in the two experiments, similar results were obtained in terms of pest incidence. The number of eggs laid by *Pieris rapae* was not significantly altered, but the number laid by cabbage root fly was greatly reduced by the clover. The number of alate *Brevicoryne* entering the crop was also greatly reduced by the presence of clover. These findings for *Pieris* and *Brevicoryne* agree with earlier findings on the effects of weed in the crop (Dempster 1969, Smith 1969). The numbers of all three pests were reduced by the clover. In the case of *Pieris* this was due to an increased number of predators, particularly *Harpalus rufipes*, which increased early larval mortality on the clover plots. The most important predators of cabbage root fly are carabids and staphylinids (Coaker & Williams 1963) and while the numbers of some of the former were reduced by the clover, the larger species of carabid were increased. Added to this there appeared to be an increased effectiveness of the chlorfenvinphos root drench on the clover plots.

These experiments do not provide any evidence on the role of diversity on the stability of pest populations. Observations over many years would be required for this. They do, however, suggest that the numbers of *Pieris*, *Brevicoryne* and *Erioischia* can be reduced by adding diversity of this sort to the crop. The clover was grown under the crop plants to provide shelter for ground-living predators and to reduce the contrast between the crop plants and the bare soil. The larger number of predators in the more diverse plots reduced survival of *Pieris* caterpillars, but with *Brevicoryne* and *Erioischia* any changes in survival could not be separated from those of immigration. In the case of the aphid, the larger number on the hoed plots was probably mainly due to larger numbers of immigrants on those plots. There may, however, have been some difference in the survival and/or reproduction, since there were bigger differences in the numbers per plant than in the numbers caught in the water traps. The smaller number of *Erioischia* on the clover plots could have been due to a higher rate of predation on these plots, but more information is required to show what effect clover has on the ovipositing flies.

The main difficulty to be overcome is the competition between clover and the crop. In this respect, the lower-growing white clover has clear advantages over red. On the other hand, the performance of white clover on different soils and in different years needs to be studied before the value of this approach to pest control can really be assessed.

References

COAKER T.H. & D.A. WILLIAMS (1963) The importance of some Carabidae and Staphylinidae as predators of the cabbage root fly *Erioischia brassicae* (Bouché). *Entomologia exp. appl.* 6, 156–64.

COLLYER E. (1953) Insect population balance and chemical control of pests: Predators of the fruit tree and red spider mite. *Chem. & Ind.* 1044–6.

DEMPSTER J.P. (1967) The control of *Pieris rapae* with DDT. I. The natural mortality of the young stages of *Pieris*. *J. appl. Ecol.* 4, 485–500.

DEMPSTER J.P. (1968) The control of *Pieris rapae* with DDT. II. Survival of the young stages of *Pieris* after spraying. *J. appl. Ecol.* 5, 451–62.

DEMPSTER J.P. (1969) Some effects of weed control on the numbers of the small cabbage white (*Pieris rapae* L.) on Brussels sprouts. *J. appl. Ecol.* 6, 339–45.

DUNNING R.A. (1971) Changes in sugar beet husbandry and some effects on pests and their damage. *Proc. 6th Br. Insectic. Fungic. Conf.* 1–8.

ELTON C.S. (1958) *The Ecology of Invasions by Animals and Plants.* Methuen, London.

VAN EMDEN H.F. (1965) The role of uncultivated land in the biology of crop pests and beneficial insects. *Scient. Hort.* 17, 126–36.

HUGHES R.D. & D.D. SALTER (1959) Natural mortality of *Erioischia brassicae* (Bouché) (Diptera Anthomyiidae) during the immature stages of the first generation. *J. Anim. Ecol.* 28, 231–41.

LEWIS T. (1965) The effects of shelter on the distribution of insect pests. *Scient. Hort.* 17, 74–84.

MACARTHUR R.H. (1955) Fluctuations of animal populations and a measure of community stability. *Ecology* 35, 533–6.

PIMENTEL D. (1961) The influence of plant spatial patterns on insect populations. *Ecology* 54, 61–9.

POLLARD E. (1971) Hedges VI. Habitat diversity and crop pests: A study of *Brevicoryne brassicae* and its syrphid predators. *J. appl. Ecol.* 8, 751–80.

ROLFE S.W.H. (1969) Co-ordinated insecticide evaluation for cabbage root fly control. *Proc. 5th Br. Insectic. Fungic. Conf.* 238–43.

SMALLMAN B.N. (1965) Integrated insect control. *Aust. J. Sci.* 28, 230–4.

SMITH J.G. (1969) Some effects of crop background on populations of aphids and their natural enemies on Brussels sprouts. *Ann. appl. Biol.* 63, 326–30.

SOUTHWOOD T.R.E. & M.J. WAY (1970) Ecological background to pest management, pp. 6–28, in *Concepts of Pest Management*, eds. R.L. Rabb and F.E. Guthrie. University of North Carolina Press, Chapel Hill.

TURNBULL A.T. & D.A. CHANT (1961) The practice and theory of biological control of insects in Canada. *Can. J. Zool.* 39, 697–753.

VOÛTE A.D. (1946) Regulation of the density of the insect populations in virgin forests and cultivated woods. *Arch. neerl. Zool.* 7, 435–70.

WATT K.E.F. (1964) Comments on fluctuations of animal populations and measures of community stability. *Can. Ent.* 96, 1434–42.

WATT K.E.F. (1965) Community stability and the strategy of biological control. *Can. Ent.* 97, 887–95.

ZWÖLFER H. (1963) Untersuchungen uber die Struktur von Parasitenkomplexen bei einigen Lepidopteren. *Z. angew. Ent.* 51, 346–57.

The use of pathogens in the control of insect pests

T. W. TINSLEY AND P. F. ENTWISTLE
Natural Environmental Research Council, Unit of Invertebrate Virology, University of Oxford

Biological control, involving the use of parasitic and predatory organisms to control pest species, has, until recently, been based largely on the exploitation of entomophagous insects. A range of different pathogenic micro-organisms is now available and this has prompted Burges & Hussey (1971) to suggest that conventional biological control could be divided into macrobiological control (parasites and predators) and microbiological control (pathogens). This paper is concerned with the second category, where pathogenic agents are used in biological control programmes. Size largely divides macro- and microbiological agents, but functionally the first group has host-searching abilities whereas the association of pathogens with their hosts often occurs by chance. Nematodes are usually included with the microbiological agents partly because of their small size, and partly because of their limited free movement and host-searching ability.

Microbial control agents

Nematodes

Nematode infections of insects are widespread and, though lethal effects are common, some non-lethal associations can be effective as they lead to the sterilization of the host. A specific symbiotic relationship has been recorded between *Neoplectana carpocapsae* and a bacterium, *Achromobacter nematophilus*, whereby the nematode moves from the gut into the host haemocoele, and then releases bacteria which, after multiplication, provide conditions suitable for the further development of the nematode. Many nematodes have free-living stages but they do not constitute a major natural mortality factor despite such survival mechanisms. The potential of nematodes in practical applications has been little investigated.

Fungi

The entomophagous habit is widespread in the fungi, notably in the Hyphomycetes (Fungi Imperfecti), in such genera as *Beauveria*, *Metarrhizium*, *Aspergillus*, *Spicaria* and *Isaria*. The importance of these fungi as mortality factors in the field has been demonstrated many times but the effects are sporadic, and in practice their use is limited because most require a humidity of 80% or more for spore germination, and the development of epizootics is thus very dependent on climatic factors. On the other hand, fungi are amenable to cultivation on artificial media and to mass production.

Protozoa

In general, protozoan infections in insects develop slowly and though certain species can produce severe effects on their hosts most do not. Protozoan infections are known to reduce host fertility and longevity and have the additional advantage of being transmitted with the egg. A measurable degree of control, natural and induced, has been described for seven species (McLaughlin 1971). Microsporidia numerically dominate the economically useful protozoa.

Bacteria

Epizootics of bacterial disease may be key factors in population regulation and often produce significant mortality effects. The use of bacteria in control programmes, and in particular of those species that produce toxins, is very firmly established. The field has been dominated by *Bacillus thuringiensis*, which is spore-forming and crystalliferous and is active against a wide spectrum of lepidopterous insects. *Bacillus popilliae* and *B. lentimorbus* have been used widely against Japanese beetle (*Popillia japonica*) in N. America.

Rickettsiae

These are obligate intracellular parasites with a general bacilliform morphology but they frequently exhibit pleiomorphism. Rickettsiae that are pathogenic in insects are very stable outside the host, unlike most of those carried by vectors to vertebrate hosts. Little work has been done with these organisms but some have been shown to exert significant control of scarabaeid beetles.

Viruses

The pathogenic viruses of insects can be conveniently divided into two groups. In the polyhedral viruses the virus particles are occluded within a proteinaceous inclusion body. These inclusions can contain from one to many virions in each polyhedron according to type. In the other important group, the 'non-occluded' viruses, the virions are not associated with inclusion bodies. DNA and RNA viruses are represented in both groups.

Occluded viruses are often very stable outside the host and may have a 'shelf life' of many years. They also have considerable field persistence in protected situations, such as the soil. On the other hand, the non-occluded viruses are generally considered to be less stable in natural situations. Some hundreds of possibly different viruses have been isolated, though regrettably few have been adequately characterized. Present experience suggests that the natural host range of these viruses is fairly restricted, and indeed some appear to be practically host-specific. This is in marked contrast to the fungi where the host range can include a wide range of insect orders and indeed, on occasion, other animal phyla. This host-specificity of viruses may be an important feature in field use. A disadvantage compared to fungi or bacteria is that viruses require living hosts for replication, but it is possible that insect cell cultures may yet solve the problem of large-scale production.

We consider that viruses offer greatest potential as microbiological control agents, as their host range is restricted and it is relatively easy to obtain large quantities of highly-purified and characterized virus. This is of vital importance for field applications, as the presence of other agents of unknown pathogenicity is highly undesirable.

Hazards in microbiological control

To what extent cross transmission of pathogens occurs between (a) unrelated insects, and (b) between insects and other animals is uncertain. Certain entomophagous fungi, e.g. some *Aspergillus* and *Beauveria* species and *Entomophthora coronata*, are known to infect vertebrates. Allergic responses have been recorded after exposure to the spores of *Beauveria bassiana*. Microsporidal infections are common to vertebrates and arthropods, including insects, and Heimpel (1971) comments 'It is highly doubtful whether species identification is sufficiently reliable to forego thorough testing of insect microsporidal pathogens before exposing vertebrates to these organisms.' Certain bacteria, e.g. *Pseudomonas aeruginosa* and *Serratia marescens*, and Rickettsiae, are certainly cross infective to vertebrates and could probably never be used with complete safety. The host range of viruses, however, as mentioned already, is probably very much more restricted. For instance, there is only slight evidence of cross-transmission of viruses from a phytophagous host to a predator (Sidor 1960, Smith 1967). There are no records of infection of vertebrates by insect viruses, despite extensive injection of many viruses into a variety of test animals (e.g. guinea-pigs, rabbits, mice, chickens and frogs) by a variety of routes (intraperitoneal, intravenous, subcutaneous and intramuscular) (Heimpel 1971). However, one cannot say that all viruses can be considered safe in all respects. Many morphological similarities exist within groups of viruses found in invertebrate and vertebrate animals as the pox-viruses and the rhabdoviruses exemplify. As some of the individual viruses in these groups are highly pathogenic to man and other vertebrates, their similarity in size and shape to certain viruses isolated from insects should dispel any attitude of complacency that may exist. We feel strongly that any virus candidate for use in microbiological control should be thoroughly characterized for pathogenicity to other animals. It also follows that its issue for general application should be subject to conformity to pre-agreed criteria. Precisely what these criteria should be has yet to be established for the U.K. but high on the list will certainly be a stipulation of purity: a microbiological preparation cannot be considered for field application until it is shown to be free from all other, possibly dangerous, contaminating pathogens.

Modes of use of pathogenic organisms

Where ecological circumstances are conducive to a rapid natural rate of spread of a particular pathogen, control may be achievable simply by releasing the pathogen, perhaps as a locally applied spray, perhaps by release of hosts infected in the laboratory. This will be discussed more fully below. When an organism is *virulent* in the sense that its biomass/infectivity ratio is favourable, so that a relatively few individuals produce the desired pathogenic effect, but its *infectivity* or spread capacity is limited, then widespread dissemination by one of the methods used with insecticides is called for, and large amounts of inoculum are necessary. Such overall spraying methods may also be required when a more rapid effect is required than can be achieved by introduction alone.

Knowledge of microbial toxins is limited to the crystalliferous toxins of *Bacillus thuringiensis* and toxins of certain fungi, e.g. *Asperigillus flavus*, *Metarrhizium anisopliae* and *Beauveria bassiana*. Commercial preparations of *B. thuringiensis* consist of endospores plus crystalliferous toxins. The separation of the toxins would be commercially uneconomic but spores could probably be inactivated by exposure to UV, leaving only toxin activity. This could be an advantage where it is desirable there should be either no pathogenic effect or no carry over of activity beyond natural toxin persistence on the host's food.

Many insects have symbiotic micro-organisms and it has been suggested from time to time that control could be achieved by their elimination from the host. As far as we know this has not been put into practice. Perhaps use could also be made of the gut microflora. There is some evidence to suggest this may assume pathogenic properties at a temperature above that which is optimal for host survival. Can habitat manipulation (e.g. removal of shade) help to produce such temperatures, or can chemical substances be used to facilitate invasion of the haemocoele by the gut flora?

There is evidence to suggest occult persistence of some viruses in insect populations and this may even prove to be a general phenomenon. These latent infections are a subject of some contention among insect virologists principally as regards the nature of the infection. This provides a mechanism whereby a virus may persist through successive generations until, for reasons which are generally obscure, it becomes manifest as an overt disease. The ability to induce the active expression of a latent virus infection would be of great interest to economic entomolgists.

The two modes of use of pathogenic organisms, 'introduction' and 'microbiological insecticides' concern two distinct concepts in their application. One is the idea of *virulence*, which involves the dosage level required to induce the pathogenic effect characteristic of the organism in question. The other is the tendency of the host and pathogen to meet. This is the expression of many aspects of host biology, including environmental factors, such as dispersal mechanisms, and qualities of the virus, such as stability and adhesion to plant surfaces; it has been summed up by Tanada (1963) as *infectivity* of the pathogen. Thus, virulence is a host/pathogen interaction whilst infectivity is an ecological concept. The terms are for convenience only and are not mutually exclusive. The resistance of the host to the pathogen may, within certain limits, be a function of environmental resistance to the host, so that virulence may be affected by general ecological circumstance. Later, we shall attempt to relate virulence and some aspects of infectivity to the plant pathologist's concept of inoculum potential and more particularly to inoculum capacity. Infectivity, like virulence, is a somewhat variable quality: changes in the ecosystem of a particular host/pathogen may substantially affect the level of infectivity by tending to aid or suppress not just the degree of distribution of the pathogen but more particularly the distribution in relation to the host individuals.

Epizootiology

Three features of insect viruses are of special importance in relation to their success as agents in insect control, survival, dispersal (which are particular aspects of infectivity as we have defined it) and virulence.

Survival

The survival capacity of an occluded virus is often very great in the laboratory and in certain circumstances also in the field. Thus *in vitro* the nuclear polyhedrosis virus (NPV) of *Bombyx mori*, when kept under refrigeration, was still infective after 21 years (Steinhaus 1960), the NPV of *Gilpinia hercyniae* after 11 years (Neilson & Elgee 1960) and in the field the NPV of *Trichoplusia ni* may still be active in the soil for at least four years, almost certainly longer (Jaques 1969). The persistence of non-occluded viruses is less well known, but some are certainly very unstable outside the host, e.g. that of *Panonychus ulmi* (Putman 1970) and mosquito iridescent virus (MIV) (Linley & Nielsen 1968). Ultraviolet light seems the most important degrading factor for all insect viruses, but other conditions, such as pH, may be just as important with non-occluded viruses and at least some occluded viruses (Falcon 1971, Gudauskas & Canerday 1968).

Though a high degree of persistence of viruses outside the host enhances the epizootic potential and also aids virus continuance at an enzootic level, many viruses are able to persist for much of the year within the host. Such persistence may be in an actively multiplying state (as with some cytoplasmic polyhedrosis viruses (CPV) of low pathogenicity) or in a non-lethal condition (as with latent infection), or in an alternation of high and low activity levels. We may contrast two examples. An NPV of the spruce sawfly (*Gilpinia hercyniae*) multiplies only in digestive cells of midgut tissues. It is always active in the host larvae which are killed 5–10 days after ingestion. Larvae that acquire the virus within three days from the end of the last feeding instar survive to the prepupal stage. Here the gut cells infected in the larvae are sloughed off and come to lie, with their virus burden, in the lumen of a gut reformed from nidi of small undifferentiated cells which previously lay only on the basal membrane between the original digestive cells. The prepupal gut of undifferentiated 'embryonic cells' is resistant to infection. The prepupa is the diapause stage and in it virus may be carried through the winter. Virus multiplication recommences in digestive cells now developed from some of the prepupal gut 'embryonic cells'. This proceeds further in the adult which may thus carry a heavy virus load (Bird 1953). Adults contaminate foliage, especially during oviposition, and the virus survives long enough to be acquired by the emergent larvae (Neilson & Elgee 1968). Larvae of *Aedes taeniorhynchus* (Wied.) may be killed by a general systemic infection of the non-occluded mosquito iridescent virus (MIV). This virus seems unstable in water and appears to be acquired only when healthy larvae feed on those that have died from virus infection. Larvae thus infected complete their development and give rise to adults which lay eggs containing virus. The resulting larvae die of the infection in about the 4th instar (Linley & Nielsen 1968). The state of MIV in the adult is unknown, but it evidently does not constitute a lethal infection, and may well have entered gonad tissue during its multiplication phase in the larval stage. In *Gilpinia* and *Aedes* then, an alternation of pathogenic and non-pathogenic states of the virus is correlated with the host life cycle. This, in the mosquito, gives survival totally within the host, and in the sawfly, persistence in the host essentially broken only by a limited period of exposure between oviposition and ingestion by first instar larvae, and subsequently between this and perhaps two further recycles before again entering a late final instar larva and persisting through the eonymph. *Gilpinia* eonymphs may remain in diapause for up to six winters (Balch 1939) and this gives the possibility of very prolonged survival of the virus, perhaps helping its survival through periods when, owing to low host population density, it might be reduced to a very low level or even to local extinction.

CPVs in Lepidoptera commonly survive through the host pupal stage to the adult to infect the next larval generation by external contamination of the egg; but some NPVs and CPVs persist outside the host, at least over the winter. Thus the NPV of *Malacosoma fragile* may do so on the bark of trees contaminated by larvae infected in the previous summer and, during the winter, polyhedra may be washed onto overwintering eggs laid on the bark, to be ingested during hatching of larvae in the spring (Clark 1956). In *Lymantria dispar* NPV persists on the surface of the eggs through the winter (Doane 1969).

The inevitable course of all virus is downwards, and virus which reaches the soil before being inactivated by abiotic factors may survive long periods. Rain splash carrying soil onto the lower leaves of plants has been shown to account for the recycling of some such viruses in new host generations (Jaques 1964).

There is some evidence of virus persistence in a latent form in insects, e.g. Sigma virus of *Drosophila* (Seecof 1968), CPV of *Antheraea eucalypti* (Grace 1962), granulosis virus (GV) of *Pieris brassicae cheiranthi* (David *et al.* 1968). These examples indicate how the phenomenon occurs in several types of insect viruses; perhaps it occurs in all types. The nature of latent infections has not been fully investigated. The many *ad hoc* experiments on the activation of such conditions indicate that the extent of latency in populations may be demonstrated by the induction of overt infections. The presence of latent infection has been demonstrated by means of a wide range of chemical stressors (e.g. thioglycolic acid, sodium fluoride, boric acid), by some other stressors (cold exposure, starvation, change of food plant) and by feeding insect viruses isolated from other insect species. The conclusion that has emerged is that there are no generally applicable stressors; what activates latent infection in one host is usually not effective for another. Lack of basic understanding need not discourage the use of latency termination as a method of virus control of insects. Successful investigations, however, might give us the means of obtaining predictable results, perhaps by leading to the discovery of generally applicable stressors.

Dispersal

Two main directions of dispersal exist. *Vertical* dispersal, or transmission of virus between successive generations, and *horizontal* dispersal, or transmission where spread of virus occurs between host individuals of the same generation; the latter is essentially a spatial concept. The two are not necessarily mutually exclusive for there may be a strong spatial element involved in vertical transmission, as when infected adults oviposit following migration. For instance, the great *Pieris brassicae* migration of 1955 is thought to have been responsible for the appearance here of its GV (Smith & Rivers 1956), for the disease was prevalent again in France in 1954 after nearly 30 years' eclipse (Paillot 1926). On a less spectacular scale, the spread of virus by viruliferous adults is commonplace, and accounts for local spread and some jump spread. Adults of *Gilpinia hercyniae* may fly at least one mile before oviposition, and so disperse their virus burden as widely as their eggs (Balch 1939). This capacity has been subject to exploitation with *Colias philodoce eurytheme*, the alfalfa caterpillar; a paste containing NPV was applied to adult female genitalia, and as a result the virus became effectively dispersed in larval populations in the area of adult release (Martignioni & Milstead 1962). The introduction of cocoons derived from larvae sublethally infected with an NPV proved an efficient way of seeding populations of *Neodiprion swainei* and of obtaining control (Smirnoff 1962).

Oviposition habits influence the efficiency of transovarial spread of viruses of sawflies. *Neodiprion sertifer* lays eggs in groups, whilst *Gilpinia hercyniae* lays them singly, thus creating a greater number of infection foci. This has been used as a partial explanation of the greater spread of *Gilpinia* NPV (Bird 1961). Both these sawflies became established in eastern North America and flourished in the absence of natural enemies (including virus). For instance, *Gilpinia*, which had infested 2,000 square miles when discovered in the Gaspé Peninsula of Canada in 1930, was heavily infesting 12,000 square miles by 1938. Accidentally introduced NPV, first noted in the field in 1938, spread unassisted from south to north within a period of four years and was the main factor in controlling the host species at economic levels over most of its range (Balch & Bird 1944). On the other hand the NPV of *N. sertifer* was introduced from Sweden but did not keep pace with the spread of its host.

The rate of virus spread in *N. sertifer* did not for instance compare with that in the indigenous *Neodiprion lecontei*, and this has been attributed to the fact that *N. lecontei* has a well-established complement of natural enemies which would help the tree-to-tree spread of virus, whilst the alien *N. sertifer* had few associated parasites (Bird 1961). There is evidence of maintenance of virus viability after passage through the gut of both bird predators (e.g. NPV of *Trichoplusia ni* through *Passer domesticus* (Hostetter & Biever 1970) and the NPV of *N. sertifer* through the cedar waxwing (Bird 1955)), and invertebrate predators such as *Rhinocoris annulatus* (Franz *et al.* 1955), crickets and opilionids. It may be safe to predict that this is a pH effect, as polyhedra can be dissolved only in weak alkali in the range pH 10–11, thus enabling infectious but unstable virus particles to be released. The gut of vertebrates, however, is acid and that of carnivorous insects is, with some exceptions, generally more acid than in phytophagous species. For instance, the midgut pH of Lepidoptera varies between 9 and 10, whereas most predacious Coleoptera have weakly acid gut juices. Hence virus may be dispersed by being carried in the guts of predators of many kinds.

Dispersal of virus by parasites may occur by contamination of the ovipositor although, with both parasites and predators, virus may be transmitted passively as an external contaminant. Vertical passage of virus may also be achieved through parasites as in the case of the tachinid *Voria ruralis* developing within NPV-infected larvae of *Trichoplusia ni*; virus persists on the outside of the parasite puparium and may be picked up by the emerging adult (Soo Hoo *et al.* 1971).

Abiotic factors may also be responsible for virus dispersal. The influence of wind remains unassessed, but rain certainly disperses virus downwards on trees, as has been demonstrated with *N. sertifer* and its NPV (Bird 1961). There is also the possibility of rain splash on contaminated foliage creating aerosols which are then dispersed by wind drift. These routes are aided by the disruptive effect which infection may have on host behaviour. Thus, infected larvae of *Lymantria monacha*, the nun moth, move to the top of the trees, whilst those of other species become restless and disperse, a feature especially useful in achieving virus spread between groups of insects normally feeding colonially. As virus may survive in soil for long periods, it may be dispersed under dry conditions by wind-blown soil particles, and also in water during irrigation and run-off. Both mechanisms have been invoked for the spread of the NPV of *Colias eurytheme* (Steinhaus 1941, Thompson & Steinhaus 1950).

Virulence

The susceptibility of an insect to infection varies with age of host and natural resistance. In Lepidoptera, early instars may be very much more susceptible than later ones which, on occasions, can be impossible to infect. For instance, the relative susceptibilities of first, third and fourth instars of *Malacosoma disstria* larvae to NPV were 68,000:64:1. It has been calculated that some first-instar larvae succumb to one polyhedron and some fourth instar larvae survive two billion; the total range of variation is, therefore, one polyhedron to two billion (Stairs 1965). On the other hand, this disparity is not seen in the sawfly *Gilpinia hercyniae* (Bird 1949). The concept of virus selecting for greater host resistance and reduced genetic heterogeneity has been raised in relation to insects and their viruses. It was postulated by Martignioni & Schmid (1961), for *Phryganidia californica*, that a virus epizootic reduces an insect population to a lower density, whilst simultaneously the population's resistance to the virus increases and its heterogeneity in this respect decreases. At lower host densities when virus spread is reduced and during the post-epizootic phase, the disease ceases to act as a selective sieve. With reduced selection there is a return to susceptibility, provided all susceptible individuals have not been eliminated from the population. Thus, during the low ebb of the population, heterogeneity gradually increases and so resistance to virus again declines. This theory fits reasonably well other observations on *Pieris brassicae cheiranthi* and GV (David *et al.* 1968) and on *Zeiraphera diniana* and GV (Martignioni 1957).

Inoculum potential—a unifying and analytic concept?

Many of the possible mechanisms of pathogen dispersal have been identified and studied, albeit often in qualitative terms. To understand how an epizootic follows a particular course or how that course may be most easily modified for the benefit of economic control, requires some unifying concept between (a) virulence and infectivity, and (b) the quantity of inoculum involved.

Such a concept has already been well worked out for plant diseases in the idea of inoculum potential. As defined by Dimond & Horsfall (1960), this is the 'number of independent infections that are likely to occur in a given situation in a population of susceptible healthy tissue'. The operation of this principle has been explained by analogy to potential energy, the magnitude of which is a product of intensity and capacity factors. The quantity of inoculum is the intensity factor whilst the capacity factor is the tendency of the environment to encourage or discourage disease. The capacity factor therefore completely embraces our previous concept of virulence as a factor which varies with respect to environmental pressures on the host insect and partially includes the idea of infectivity so far as this embraces the idea of virus attrition, mainly by abiotic factors.

Can the applicability of this concept be tested in relation to the spread of pathogens introduced into insect populations? Pathogens introduced at a point source will spread from the infection centre and, initially, there will be a gradient of disease. The gradient of disease is defined in plant disease studies as the slope of the line when the logarithm of incidence of disease is plotted against the logarithm of the distance from the disease source. With plant disease, it has been suggested, the gradient of disease can give information on the nature of transmission. Thus a gradient of -1 suggests two dimensional spread, such as by mycelial growth or nematode migration in soil, whilst one of -2, where the strength of the inoculum varies inversely with the square of the distance from its source, indicates three-dimensional spread, as might occur when spread is mainly by air-borne spores. With insect viruses, effective spread is thought to be mostly achieved not by active pathogen movement nor by movement under the influence of abiotic factors, but by biotic agencies such as parasites and predators, and also by the juvenile and adult stages of the host insect and other susceptible species. If so, it is strictly neither two- nor three-dimensional. However, the method by which inoculum is dispersed does not alter the fact that its incidence is proportional to the amount produced at the infection centre, but it does affect the interpretation of the power of the disease gradient. With many plant diseases, the gradient of disease is to a higher power than -2, and the disease spread is then limited, no matter how strong the intensity of the inoculum. We thus see dispersal, covered in some aspects of the previously mentioned concept of infectivity, as a separate matter from inoculum capacity. Dispersal, in fact, activates both inoculum capacity and intensity.

Practical implications arise from quantitative knowledge of the gradient of disease under particular circumstances. It would indicate the optimal distance between point source introductions for rapid diffusion of disease through a pest population. This would be of some economic value both in avoiding wastage of pathogenic material and in obtaining the swiftest protection of the crop possible, except by overall spraying. The method of dispersal of the pathogen from the infection centre is all-important, for the more efficient this is, the fewer disease-producing units need be introduced. The effect of introducing further dispersal factors, perhaps as entomophagous insects, can now be inspected experimentally and asssesed quantitatively by comparing gradients of disease. On the other hand, considering the virulence of the disease to the host, any method of increasing host susceptibility or enhancing the survival of virus viability, i.e. of increasing the value of the capacity factor, increases the value of the inoculum for a given intensity.

We should probably not be tempted into further speculative discussion until an adequate body of data is available for inspection. Preliminary data on the spread of NPV of *Gilpinia hercyniae* from small epizootic centres in populations of this sawfly in Wales and Canada suggest conformity with the concept of gradient of disease as familiar to plant pathologists, and gives us reason to hope that we may be able to consider further applications of the inoculum potential concept to the spread of insect disease.

The scope for future applied studies

There is no doubt that there are many insect pest problems in the world that may be solved, at least partially, by microbiological control. But are there any problems here in the United Kingdom?

There are two ways of looking at this question. Firstly, we can decide which of the insect pests so far unsuitable for chemical control are likely to be amenable to microbiological methods. Pests of forest trees, where the cost of repeated insecticide treatment is generally uneconomic, come readily to mind, and we have two good well-established examples among the sawflies of Coniferae, viz. *Neodiprion sertifer* on *Pinus* spp. (Rivers 1962) and *Gilpinia hercyniae* on spruces (Entwistle 1971) with the possibility of control also of *Bupalus piniarius* and *Zeiraphera diniana* yet to be explored. Then, secondly, we may ask which of the pests currently subject to insecticidal control could be as satisfactorily controlled by biological systems that conceivably could be self-perpetuating. The field now really becomes very wide. In field and market garden crops possibilities exist among the cutworm complex, hepialids, pea moth (*Laspeyresia nigricana*), pests of brassicas, such as *Mamestra brassicae*, *Plutella xylostella* and the pierids. Among soft fruits, magpie moth (*Abraxas grossulariata*), various sawflies and hepialids require consideration; whilst amongst orchard fruits, codling moth (*Laspeyresia* (*Carpocapsa*) *pomonella*), various other caterpillars and perhaps spider mites offer possibilities.

These are very real possibilities. Already, outside the UK, promising work is under way on the control of cotton bollworm (*Heliothis* spp.), of cabbage looper (*Trichoplusia ni*) with NPV, of codling moth with a GV (Falcon 1971) and of spider mites with non-occluded virus (Putman 1970).

Microbiological control is on the threshold of an exciting and useful future, one which could see it taking over much that is currently dominated by the chemical approach. After all, the role of biological control is not to deal just with the problems chemists find intractable, but rather to provide alternative and more discriminatory methods. However, to achieve this, insect microbiology must progress more assertively in the field of ecology and epizootiology.

References

BALCH R.E. (1939) The outbreak of the European spruce sawfly in Canada and some important features of its bionomics. *J. econ. Ent.* **32**, 412–18.

BALCH R.E. & R.E. BIRD (1944) A disease of the European spruce sawfly *Gilpinia hercyniae* (Htg.), and its place in natural control. *Scient. Agric.* **25**, 65–80.

BIRD F.T. (1949) *A virus (polyhedral) disease of the European spruce sawfly*. Ph.D. Thesis, McGill University, Montreal, Quebec, Canada.

BIRD F.T. (1953) The effect of metamorphosis on the multiplication of an insect virus. *Can. J. Zool.* **31**, 300–3.

BIRD F.T. (1955) Virus diseases of sawflies. *Can. Ent.* 87, 124–7.

BIRD F.T. (1961) Transmission of some insect viruses with particular reference to ovarial transmission and its importance in the development of epizootics. *J. Insect Pathol.* 3, 352–80.

BURGES, H.D. & N.W. HUSSEY, eds. (1971) *Microbial Control of Insects and Mites*. Academic Press, New York & London, 861 pp.

CLARK E.C. (1956) Survival and transmission of a virus causing polyhedrosis in *Malacosoma fragile*. *Ecology* 37, 728–32.

DAVID W.A.L., B.O.C. GARDINER & S.E. CLOTHIER (1968) Laboratory breeding of *Pieris brassicae* transmitting a granulosis virus. *J. Invertebr. Pathol.* **12**, 238–44.

DIMOND A.E. & J.G. HORSFALL (1960) Inoculum and the diseased population. In *Plant Pathology, an advanced treatise*. Vol. III, eds. J.G. Horsfall and A.E. Dimond. Academic Press, New York & London, pp. 1–22.

DOANE C.C. (1969) Transovum transmission of a nuclear polyhedrosis virus in the gipsy moth and the inducement of virus susceptibility. *J. Invertebr. Pathol.* **14**, 199–210.

ENTWISTLE P.F. (1971) Possibilities of control of a British outbreak of spruce sawfly by a virus disease. *Proc. 6th Insectic. Fungi. Conf. (1971)* Vol. 2, 475–9.

FALCON L.A. (1971) Microbial control as a tool in integrated control programs. In *Biological Control. Proc. A.A.A.S. Symposium on Biological Control, Boston, Mass.* 1969, ed. C.B. Huffaker, 346–63.

FRANZ J., A. KRIEG & R. LANGENBUCH (1955) Untersuchungen über den Einfluss der Passage durch den Darm von Raubinsecten und Vögeln auf die Infecktiosität von Insekten pathogener Viren. *Z. PflKrankh. PflPath. PflSchutz.* **62**, 721–6.

GRACE T.D.C. (1962) Development of a cytoplasmic polyhedrosis in insect cells grown *in vitro*. *Virology* **18**, 33–43.

GUDAUSKAS R.T. & D. CANERDAY (1968) The effect of heat, buffer, salt and H-ion concentration, and ultraviolet light on the infectivity of *Heliothis* and *Trichoplusia* nuclear-polyhedrosis viruses. *J. Invertebr. Pathol.* **12**, 405–11.

HEIMPEL A.M. (1971) Safety of insect pathogens for man and vertebrates. In *Microbial*

Control of Insects and Mites, eds. H.D. Burges and N.W. Hussey. Academic Press, New York & London, 469–89.

Hostetter D.L. & K.D. Biever (1970) The recovery of virulent nuclear polyhedrosis virus of the cabbage looper, *Trichoplusia ni*, from feces of birds. *J. Invertebr. Pathol.* **15**, 173–6.

Jaques R.P. (1964) The persistence of a nuclear-polyhedrosis virus in soil. *J. Insect. Pathol.* **6**, 251–4.

Jaques R.P. (1969) Leaching of the nuclear-polyhedrosis virus of *Trichoplusia ni* from soil. *J. Invertebr. Pathol.* **13**, 256–63.

Linley J.R. & H.T. Nielsen (1968) Transmission of a mosquito iridescent virus in *Aedes taeniorhynchus*. II. Experiments related to transmission in nature. *J. Invertebr. Pathol.* **12**, 17–24.

Martignioni M.E. (1957) Contributo alla conoscenza di una granulosi di *Eucosma griseana* (Hübner) (Tortricidae, Lepidoptera) quale fettore limitante il pullulamento dell' insetto nella Engadina alta. *Mitt. schweiz. Anst. forstl. VersWes.* **32**, 371–418.

Martignioni M.E. & P. Schmid (1961) Studies on resistance to virus infections in natural conditions. *J. Insect. Pathol.* **3**, 62–74.

Martignioni M.E. & J.E. Milstead (1962) Transovum transmission of the nuclear-polyhedrosis virus of *Colias eurytheme* Boisduval through contamination of the female genitalia. *J. Insect Pathol.* **4**, 113–21.

McLaughlin R.E. (1971) Use of protozoans for microbial control of insects. In *Microbial Control of Insects and Mites*, eds. H.D. Burges and N.W. Hussey. Academic Press, New York & London, 151–72.

Neilson M.M. & D.E. Elgee (1960) The effect of storage on the virulence of a polyhedrosis virus. *J. Insect Pathol.* **2**, 165–71.

Neilson M.M. & D.E. Elgee (1968) The method and role of vertical transmission of a nucleopolyhedrosis virus in the European spruce sawfly, *Diprion hercyniae*. *J. Invertebr. Pathol.* **12**, 132–9.

Paillot A. (1927) Sur un nouvelle maladie du noyau ou grasserie des chenilles de *P. brassicae* et un nouveau groupe de micro-organismes parasites. *C.r. hebd. Seanc. Acad. Sci., Paris*, **182**, 180–2.

Putman W.L. (1970) Occurrence and transmission of a virus disease of the European Red Mite, *Panonychus ulmi*. *Can. Ent.* **102**, 305–21.

Rivers C.F. (1962) The use of a polyhedral virus disease in the control of the Pine Sawfly *Neodiprion sertifer* Geoffr. in North West Scotland. *Int. Pathol. Insectes, Paris*, 476–80.

Seecof R. (1968) The sigma virus infection of *Drosophila melanogaster*. *Current Topics in Micro. and Immun.* **42**, 59–93.

Sidor C. (1960) A polyhedral virus disease of *Chrysopa perla*. *Virology* **10**, 551–2.

Smirnoff W.A. (1962) Trans-ovum transmission of virus of *Neodiprion swainei* Middleton. (Hymenoptera, Tentheridinidae). *J. Insect Pathol.* **4**, 192–200.

Smith K.M. (1967) *Insect Virology*. Academic Press, New York & London, pp. 256.

Smith K.M. & C.F. Rivers (1956) Some viruses affecting insects of economic importance. *Parasitology* **46**, 235–42.

Soo Hoo C.F., R.S. Seay & P.V. Vial (1971) Surface disinfection of a nucleopolyhedrosis virus contaminant of *Voria ruralis* using sodium hypochlorite. *J. econ. Ent.* **64**, 988–9.

Stairs G.R. (1965) Quantitative differences in susceptibility to nuclear polyhedrosis virus among larval instars of the Forest Tent caterpillar, *Malacosoma disstria* (Hübner). *J. Invertebr. Pathol.* **7**, 427–9.

STEINHAUS E.A. (1948) Polyhedrosis ('wilt disease') of the alfalfa caterpillar. *J. econ. Ent.* 41, 859–65.

TANADA Y. (1963) Epizootiology of infectious diseases. In *Insect Pathology, an advanced treatise*, ed. E.A. Steinhaus. Academic Press, New York & London, Vol. 2, 423–75.

THOMPSON C.G. & E.A. STEINHAUS (1950) Further tests using a polyhedrosis virus to control the alfalfa caterpillar. *Hilgardia* 19, 412–45.

Control of insects by exploiting their behaviour

J. M. CHERRETT *Department of Applied Zoology, University College of North Wales, Bangor, Caernarvonshire*
T. LEWIS *Department of Entomology, Rothamsted Experimental Station, Harpenden, Hertfordshire*

Various aspects of insect behaviour can be exploited for control purposes. Mating behaviour causes some male insects to be caught in traps baited with female sex pheromones, and attempts to inhibit mating by saturating the environment with sex pheromone to confuse the males have also been tried (Shorey 1970). Some pests can be encouraged to lay eggs on trap crops which are subsequently destroyed (Fernald 1935). Alarm behaviour has also been exploited to reduce the incidence of oviposition, for example by simulating the sounds produced by echo-locating bats to repel the European corn borer *Ostrinia nubilalis* (Belton & Kempster 1962). Knowledge of migratory behaviour and of the responses of the migrating animal to environmental factors have enabled locust swarms to be tracked and destroyed (Hemming 1968), the patterns of settlement of windborne arthropods to be predicted and manipulated (Lewis 1969, Lewis & Stephenson 1966), and the alighting of aphids and thrips on crops to be greatly reduced (Smith *et al.* 1964, Ota & Smith 1968). The thigmotactic behaviour of lepidoptera when seeking pupation sites permits the use of pupation traps to catch codling moth on apple trees (Putman 1963), and brood-tending behaviour in ants might be used to introduce toxicants into the nest in simulated brood (Glancey *et al.* 1970).

However, the most important aspect of insect behaviour exploited in their control is feeding behaviour. Food sources can be made unattractive by chemical repellents, a principle much used in protecting Man from blood-sucking insects (Dethier 1956), and even after an insect has been attracted to a suitable host it can be deterred from feeding by the use of antifeedants sprayed on the crops (Munakata 1970). More commonly, feeding behaviour is used to lure insects into traps, or to encourage them to consume a lethal dose of toxicant, and it is this latter technique, the use of poison baits, with which we are concerned in this paper.

Most poison baits consist of four components.

1 *A carrier* of relatively inert material providing the structure of the bait. In the case of 'poison solutions' (Peacock *et al.* 1950), the blotting paper in which the solution was taken up, or even the water itself, fulfils this role.

2 *Attractants*, which may be an integral part of the carrier, as in some leaf-cutting ant baits where an insecticide, aldrin or 'Mirex', is mixed dry with citrus pulp (Cherrett 1969a), or which are added to the carrier, e.g. sugar and molasses (Stapley 1948, Blanche 1965) or vegetable oils (Hays & Arant 1960, Tyler 1964).

3 *A toxicant*, which has hitherto usually been a stomach poison. More recently, contact insecticides such as aldrin have been used.

4 *Other materials* added for formulation purposes, such as waterproofing agents (Webley 1966) and preservatives (Travis 1939).

Baiting is a standard control technique for vertebrates and terrestrial molluscs, and was widely used with inorganic stomach poisons such as sodium arsenite, lead arsenate, sodium fluoride and thallium sulphate for a variety of insect pests. However, when the chlorinated hydrocarbon insecticides were introduced the use of baits against cockroaches, grasshoppers, cutworms, silverfish and flies was largely discontinued because they were more expensive to apply (Beatson 1968).

Poison baits have always been particularly valuable for controlling social insects. The aggregation of individuals into nests renders general spraying inefficient (Wagner & Reierson 1969), the nests are often difficult to locate, and subterranean nests of ants and termites are difficult to kill completely by the direct application of poison. Most social insects carry food back to their nest, and many have a system of trophallaxis by which food is shared among adults, and fed to larvae (Eisner & Brown 1956), a pattern of behaviour exploited when baits are used. Even amongst non-social insects, increasing resistance to contact insecticides, and the environmental hazards associated with the general application of large doses of broad-spectrum pesticides, have renewed interest in stomach poisons formulated as baits to attract a limited range of pest species (Tyler 1964).

The peculiar advantages of using bait for the control of social insects are well illustrated by the campaign against the imported fire ant *Solenopsis saevissima* v. *richteri* in the southern United States, briefly outlined below.

Control of the imported fire ant in the United States—A case history

A dark-coloured, southern variety of the South American fire ant *Solenopsis saevissima* was accidentally introduced into Mobile, Alabama around 1918, and although it became established, it hardly spread. About 1930, however, a lighter coloured, more northerly variant was again accidentally introduced into the same area and began to spread rapidly (Wilson & Brown 1958). During the period 1949–57, the original population expanded at a rate of 5 miles per year, whilst many secondary infestations, often centred on nurseries, became established in ten states bordered by Texas, Arkansas, North Carolina and Florida. The ant was considered a pest because its mounds hindered harvesting and damaged agricultural machinery, and because of its painful stings. Consequently in 1957 the United States Department of Agriculture embarked upon a campaign to treat some 20,000,000 acres (8,000,000 ha) of infested land in ten of the southern states. The method initially adopted was aerial spraying of the area with dieldrin or heptachlor at a rate of 2 lb active ingredient per acre (2·24 kg/ha). This was eventually reduced to two applications of 0·28 kg (0·25 lb).

In the event, the important fire ant was not eradicated, nor was its spread appreciably checked, but some secondary pest outbreaks did result from the treatment, in particular increased damage by the sugar cane borer (Hensley *et al.* 1961), and heavier invasions of houses by earwigs (Gross & Spink 1969). Deaths of livestock and poultry were reported, and severe destruction of wildlife claimed, although some of these claims have been disputed by the United States Department of Agriculture. As a result, the fire ant eradication campaign has been widely condemned by conservationists as an illustration of inadequately researched, crude control measures entered upon with little knowledge of, or regard for, the biology of the insect and the ecology of its habitat (Brown 1961, Carson 1963, Rudd 1965, Ehrlich & Ehrlich 1970). Indeed Rudd (loc. cit.) wrote 'I believe that the failure of this program marks the end of eradication attempts by "blitz" methods'.

As early as 1939, Travis had carried out extensive tests on poison baits against the native fire ant *Solenopsis geminata*. The common inorganic insecticides rendered his sugar bait unattractive, but thallium acetate gave satisfactory results. Green (1952) however found baits containing DDT, dieldrin or thallium sulphate ineffective in the field against the imported fire ant, and interest in baiting lapsed in the face of mass spraying of dieldrin and heptachlor. Hays & Arant (1960) returned to the possibilities of control by baiting, and tested 400 formulations, concluding that peanut butter with 0·125% Kepone (decachlorooctahydro-1,3,4-metheno-2H-cyclobuta [cd] pentalen-2-one) as toxicant produced the best results, giving 100% field control at a rate of 4–6 lb per acre (4·5–6·7 kg/ha). Further research to discover attractive bait material followed (Bartlett & Lofgren 1961, Lofgren *et al.* 1961) and Lofgren *et al.* (1962) reported a promising new toxicant, an analogue of Kepone, eventually given the trade name of 'Mirex' (dodecachlorooctahydro-1,3,4-metheno-2H-cyclobuta [cd] pentalene). In the search for an ideal toxicant for fire ants, Stringer *et al.* (1964) emphasized that delayed intoxication was essential if the toxicant was to be carried back to the nest and passed trophallactically around the colony before mortality began to disrupt foraging and social life. Accordingly, they stipulated three requirements of such a toxicant:

1 delayed killing action over at least a tenfold, and preferably a hundredfold dose range;

2 ready trophallactic transfer from one ant to another, resulting in the death of the recipient;

3 no repellency to the ants at effective concentrations.

To these might be added the environmental requirements of specificity (a low toxicity to other groups of animals), and non-persistence, so that the toxicant does not build up in the environment or accumulate along food chains. Table 1 compares five toxicants used in fire ant control and shows the superiority of 'Mirex'.

Table 1. A comparison of the properties of 5 toxicants used in fire ant control

Toxicant	Mode of action	Speed of action[1]	Mammalian toxicity[2]	Comments
Aldrin	contact	Delayed at concentrations 0·0025–0·01%	67	Highly persistent
Dieldrin	contact	Too rapid at high concs., insufficient kill at low	46	Highly persistent
Heptachlor	contact and some fumigant	Delayed at concentrations 0·0025–0·01%	male, 100 female, 162	Epoxide highly persistent
Kepone	stomach	Delayed at ×10–×99 dosage range	male, 132 female, 126	
Mirex	stomach	Delayed at >×100 dosage range	male, 312 female, 700	No accumulation in cows or vegetables[3]

[1] From Lofgren *et al.* (1967). Delayed action defined as causing <15% mortality 24 h after treatment, but >89% mortality by the end of the test.

[2] LD_{50} mg/kg for rats.

[3] Lofgren *et al.* (1964).

In 1961, trials were begun on the aerial application of 'Mirex' baits, containing 85% corncob grits, about 14% soya bean oil, and from 0·075 to 0·3% 'Mirex' (Lofgren *et al.* 1964). Excellent control was obtained with baiting rates which varied from 3 lb to 12·5 lb per acre (3·4 to 14·0 kg/ha), although in some tests it took 16 weeks for all ants to die after a single application. Substitution of 'Mirex' for dieldrin and heptachlor in control campaigns began in 1963. By 1970, some 3,000,000 acres (1,213, 800 ha) had been treated in Georgia, Florida and Mississippi with a 'Mirex' bait made of corncob grits, soya bean oil and 0·3% 'Mirex', and scattered at the rate of 3 lb/acre (3·4 kg/ha).

The main advantages of the use of 'Mirex' bait over the general spraying of dieldrin and heptachlor are:

1 The foraging behaviour of the ants is utilized to accumulate in subterranean nests a relatively high concentration of a toxicant broadcast at a very low concentration. Thus 'Mirex' is applied at 10 g/ha active ingredient, compared with rates for dieldrin and heptachlor which ranged from 2240 g/ha to 560 g/ha.

2 The slow-acting, non-repellent nature of 'Mirex' allows the ants' trophallactic behaviour to spread the toxicant effectively throughout the colony.

3 The relative specificity of 'Mirex', as indicated by its high LD_{50}s for acute oral poisoning in mammals, minimizes the danger of harmful side effects on other fauna, especially as the bait itself is not attractive to many species.

These developments have gained the approval of some conservationists, and Ehrlich & Ehrlich (1970) wrote '. . . this program shows what improvement is possible if one pays some attention to the ecology of a situation, rather than launching vast broadcast spray programs'. Even so, the United States Department of Agriculture's recent programme for treating with bait 150 million acres (61 million ha) in 9 southern states has been attacked by conservationist groups, and is suspended pending an enquiry. Lack of long-term information on toxic side effects, possible carcinogenic action of 'Mirex' on mice, and toxicity to shrimps and crabs, together with the claim that the harmful effects of fire ants have been exaggerated, are all cited as grounds for discontinuing the campaign (Article in *Time Magazine* (1970) 96 (18), 32–3).

The development of baits and baiting schemes for leaf-cutting ant control

Leaf-cutting, fungus-growing ants of the genera *Atta* and *Acromyrmex* are amongst the most serious general insect pests in the tropical and sub-tropical regions of the New World, perhaps causing $1,000 million U.S. worth of damage annually (Cramer 1967). Indeed, in Trinidad, Cherrett & Sims

(1968) estimated losses to citrus and cocoa alone as $600,000 T.T. per annum. Traditionally, leaf-cutting ant control has involved searching for nests and then destroying them by physical or chemical means (Mariconi 1970). But as large *Atta* nests can forage up to 150 m, and land use in many of the agricultural areas is diverse with scrub and forest interspersed among many small, individually owned plots each with varying standards of ant control, the location and destruction of nests is slow, inefficient and expensive, therefore often neglected. Also, portions of large irregularly-shaped nests may be overlooked and only partly destroyed by these methods.

A considerable advance in control techniques was made with the large-scale introduction of poison baits in the early sixties (Anon. 1965, Blanche 1965, Echols 1965, Lopez 1966 and Zárate 1966). Most of these used aldrin or 'Mirex' as toxicants, and the most successful, 'Mirex 450', consisted of 91·05% citrus pulp, 8·5% once-refined soya bean oil and 0·45% 'Mirex' (Echols 1966a). This bait could be applied whenever damage occurred, and as the foraging workers took the pellets back to their nest, the nest need not be found. As with the fire ant, 'Mirex' dissolved in the soya bean oil was spread around the colony (Echols 1966b) and even large *Atta* nests, which may be 250 m^2 (Cherrett 1968), reaching to a depth of over 4 m (Moser 1963), and containing over 2 million individuals (Martin *et al.* 1967), were completely killed. For large-scale control programmes of the type sometimes attempted in the past (Anon. 1945, Carvalho 1945, Kennard 1965), and especially in difficult terrain, aerial application of bait is essential to ensure adequate cover. As 'Mirex 450' disintegrates badly when wet (Cherrett & Merrett 1969) and is expensive, research is in progress to produce a good bait for aerial distribution, and to obtain the ecological and behavioural information necessary to devise an efficient baiting scheme, requiring a minimum amount of insecticide. The work embraces the following aspects.

Distribution and density of nests

Atta cephalotes (L.) and *Acromyrmex octospinosus* (Reich) are the only two pest species of leaf-cutting ants found in Trinidad, and when tropical rain forest is cut down and replaced by agriculture, a shift in the dominant species occurs, *A. cephalotes* being replaced by *A. octospinosus*, which rapidly becomes the more serious pest (Cherrett 1968). Typical nest densities found in some crops where traditional control methods are used are given in Table 2. Knowledge of the extent of damage caused by different sized populations enables the likely profitability of control measures to be assessed. Thus, 10 *Acromyrmex* colonies per acre (25/ha) may not be worth controlling in old established citrus orchards, but a similar population would be devastating to young trees.

Table 2. Examples of nest densities of leaf-cutting ants[1] and their effect on crops in Trinidad

Crop	Mean nests/acre	Mean nests/ha	Damage
Orange (3–4 yr old)	14·3	35·3	Very serious. 50% of trees killed and most others retarded.
Grapefruit (3 yr old)	1·0	2·5	Serious. 10% of original plants killed. Many trees retarded.
Orange (6 yr old)	1·4	3·6	Trivial.
Grapefruit (30 yr old)	11·0	27·2	Unassessable, probably trivial.
Cocoa (1 yr old)	25·0	61·8	Very serious. 15% of trees dead. 30% of leaves on remaining trees attacked.
Cocoa (25 yr old)	10·0	24·7	Probably trivial.

[1] Mostly *Acromyrmex octospinosus*.

Foraging distances

Knowledge of the area foraged from each nest allows criteria to be established for the placement of bait. Uniform application by aircraft is always difficult to achieve, and an insistence on very high standards of uniformity would increase application costs, and probably involve a higher application rate. Fortunately, as long as the coverage of bait is fairly comprehensive, its evenness of distribution is not important, providing concentrations of bait fall within foraging range of all nests. A typical nest of *A. octospinosus* 0·75 m² (0·9 yd²) in extent forages 20–30 m, whilst an established *A. cephalotes* nest of 146 m² (175 yd²) would forage from 100–150 m. Thus, with a swath width of about 20 m, bait will fall within range of almost all nests (Fig. 1) (Lewis 1972). In practice, parallel swaths laid down over 3 m tall scrub, with their centres 21 m apart, poisoned 90–91% of colonies in the treated area, showing, as predicted, that this application pattern placed adequate quantities of bait within the foraging range of established nests.

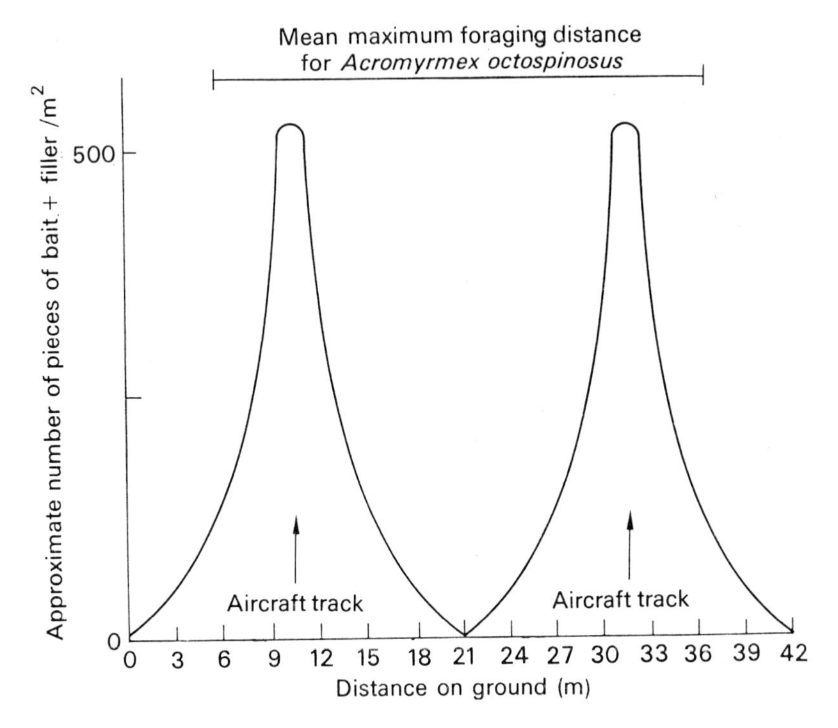

Figure 1. The approximate pattern and extent of swaths of citrus meal bait on the ground after application of 2 lb (0·91 kg) bait + 8 lb (3·63 kg) of filler by an aircraft from a height of 15 m. Most of the material is concentrated in the central 0·7 m of each 21 m swath. *Atta cephalotes* forages up to 150 m from the parent nest and *Acromyrmex octospinosus* 30 m (*cf.* uppermost line), so some bait fell within foraging range of all nests. (After Lewis 1972.)

The distance foraged by these two species depends on the size of the colony, as shown in Fig. 2 for *A. cephalotes* in relatively undisturbed rain forest and for *A. octospinosus* on cultivated ground in Trinidad. Accordingly, to obtain maximum control, it is advantageous to avoid the time of the year when very small colonies are being established.

Sexual cycles

In Trinidad, both *A. cephalotes* and *A. octospinosus* produce sexuals once a year. *A. octospinosus* has its main flight immediately after the first rains of the wet season in May/June, so earning the sexuals the local name of 'rain

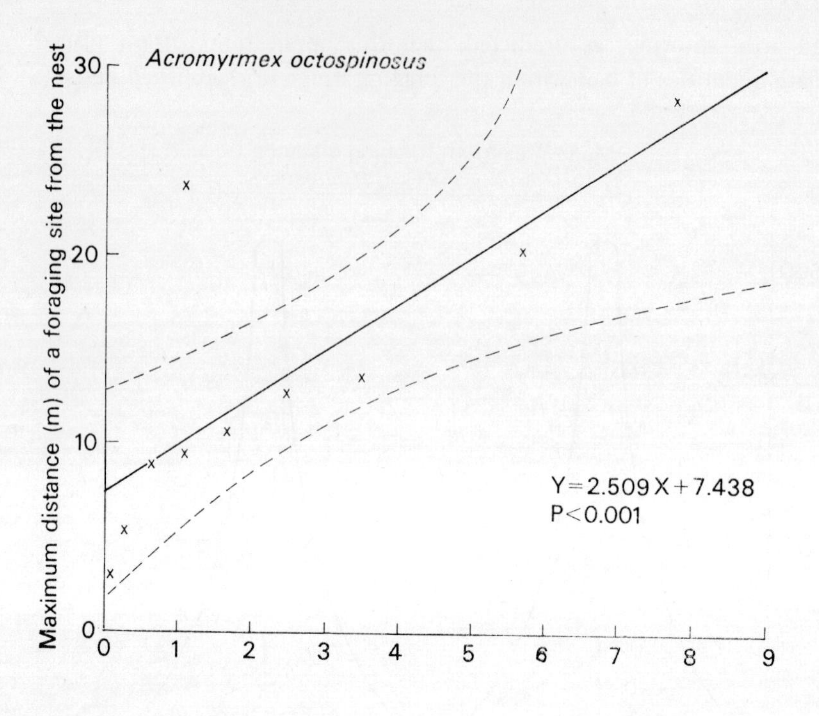

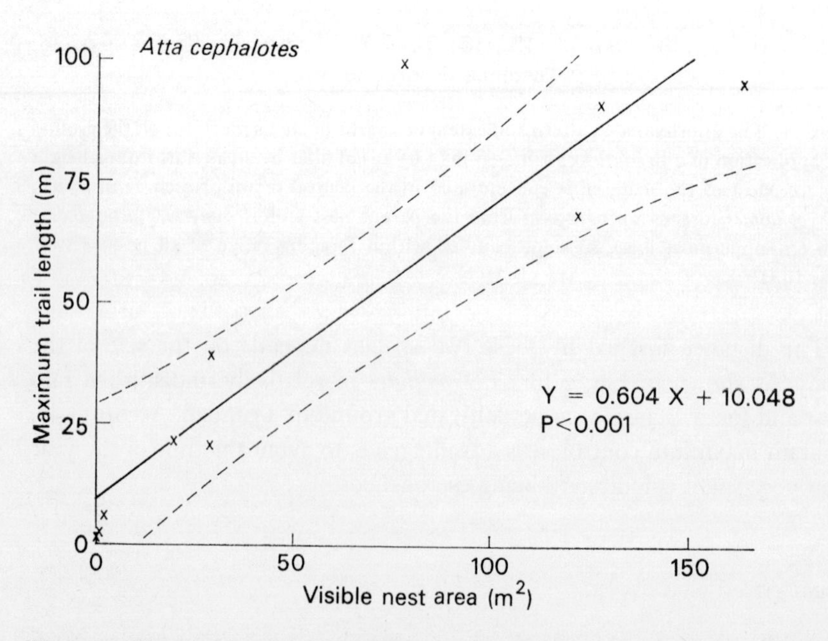

Figure 2. The nest size, judged by the area of disturbed earth, in relation to the maximum trail length for each of a number of different-sized nests visited once. Regression lines together with their 95% confidence limits are given.

flies', although a few more sexuals are released on particularly wet days during the next 2–3 months. *A. cephalotes*, by contrast, has its main flight almost 8 weeks later when the wet season is well under way (Fig. 3). Bait applied shortly before the mating flight has the double advantage of operating when the mean nest size, and hence foraging area, is greatest, and of killing

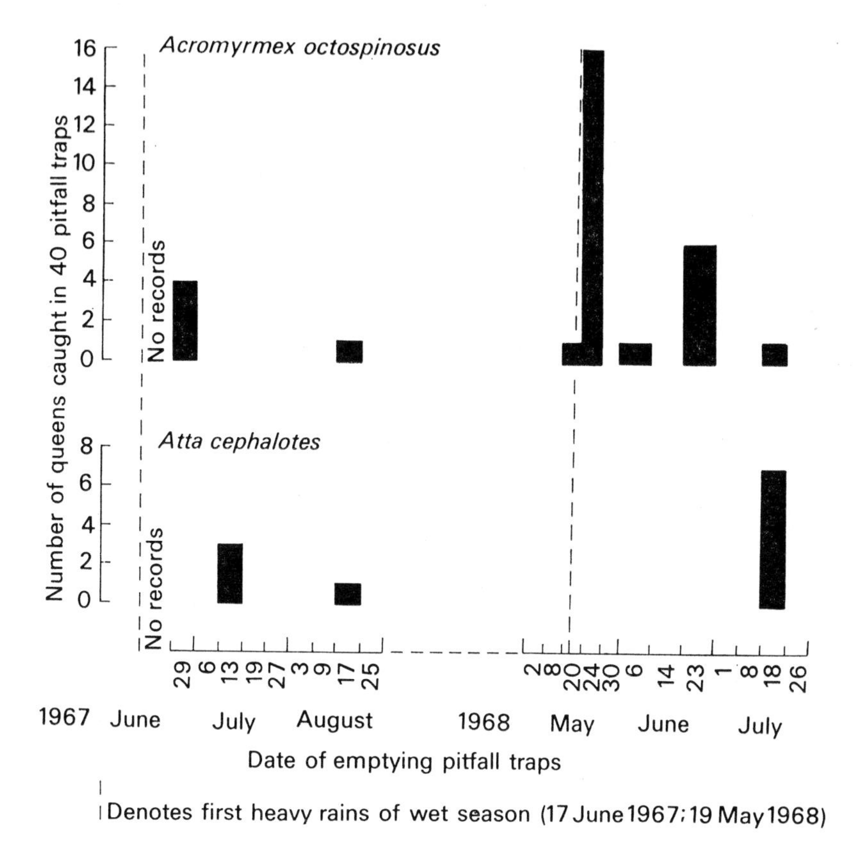

Figure 3. The numbers of queens caught in 40 pitfall traps placed in forest and nearby clearings recorded at approximately weekly intervals for 56 weeks. No queens were caught between September 1967 and April 1968.

developing sexuals, thus reducing the number of fertilized queens subsequently available for founding new colonies. The knowledge that queens of *A. cephalotes* can disperse over 6 miles (9·7 km) (Cherrett 1969b) and that the average annual rate of spread of the introduced *A. octospinosus* in Guadeloupe is 0·5 miles (0·8 km) (Jenly, personal communication) is also valuable in determining the optimum size for areas subjected to large-scale control measures. As newly-established colonies need more than a year to grow beyond a size vulnerable to baiting, re-infestation of cultivated areas by young queens from uncultivated land should present no problem when bait is applied in an annual concerted effort. Also, as long as nests within about 200 m of cultivated land are destroyed, little advantage would ensue from blanket baiting large blocks of forest, and harmful effects on wildlife could be almost completely avoided.

Bait attractants

Studies on the factors which induce workers to pick up materials and carry them back to their nests suggest that the ants stumble accidentally on the bait and then examine it (Cherrett 1972). Chemicals which induce pick-up of baits should properly be called 'arrestants' (Beck 1965) although attractants is the term more generally used. Ideally, the attractants in bait should be active enough to stimulate immediate pick-up; they should so excite workers that many are recruited for foraging bait; they should be persistent and active at low concentrations so that baits remain attractive in the field for as long as possible, preferably for several weeks; and they should be specific to pest species of leaf-cutting ants. There are three obvious sources for such chemicals:

1 *Attractants normally present in materials cut by the ants.* Citrus pulp, the base for most successful baits, has been analysed and found to contain a mixture of attractants and repellents, most of the latter evaporating during drying (Cherrett & Seaforth 1970). For both *A. cephalotes* and *A. octospinosus*, the unabsorbed sapids from a polyamide resin comprising, amongst other things, the carbohydrates, formed the most attractive of the relatively refined fractions. The attractiveness of this fraction was largely simulated by a synthetic mixture of glucose, sucrose and fructose, and Mustafa (1971) has demonstrated the synergistic action of 8 sugars in inducing feeding in both species.

2 *Chemicals present in the fungus cultivated by the ants and which induce feeding and tending behaviour.* Martin *et al.* (1969) describe 4 sugars from the fungus cultured by *Atta colombica tonsipes*, and these were included by Mustafa (loc. cit.) in the sugar mixtures he tested. No other studies on the attractiveness to ants of fungus chemicals appear to have been published.

3 *Social pheromones produced by the ants.* Preliminary experiments failed to reveal any 'food'-marking chemicals (Barrer & Cherrett 1972), so we are now looking for substances that might enable the ants to recognize their brood in the hope that simulated brood could be used as a poison bait, as the work of Glancey *et al.* (1970) on the fire ant suggests. Finally, Wilson (1962) has shown that the scent trail pheromone of the fire ant not only guides foragers along trails, but also excites them to venture from the nest.

Perhaps the addition of scent trail pheromone to a bait would enhance the amount of foraging once the first piece of bait had been brought to the nest, and this is being investigated.

Hitherto, leaf-cutting ant baits have been made from naturally occurring vegetable materials such as citrus pulp, grain and bagasse, the latter fortified with molasses (Blanche 1965), although Echols (1966a) claims that soya bean oil is the principal attractant for *Atta texana* (Buckley) in 'Mirex 450', the citrus peel acting only as a carrier. It seems possible that baits formulated to cater specifically for the known preferences of ants in general, or species in particular, would increase their attractiveness. Also, the preference for novelty in the selection of substrate brought back to the nest will help to ensure maximum bait foraging (Cherrett 1972).

The physical properties of baits

If a broadcast bait is to remain active in the field for 2–3 weeks in humid tropical conditions, rain must not cause physical disintegration as happens with the compressed pellets of 'Mirex 450' (Echols & Biesterfeldt 1966), mould should not develop to render particles unattractive, and toxicant and attractants should not be unduly leached out. Waterproofing citrus/soya bean oil/aldrin bait with a hydrophobic surface deposit of siloxane helps to prolong its life under wet conditions (Cherrett & Merrett 1969), and has the added bonus of increasing the attractiveness of freshly made bait (see Table 3). In Trinidad plain baits are adequate when applied at the end of the dry season

Table 3. The number of pieces of plain and waterproofed bait collected in standard tests by laboratory nests of *Acromyrmex octospinosus* after different periods of weathering

Period of weathering (days)	Cumulative rainfall (cm)	Number of pieces of bait collected	
		Plain	Waterproofed
0	0	102	121
2	0	51	138
4	3·68	11	99
8	4·62	32	126
12	5·13	4	81
14	5·13	13	90
18	5·21	37	128
22	5·87	27	138
26	5·94	23	124
30	7·01	17	87
	Σ	317	1132

just before the mating flights of *A. octospinosus*, and this represents a considerable saving, as plain bait costs around 5p per kg compared with 11½p for waterproofed bait.

One of the principal disadvantages of scattering conventional baits by air is that a high percentage of both the weight and the volume of the bait (probably > 80%) consists of biologically inert structural materials. As application costs commonly constitute a high proportion (about 90–95%) of the total costs of baiting, it would be advantageous if a greater area could be baited on each flight by increasing the proportion of biologically active material carried. Markin *et al.* (1969), and Markin & Hill (1971), who scattered 'Mirex' dissolved in soya bean oil and microencapsulated in a gelatin or plastic wall, claim that when the bait is scattered dry, this technique increases the carrying capacity of the aircraft fivefold. These capsules, tested against the fire ant, incorporated only 13% inert material (i.e. capsule wall) when compared with the standard formulation containing 85% of largely inert corncob grits. The capsules, less than 1 mm in diameter, were picked up by the ants, which seemed able to detect the presence of oil inside them, open them, and ingest the contents. Some capsules remained intact in the field for 40 days, during which time the oil was neither leached out, nor oxidized as occurs with corncob bait, and at a rate of 100 g of capsules per acre (247 g/ha) the material was as effective as the standard bait, although more expensive. A variety of synthetic carriers have been tested, especially vermiculite and clay (Lofgren *et al.* 1963) in fire ant baits, and 'Oasis' in slug baits (Smith & Boswell 1970). Mustafa (1971) tested expanded polystyrene and 'Ufoam', and although neither of these would take up adequate quantities of oils or sugars in their expanded state, the incorporation of attractants and toxicants in rapidly setting foams is an attractive possibility. This is especially so if the foams could be produced from liquid precursors by scattering devices on aircraft, and provided the foam eventually decomposed.

Conclusions

Against some insect pests the use of insecticides formulated as toxic baits has certain distinct advantages over general application by spraying or dusting, namely:

1 Active ingredient can be spread sparsely in the environment, leaving the pest's food-seeking behaviour to concentrate toxicants to lethal doses.

2 Stomach poisons are particularly appropriate for use in baits, and together with suitable attractants they provide some degree of specificity in the organisms killed.

3 Modern formulation techniques could be used to prolong the effective life of the bait.
4 Baits require less critical placement than non-persistent contact insecticides so may often be cheaper to use.

Baits are particularly valuable for the control of social insects which are often difficult to kill by direct spraying or dusting. The practical value of this approach, which requires a detailed knowledge of pest biology and behaviour, is illustrated by fire ant control where baiting has greatly reduced environmental damage with no loss of control efficiency. Work to extend this approach to leaf-cutting ant control is currently being undertaken.

Acknowledgements

These studies have been supported by grants from the Overseas Development Administration of the Foreign and Commonwealth Office.

References

Anon. (1945) Campaña de exterminio de la plaga del zompopo, iniciada por el Departamento de Defensa Agricola del Ministerio de Agricultura. *Revta. agric. Guatem.* **1**, 787–92.
Anon. (1965) Tatuzinho resolve um grande problema. *Boletim do Campo.* **192**, 18–21.
Barrer P.M. & J.M. Cherrett (1972) Some factors affecting the site and pattern of leaf-cutting activity in *Atta cephalotes* (L.) (Hymenoptera: Formicidae). *J. Ent.* (A). 47, 15–27.
Bartlett F.J. & C.S. Lofgren (1961) Field studies with baits against *Solenopsis saevissima* v. *richteri*, the imported fire ant. *J. econ. Ent.* 54, 70–3.
Beatson S.H. (1968) Eradication of Pharaoh's ants and crickets using chlordecone baits. *Int. Pest Cont.* **10**, 8–10.
Beck S.D. (1965) Resistance of plants to insects. *A. Rev. Ent.* **10**, 207–32.
Belton P. & R.H. Kempster (1962) A field test on the use of sound to repel the European Corn Borer. *Entomologia exp. appl.* 5, 281–8.
Blanche D. (1965) Appats empoisonnés contre la Fourmi-manioc á la Guadeloupe. *Congrès de la protection des cultures tropicales. Compte rendu des travaux*, Marseille, 449–54.
Brown W.L. (1961) Mass insect control programs: four case histories. *Psyche* 68, 74–108.
Carson R. (1963) *Silent Spring*. Hamish Hamilton, London.
Carvalho J.C. (1945) O combate às formigas. *Biológico* **11**, 227–31.
Cherrett J.M. (1968) Some aspects of the distribution of pest species of leaf-cutting ants in the Caribbean. *Proc. Am. Soc. hort. Sci. Trop. Region* **12**, 295–310.
Cherrett J.M. (1969a) Baits for the control of Leaf-cutting Ants. 1. Formulation. *Trop. Agric. Trin.* 46, 81–90.
Cherrett J.M. (1969b) A flight record for queens of *Atta cephalotes* L. (Hymenoptera: Formicidae). *Entomologist's mon. Mag.* **104**, 255–6.

CHERRETT J.M. (1972) Chemical aspects of plant attack by leaf-cutting ants. In *Phytochemical Ecology*, ed. J.B. Harborne. Academic Press, New York & London, pp. 13–24.

CHERRETT J.M. & M.R. MERRETT (1969) Baits for the control of Leaf-cutting Ants. 3 Waterproofing for General Broadcasting. *Trop. Agric., Trin.* 46, 221–31.

CHERRETT J.M. & C.E. SEAFORTH (1970) Phytochemical arrestants for the leaf-cutting ants, *Atta cephalotes* (L.) and *Acromyrmex octospinosus* (Reich), with some notes on the ants' response. *Bull. ent. Res.* 59, 615–25.

CHERRETT J.M. & B.G. SIMS (1968) Some costings for leaf-cutting ant damage in Trinidad. *J. agric. Soc. Trin.* 68, 313–24.

CRAMER H.H. (1967) *Plant Protection and World Crop Production.* 'Bayer' Pflanzenschutz, Leverkusen.

DETHIER V.G. (1956) Repellents. *A. Rev. Ent.* 1, 181–202.

ECHOLS H.W. (1965) Town ants controlled with Mirex baits. U.S. Forest Service Research Notes, SO-18.

ECHOLS H.W. (1966a) Texas leaf-cutting ant controlled with pelleted Mirex bait. *J. econ. Ent.* 59, 628–31.

ECHOLS H.W. (1966b) Assimilation and transfer of Mirex in colonies of Texas leaf-cutting ants. *J. econ. Ent.* 59, 1336–8.

ECHOLS H.W. & R.C. BIESTERFELDT (1966) Controlling Texas leaf-cutting ants with Mirex. Forest Service, U.S. Dept. of Agric. SO-38.

EHRLICH P.R. & A.H. EHRLICH (1970) *Population, Resources, Environment: Issues in Human Ecology.* Freeman, San Francisco.

EISN R T. & W.L. BROWN, Jr. (1956) The evolution and social significance of the ant proventriculus. *Proc. Xth Int. Congr. Ent.* 2, 503–8.

FERNALD H.T. (1935) *Applied Entomology.* McGraw-Hill, New York.

GLANCEY B.M., C.E. STRINGER, C.H. CRAIG, P.M. BISHOP & B.B. MARTIN (1970) Pheromone may induce brood tending in the fire ant, *Solenopsis saevissima. Nature, Lond.* 226, 863–4.

GREEN H.B. (1952) Biology and control of the imported fire ant in Mississippi. *J. econ. Ent.* 45, 595–7.

GROSS H.R., Jr. & W.T. SPINK (1969) Responses of striped earwigs following applications of heptachlor and Mirex, and predator-prey relationships between imported fire ants and striped earwigs. *J. econ. Ent.* 62, 686–9.

HAYS S. & F.S. ARANT (1960) Insecticidal baits for control of the imported fire ant, *Solenopsis saevissima richteri. J. econ. Ent.* 53, 188–91.

HEMMING C.F. (1968) *The Locust Menace.* Anti-Locust Research Centre, London.

HENSLEY S.D., W.H. LONG, L.R. RODDY, W.J. MCCORMICK & E.J. CONCIENNE (1961) Effects of insecticides on the predatory arthropod fauna of Louisiana sugarcane fields. *J. econ. Ent.* 54, 146–9.

KENNARD C.P. (1965) Control of leaf-cutting ants (*Atta* spp.) by fogging. *Expl. Agric.* 1, 237–40.

LEWIS T. (1969) The distribution of flying insects near a low hedgerow. *J. appl. Ecol.* 6, 443–52.

LEWIS T. (1972) Aerial baiting to control leaf-cutting ants. *Pestic. Abstr.* 18, 71–4.

LEWIS T. & J.W. STEPHENSON (1966) The permeability of artificial windbreaks and the distribution of insects in the leeward sheltered zone. *Ann. appl. Biol.* 58, 355–63.

LOFGREN, C.S., F.J. BARTLETT & C.E. STRINGER (1963) Imported fire ant toxic bait studies: evaluation of carriers for oil baits. *J. econ. Ent.* 56, 62–6.

LOFGREN C.S., F.J. BARTLETT, C.E. STRINGER & W.A. BANKS (1964) Imported fire ant toxic bait studies. Further tests with granulated Mirex-Soybean oil bait. *J. econ. Ent.* 57, 695–8.

LOFGREN C.S., C.E. STRINGER, W.A. BANKS & P.M. BISHOP (1967) Laboratory tests with candidate bait toxicants against the imported fire ant. U.S.D.A. Agricultural Research Service. ARS 81–14, 1–25.

LOFGREN C.S., C.E. STRINGER & F.J. BARTLETT (1961) Imported fire ant toxic bait studies: The evaluation of various food materials. *J. econ. Ent.* 54, 1096–1100.

LOFGREN C.S., C.E. STRINGER & F.J. BARTLETT (1962) Imported fire ant toxic bait studies: G.C.—1283, a promising toxicant. *J. econ. Ent.* 55, 405–7.

LOPEZ R.E.D. (1966) Resultados preliminares con el Mirex 450, en el control del Zompopo. Min. de Agric. El Salvador Sección de Parasitologia Vegetal. Circular No. 69.

MARICONI F.A.M. (1970) As Saúvas. Editôra Agronômica 'Ceres'. São Paulo.

MARKIN G.P. & S.O. HILL (1971) Microencapsulated oil bait for control of the imported fire ant. *J. econ. Ent.* 64, 193–6.

MARKIN G.P., C.J. MAUFFRAY & D.J. ADAMS (1969) A granular applicator for very low volumes of micro-encapsulated insect bait or other materials. U.S. Dept. Agr. ARS (Ser) 81–34, 4 pp.

MARTIN M.M., G.A. CARLS, R.F.N. HUTCHINS, J.G. MACCONNELL, J.S. MARTIN & O.D. STEINER (1967) Observations on *Atta colombica tonsipes* (Hymenoptera: Formicidae). *Ann. ent. Soc. Am.* 60, 1329–30.

MARTIN M.M., R.M. CARMAN & MACCONNELL J.G. (1969) Nutrients derived from the fungus cultured by the fungus-growing ant, *Atta colombica tonsipes. Ann. ent. Soc. Am.* 62, 11–13.

MOSER J.C. (1963) Contents and structure of *Atta texana* nest in summer. *Ann. ent. Soc. Am.* 56, 286–91.

MUNAKATA K. (1970) Insect antifeedants in plants. In *Control of Insect Behaviour by Natural Products*, ed. D.L. Wood *et al.* Academic Press, New York & London.

MUSTAFA A.H.I. (1971) Synthetic baits for leaf-cutting ants. Unpublished M.Sc. thesis, University of Wales.

OTA A.K. & F.F. SMITH (1968) Aluminium foil—thrips repellant. *Am. Rose A.* 53, 135–9.

PEACOCK A.D., D.W. HALL, I.C. SMITH & A. GOODFELLOW (1950) The biology and control of the ant pest *Monomorium pharaonis* (L.). *Scott. Dept. Agric. Misc. Publ.* 17, 51 pp.

PUTMAN W.L. (1963) The Codling Moth, *Carpocapsa pomonella* (L.) (Lepidoptera: Tortricidae): A review with special reference to Ontario. *Proc. ent. Soc. Ont.* 93, 22–60.

RUDD R.L. (1965) *Pesticides and the Living Landscape.* Faber & Faber, London.

SHOREY H.H. (1970) Sex pheromones of Lepidoptera. In *Control of Insect Behaviour by Natural Products*, ed. D.L. Wood *et al.* Academic Press, New York & London, pp. 249–84.

SMITH F.F. & A.L. BOSWELL (1970) New baits and attractants for slugs. *J. econ. Ent.* 63, 1919–22.

SMITH F.F., G.V. JOHNSON, R.P. KAHN & A. BING (1964) Repellency of reflective aluminium to transient virus-vectors. *Phytopathology* 54, 748.

STAPLEY J.H. (1948) *Pests of Farm Crops.* E. & F.N. Spon, London.

STRINGER C.E., C.S. LOFGREN & F.J. BARTLETT (1964) Imported fire ant toxic bait studies: Evaluation of toxicants. *J. econ. Ent.* 57, 941–5.

TRAVIS B.V. (1939) Poisoned-bait tests against the fire ant, with special reference to thallium sulfate and thallium acetate. *J. econ. Ent.* 32, 706–13.

TYLER P.S. (1964) Kepone bait for the control of resistant German cockroaches. *Int. Pest Cont.* 6, 10–12.

WAGNER R.E. & D.A. REIERSON (1969) Yellow jacket control by baiting. 1. Influence of toxicants and attractants on bait acceptance. *J. econ. Ent.* 62, 1192–7.

WEBLEY D. (1966) Waterproofing of metaldehyde on bran baits for slug control. *Nature, Lond.* **212**, 320–1.

WILSON E.O. (1962) Chemical communication among workers of the fire ant *Solenopsis saevissima* (Fr. Smith). 1. The organization of mass foraging. *Anim. Behav.* **10**, 134–64.

WILSON E.O. & W.L. BROWN (1958) Recent changes in the introduced populations of the fire ant. *Evolution, Lancaster, Pa.* **12**, 211–18.

ZÁRATE L.L. (1966) Cebos contra la hormiga 'coqui' *Atta cephalotes* L. en Tingo Maria. *Revta peru. Ent. agric.* 7, 45–9.

Insect control by hormones

C. N. E. RUSCOE *Plant Protection Ltd., Jealott's Hill Research Station, Bracknell, Berkshire*

The control of insects by hormones—or more specifically by chemicals acting on insect endocrine systems—is often cited as a method of biological control. But it simply represents a sophisticated form of chemical control—the 'third generation pesticide' (Williams 1967). Use of hormones to control insect pests is likely to parallel in many respects (for example those of synthesis, application and assessment) the use of conventional insecticides.

Development of insecticides has, however, generally followed the pattern of random screening for active compounds, followed by a later analysis of mode of action. The discovery of the potential of hormone insecticides, on the other hand, arose initially from a study of the physiology of insect growth and development. Insect control by hormones therefore provides an excellent example of the contribution made by biological studies to pest control. The method does potentially offer some of the advantages of pure or integrated biological control—for example, specificity of action, avoiding effects on beneficial insects, wildlife and man, and also the possibility that resistance might be slow to develop.

Insect control by hormones is still a long way from being a practical proposition however. Although there have been a number of reports of successes in the field, there has been an almost complete lack of detailed information. I must therefore concentrate on describing the features of insect endocrine systems which make them suitable (or unsuitable) for insecticidal exploitation, indicating in the light of results which are available, how near we are to practical use of hormone insecticides. The most important endocrine system in this context is that which involves the insect juvenile hormone.

Juvenile hormone

This hormone (Fig. 1) has a number of functions, the most striking being the promotion of the characters of the immature insect (Roller & Bjerke 1965, Wigglesworth 1970). When the hormone is present, only larval or nymphal moults occur, but when secretion ceases the insect matures to the adult.

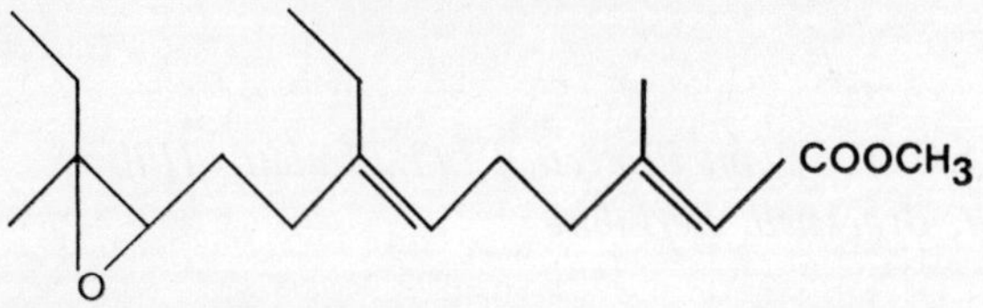

$COOCH_3$ O

Figure 1. Juvenile hormone.

The hormone is secreted again in the mature insect, where it is involved in male accessory gland function, and vitellogenesis in the female (Wigglesworth 1970). It also promotes the secretion of moulting hormone by the thoracic glands (Krishnakumaran & Schneiderman 1965).

Juvenile hormone mimics

Application of the hormone during periods of tissue growth which do not normally occur in the presence of high titres of endogenous hormone can cause disruption of maturation and reproduction. These effects can also be produced by a number of compounds which mimic the action of juvenile hormone. Although application of the hormone can cause lethal growth disruption or chemosterility in a wide spectrum of insects, at doses comparable with those of conventional insecticides, certain of the mimics appear to have similar or greater activity, and this suggests that such compounds can be used insecticidally.

The mimics are mainly aliphatic or monocyclic sesquiterpenes, or aliphatic monoterpenes attached to aromatic rings. In the first category, as well as juvenile hormone itself, are farnesol and various highly active derivatives such as DMF and DEF (the methyl and ethyl esters of farnesoic acid dihydrochloride) (Law *et al.* 1966, Slama *et al.* 1969) (Fig. 2). In the second category is juvabione, the 'paper factor' of Slama & Williams (1966), and

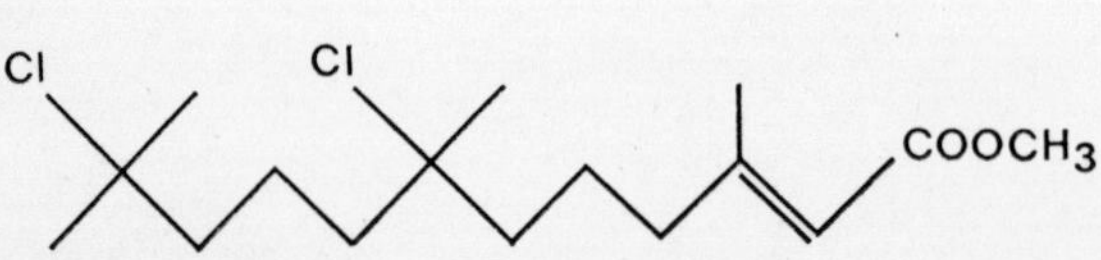

Cl Cl $COOCH_3$

Figure 2. DMF.

aromatic analogues such as para-substituted benzoates (Slama *et al.* 1968). Mimics in this category are active only on cotton stainer and linden bugs of the family Pyrrhocoridae (Fig. 3). Urethane and peptide derivatives of the two groups (Schwarz *et al.* 1970, Zaoral & Slama 1970), and compounds of generally similar structure, such as terpenoid ethers (Bowers 1969, Pallos *et al.* 1971, Cruickshank & Palmere 1971) are also potent juvenile hormone mimics.

Mimetic activity is also found in certain non-isoprenoid straight-chain compounds, such as dodecyl methyl ether (Schneiderman *et al.* 1965), and insecticide synergists such as sesoxane (Bowers 1968) (Fig. 4). Hybrid compounds, intermediate in structure between the synergist and terpenoid types, are also very active (Bowers 1969, McGovern *et al.* 1971, Walker & Bowers 1970) (Fig. 5).

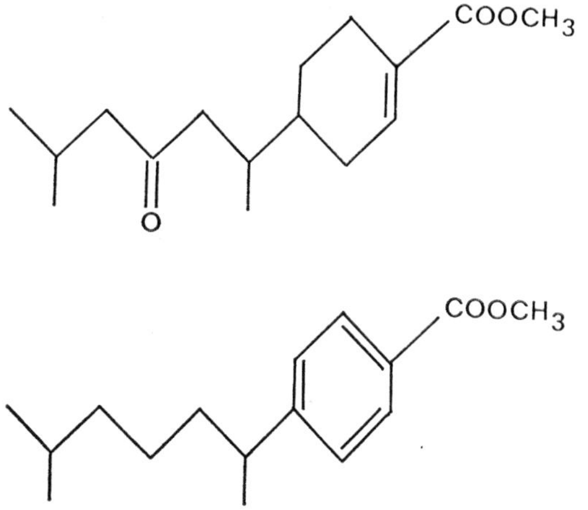

$COOCH_3$

O

$COOCH_3$

Figure 3. Above: juvabione. Below: methyl (dimethylhexyl) benzoate.

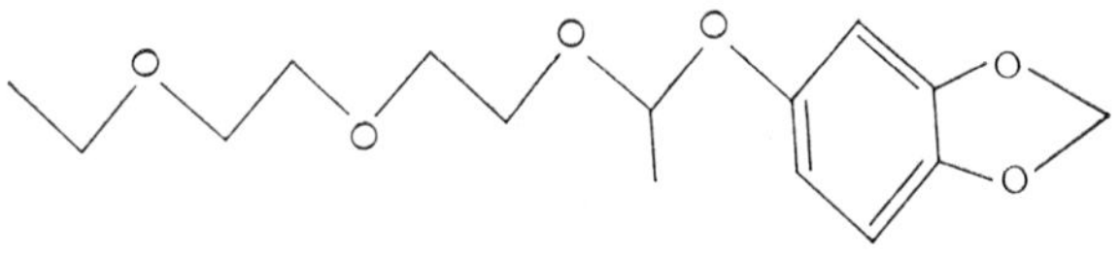

O O O O O O

Figure 4. Sesoxane.

It is likely that the mimics function by occupying juvenile hormone receptors, either at primary sites involved in the juvenile gene 'switching-on' function of the hormone (Wigglesworth 1970), or at secondary sites where the hormone is degraded. The latter possibility is attractive in view of the mimetic action of the synergists, which prevents degradation of conventional insecticides (Bowers 1968, Dyte 1969).

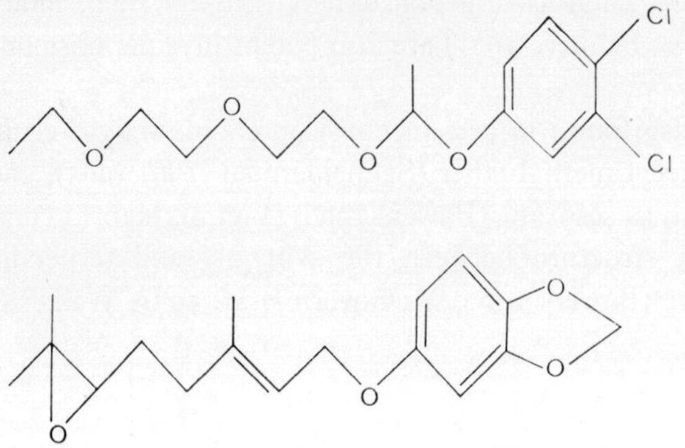

Figure 5. Hybrids.

Insecticidal effects

I would now like to describe the disruptive effects of juvenile hormone or its mimics when applied as insecticides.

In the hemimetabolous insects such as bugs, lice and cockroaches, treatment of final instar nymphs causes them to moult to sterile supernumerary nymphs or, at lower doses, to deformed sterile adults. At still smaller doses there may be no morphological effects, but the resultant adults are sterile (Slama & Williams 1966, Vinson & Williams 1967, Critchley & Campion 1971).

In the Holometabola, the beetles, butterflies and flies, treated final instar larvae moult to pupal/adult intermediates, pupal development may become blocked, or the pupal/adult moult disrupted (Wigglesworth 1969, Bowers 1969, Slama *et al.* 1969).

Insect eggs abort at blastokinesis if treated early in development (Slama & Williams 1966, Vinson & Williams 1967, Walker & Bowers 1970). Eggs may be treated directly, or may be affected while still within the body of the female. Masner and his co-workers (Masner *et al.* 1970) showed that DMF and DEF may also be passed from treated males in copulation, sterilizing the females.

Mimic applications to more developed eggs may produce subsequent disruption of larval or nymphal development (Riddiford & Williams 1967, Riddiford 1970).

Toxicology

Mimics for which data are available possess extremely low levels of mammalian toxicity. The acute oral LD_{50} (rat) of DMF is over 1600 mg/kg (Downing *et al.* 1970), that of various terpenoid ethers and amines is over 3000 mg/kg (Cruickshank & Palmere 1971, Pallos *et al.* 1971), and of the synergist sesoxane it is 2000 mg/kg. Conventional insecticides with an acute oral LD_{50} of over 500 mg/kg are considered reasonably safe, and the figures for the mimics are over an order of magnitude greater than those for organophosphates (OPs) in common use.

The juvenile hormone mimics appear to have no undue side-effects on mammals (Ellis *et al.* 1970) and there is no evidence of carcinogenicity among the sesquiterpenes, for example. It has been suggested that use of mimics as insecticides might produce unwanted effects on the *insects* however—for example the production of giant larvae. But although supernumerary Hemipteran nymphs (e.g. *Rhodnius*) have good appetites, it must be remembered that they result from large doses of hormone. So far, it is only in the laboratory that giant supernumeraries have been produced, so reports of crop destruction by large Colorado beetle larvae resulting from hormone treatment (Walton 1968) may be somewhat exaggerated.

Specificity

The fact that juvabione and a number of related mimics are active only on Pyrrhocorid bugs (Slama *et al.* 1968, Zaoral & Slama 1970), and that structural changes in chemical series of mimics can shift the activity spectrum (Romanuk 1970), has suggested that compounds could be 'tailor-made' to be selective against important natural pest groups (Williams 1967). But no mimics selective in action against, for example, Lepidoptera or aphids have yet been described. It appears that the Pyrrhocoridae are particularly sensitive to hormone mimics and growth disruptive compounds in general.

Insecticidal use

Even the non-selective juvenile hormone mimics have properties which suggest that they will never be used as broad spectrum insecticides. The lack of larvicidal effect means that pest targets must be those in which there is a relatively short life cycle, and/or where the first larval generation does little or no damage. If the pest is resistant to conventional insecticides, and lives in a situation requiring use of non-toxic chemicals, it is even more suited for control by juvenile hormone mimics.

Juvenile hormone insecticides are therefore most likely to be used against certain Lepidoptera and Diptera, aphids, and stored product Coleoptera. Target Lepidoptera are those which are multivoltine in warm regions, the first generation causing negligible damage. In this category are some of the fruit moths (for example codling and olive moths) or *Spodoptera* in N. Egypt where small winter generations develop on Berseem clover, and the subsequent summer generations migrate to cotton (Bishara 1934). In these cases, disruption at the end of the first generation should prevent significant crop damage provided there is no later immigration of pests. Certainly codling moth adults tend to remain in the orchard where they emerged (Balachowsky & Mesnil 1935). It has been reported that infestations of this pest have been controlled by applications of 20 mg of juvenile hormone per tree (Roller 1968) whereas standard compounds such as azinphos-methyl may have to be used at rates of 0·5 g per tree. Fruit tree moth control by hormone mimics may therefore be a practical proposition, but no results of large-scale trials have yet appeared.

Treatment of aphids in the laboratory with DMF is reported to disrupt growth and reduce fecundity and alate development (White 1968). But the latter effect is of little use in combating established infestations and relatively high doses of the mimic are required to produce sufficient disruption of growth and reproduction to act as a control measure. Applications of approximately 1000 times the dose required to disrupt the growth of Lepidoptera are needed (White & Lamb 1968). Obviously mimics much more active than DMF are required as hormone aphicides.

Stored product Coleoptera can in theory be controlled by treatment of stores or grain with mimics, which should prevent reproduction of invading beetles. Many of the species concerned are organophosphorus-resistant, and an insecticide of low mammalian toxicity is required. A number of juvenile hormone mimics are active on stored product Coleoptera as growth disruptors and chemosterilants, and certain of the hybrid mimics have contact and fumigant ovicidal action, at rates equivalent to organophosphorus compounds such as parathion (Walker & Bowers 1970, McGovern *et al.* 1971). However, grain admixture experiments with DMF have shown that even this active mimic has to be used at rates one hundred times greater than those of standard organophosphorus compounds to give similar control, even of resistant stored product beetles (Thomas & Bhatnagar-Thomas 1968).

Providing that mimics stable to hydrolysis can be found (Pallos *et al.* 1971), a juvenile hormone mosquito 'larvicide' seems to be one of the most likely commercial uses for mimics. Obviously the lack of larvicidal effect is unimportant in this case if adult emergence can be prevented, and *Aedes aegypti* larvae reared in a 2·5 ppm solution of DMF are unable to become adult (Spielman & Williams 1966, Spielman & Skaff 1967). Although conventional larvicides are active at lower doses, the development of resistance is decreasing the gap and increasing toxicity problems. The Monsanto compound MON 0585 (Fig. 6), a tertiary butylated phenol derivative, is an antoxidant rather than a true juvenile hormone mimic, but like the hormone it causes a block at metamorphosis (Sacher 1971). The compound is cheap, is of low mammalian and fish toxicity (1890 mg/kg, rat, acute oral) and gives control of resistant mosquitoes at rates of 1–2 kg/ha. This rate, however, is still 40–80 times the recommended rates of larvicides such as 'Dursban'.

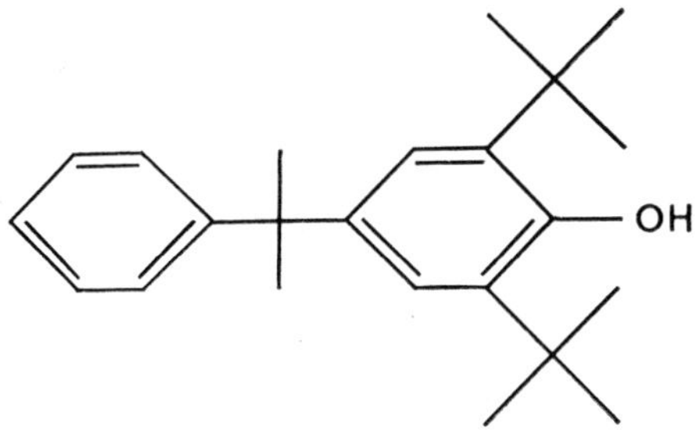

OH

Figure 6. MON 0585.

Juvenile hormone mimics may also find an outlet in control techniques using sterile male release. The low toxicity of the mimics and lack of effect on mating competitiveness at low concentrations may prove an advantage over conventional chemical or irradiation procedures, especially if the 'venereal disease' effect of mimic transfer in copulation can be utilized (Masner *et al.* 1970). Once again suitable mimics will have to be cheap and active to avoid making an already expensive technique more so.

The low toxicity of the mimics also suggests that they could be used in medical or veterinary outlets where conventional insecticides are too toxic. Insect parasites may not be the only pests to be controlled; farnesol derivatives have disruptive effects on mammalian gut nematodes of the genus *Trichinella* (Shanta & Meerovitch 1970) and juvenile hormone inhibits the growth of trypanosomes (Mors *et al.* 1967, Ilan *et al.* 1969).

Requirements

Juvenile hormone mimics of greater intrinsic activity are obviously required for insecticidal use, but probably their most important failing lies in their lack of persistence. The mode of action of the mimics means that each insect has to be treated when it reaches the sensitive stage in the final instar nymph or larva. This means that under field conditions a greater degree of persistence than that of most organophosphorus compounds or carbamates is required. Juvenile hormone and many terpenoid mimics are somewhat unstable however, due partly to the presence of epoxy and ester groups. Diversion from the close analogues of juvenile hormone may provide more stable mimics (Pallos *et al.* 1971) and it is compounds such as geranyl ethers and synergist hybrids that appear most promising from this point of view.

Moulting hormone

The hormone system governing insect ecdysis is another aspect of insect endocrine physiology that appears to have insecticide potential. The moulting hormones of insects are steroids, and include α-ecdysone and its 20-hydroxy derivative β-ecdysone (Butenandt & Karlson 1954, Karlson 1956) (Fig. 7).

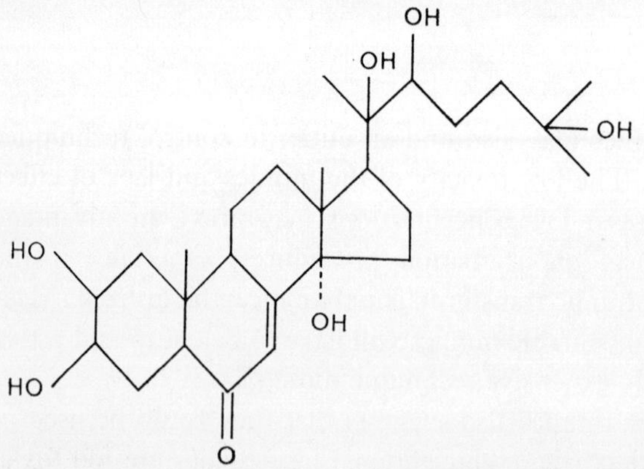

Figure 7. β-Ecdysone.

Both stimulate ecdysis when injected, although it appears that β-ecdysone, as well as being identical with crustecdysone, the moulting hormone of crustacea, is the predominant hormone in insects. α-ecdysone may form an intermediate in its synthesis (Wier 1970, Willig *et al.* 1971).

Most relevant to insecticidal use are the results of investigations involving contact or dietary application of ecdysones and their analogues. Both α- and β-ecdysone inhibit growth and cause insect sterility when supplied at low concentrations in the diet of Diptera (*Stomoxys*, *Musca*) and Coleoptera (*Tribolium*) (Robbins *et al.* 1968, 1970, Wright & Kaplanis 1970).

Moulting hormone mimics: insecticidal effects

A number of steroids of plant origin and various synthetic moulting hormone analogues have disruptive activity as great as or greater than that of the insect ecdysones when ingested (Slama 1969, Berkoff 1969, Ellis *et al.* 1970). Cyasterone and ponasterone are among the most active phytoecdysones (Fig. 8), and the synthetic compound 'Triol' (essentially ecdysone with a chloresterol side chain) is a potent growth inhibitor and chemosterilant for Coleoptera (*Anthonomus*, *Tribolium*), Diptera (*Musca*), Lepidoptera (*Manduca*) and Orthoptera (*Blattella*) (Robbins *et al.* 1968, 1970).

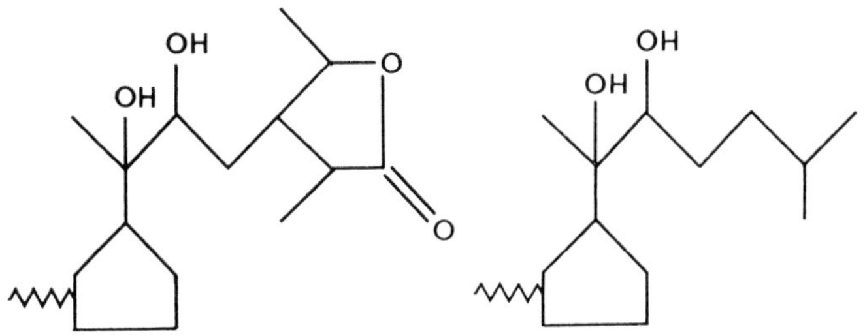

Figure 8. Left: Cyasterone. Right: Ponasterone. Only the side-chain structure is given, the five-membered ring being identical with that of β-ecdysone.

The effects of ingestion of the ecdysone analogues are mortality during the larval instars, especially at ecdysis and metamorphosis, and sterility of females, probably due to inhibition of vitellogenesis (Robbins *et al.* 1968, 1970, Earle *et al.* 1970, Wright *et al.* 1971). Despite reports that large doses of ecdysones and phytoecdysones are active when applied topically in suitable solvents (Sato *et al.* 1968, Kaplanis *et al.* 1970), it seems likely that the compounds known at present are much more active when ingested.

Insect moulting hormones are synthesized from chloresterol (Willig *et al.* 1971), and 'Triol' can form an intermediate in this synthesis, which may explain its high activity (Kaplanis *et al.* 1969). Ingestion of nitrogen-containing chloresterol analogues (azochloresterols) blocks larval tobacco hornworm (*Manduca*) and boll weevil (*Anthonomus*) development by preventing formation of chloresterol from other dietary steroids (Svoboda & Robbins 1967, Earle *et al.* 1967). Certain compounds are known which, although producing effects on insects similar to those of the ecdysone analogues, have rather less structural relationship with the moulting hormones. Azadirachtin is an oxidized triterpene which is a systemic antifeedant against locusts (Butterworth & Morgan 1971, Gill & Lewis 1971). In other insects, including Lepidoptera, Coleoptera and Hemiptera, it is a potent inhibitor of larval growth and at lower doses disrupts metamorphosis (Ruscoe 1972). Although azadirachtin is probably a metabolic poison rather than a hormone mimic, it shows that compounds which inhibit growth and moulting at low doses need not necessarily be steroids (Berkoff 1969).

Insecticidal use

The disruption of larval growth by ecdysone analogues has obvious advantages over the use of juvenile hormone mimics for pest control. The low contact/high ingestion activity of the compounds suggests, however, that they will act selectively against gross feeders such as Lepidoptera, Coleoptera and Acrididae, and they will probably have little effect on insect predators. It is also possible that the ecdysone analogues have plant systemic action, in view of their water-solubility and the natural occurrence of phytoecdysones. Many of the compounds appear to have the necessary intrinsic activity for insecticidal use. The synthetic ecdysones and azochloresterols produce lethal growth disruption in the parts per million range in the diet (Svoboda & Robbins 1967, Robbins *et al.* 1970, Earle *et al.* 1970) while a spray deposit of azadirachtin on a leaf surface causes death of tobacco budworm (*Heliothis*) larvae at rates much lower than those used for standard organophosphorus compounds. But it is vital that candidate moulting hormone insecticides be highly active, since they are even more costly to produce than juvenile hormone mimics. The absence of published field results may reflect this high price of production.

Toxicology

Although ecdysone analogues are relatively stable, and may have systemic activity, both these factors intensify the toxicological disadvantages of moulting hormone mimics. Admittedly, dosing experiments on mammals with ecdysones have generally produced no ill-effects (Ellis *et al.* 1970, Berkoff 1969), and ecdysone, far from being carcinogenic, seems to cause regression of sarcoma tumours in mice (Burdette 1964). But azochloresterols and β-ecdysone induce low chloresterol levels in mammals (Lupien *et al.* 1969), and the fact that steroids form the basis of vertebrate endocrine systems (Fieser & Fieser, 1959) means that the long-term mammalian effects of candidate moulting hormone insecticides will have to be examined closely.

There will be great difficulties in marketing a compound both as an insecticide and as a contraceptive pill.

β-Ecdysone is also the moulting hormone of Crustacea and arachnids (Krishnakumaran & Schneiderman 1969), and ecdysones also have growth effects on symbiotic Protozoa (Cleveland *et al.* 1960), trematodes (Muftic 1969) and plants (Carlisle *et al.* 1963). This also suggests that the environmental problems arising from the use of moulting hormone insecticides may be far greater than the already formidable difficulties associated with the use of conventional compounds.

Other hormones

Finally, it is worth mentioning the hormones involved in excretion and salivation in relation to insect control. A number of conventional insecticides induce extensive diuresis in *Rhodnius*, disturbing the blood ionic equilibrium and leading to desiccation (Casida & Maddrell 1971). Desiccation will also occur if extensive salivation is induced, while prevention of the secretion may lead to starvation. 5-Hydroxytryptamine stimulates the function both of Malpighian tubules and of salivary glands and a number of analogues and some less closely related compounds have stimulatory or inhibitory effects on the glands *in vitro* (Maddrell *et al.* 1971, Berridge 1972). There is therefore a possibility of insecticidal use of diuretic or salivary hormone mimics.

Perhaps the most important aspect of these studies is the possibility that other insect endocrine systems may be used as targets for insecticides. The endocrine and neurosecretory systems specific to insects—for example those involved in cuticle plasticization or hardening, and in blood sugar regulation—obviously all deserve study from this angle.

Conclusions

In summary, we have seen that hormone mimics are known which have lethal growth-disruptive effects on insects at rates approaching those of organophosphorus compounds and carbamates. But the majority of these compounds are expensive—even the most optimistic estimates of production prices are over twice those of projected new conventional insecticides. A high level of activity and hence a low rate of application is therefore a vital economic requirement for a hormone insecticide. Furthermore, many of the juvenile hormone mimics are somewhat unstable, yet it is these compounds especially which must have a long residual life for insecticidal use. This situation may be improved by formulation—for example by ultra-violet protection or encapsulation, but this will further increase the price. Price considerations will also limit the development of mixed hormone and conventional insecticides (Critchley & Campion 1971).

It has been predicted that hormone insecticides might not suffer from the development of pest resistance (Williams 1967). But resistance will develop to any form of control which leaves survivors; even constant use of the fly-swat will select for flies with quicker reactions. Hormone degradation systems are present in insects, and could be adapted for the breakdown of exogenously applied hormones (Mordue 1969). It is possible, however, that these breakdown systems are themselves being blocked competitively by the mimics (Dyte 1969) and this may tend to inhibit the development of resistance.

In this context it should be noted that, despite much use, there is negligible plant resistance to hormone herbicides, and it seems possible that insect hormone mimics have provided a defence measure for certain plants for a great length of time (Slama 1969, Schneiderman *et al.* 1970). Furthermore, hormone insecticides may provide a method of combating organophosphorus-resistant strains and could also overcome the phenomenon of cross-resistance, where insects treated with, for example, a particular organophosphorus compound, develop a general resistance to organophosphorus compounds and carbamates (Grayson & Cochran 1968).

Perhaps the most important result of the consideration of insect control by hormone techniques is that major insecticide companies now screen chemicals for delayed growth disruption effects as well as for quick kill. This procedure brings to light not only juvenile or moulting hormone mimics, but also compounds which disrupt insect growth for less apparent reasons, perhaps by a more indirect effect on general hormone balance or hormone regulated processes. It seems probable that the successful 'hormone' insecticide will be a compound of this type—one which has effects similar to the hormone mimics, but is also very active and inexpensive. Development of such chemicals should extend the concept of insect control by 'hormones' from the theoretical to the practical plane.

References

BALACHOWSKY A. & L. MESNIL (1935) *Les insectes nuisibles aux plantes cultivées.* Vol 1, Etabl. Busson, Paris.

BERKOFF C.E. (1969) The chemistry and biochemistry of insect hormones. *Q. Rev. chem. Soc.* 23, 372–91.

BERRIDGE M.J. (1972) The mode of action of 5-hydroxytryptamine. *J. exp. Biol.* 56, 311–22.

BISHARA I. (1934) The cotton worm (*Prodenia litura* F.) in Egypt. *Bull. Soc. R. ent. Égypte.* 18, 288–420.

Bowers W.S. (1968) Juvenile hormone: activity of natural and synthetic synergists. *Science* **161**, 895–7.

Bowers W.S. (1969) Juvenile hormone: activity of aromatic terpenoid ethers. *Science* **164**, 323–5.

Burdette W.J. (1964) Hormonal heterophily and the control of growth. *Acta Un. int. Cancr.* (1964) **20**, 1531–3.

Butenandt A. & P. Karlson (1954) Uber die Isolierung eines Metamorphose-Hormons der Insekten in Kristallisierter Form. *Naturf.* (B) **9**, 389–91.

Butterworth J.H. & E.D. Morgan (1971) Investigation of the locust feeding inhibition of the seeds of the Neem tree, *Azadirachta indica*. *J. Insect. Physiol.* **17**, 969–77.

Carlisle D.B., D.J. Osborne, P.E. Ellis & J.E. Moorhouse (1963) Reciprocal effects of insect and plant growth substances. *Nature, Lond.* **200**, 1230–1.

Casida J.E. & S.H.P. Maddrell (1971) Diuretic hormone release on poisoning *Rhodnius* with insecticide chemicals. *Pestic. Biochem. Physiol.* **1**, 71–83.

Critchley B.R. & D.G. Campion (1971) Effects of a juvenile hormone analogue on growth and reproduction in the cotton stainer *Dysdercus fasciatus* Say. *Bull. ent. Res.* **61**, 49–53.

Cleveland L.R., A.W. Burke, Jr. & P. Karlson (1960) Ecdysone induced modifications in the sexual cycle of the protozoa of *Cryptocercus*. *J. Protozool.* **7**, 229–39.

Cruickshank P.A. & R.M. Palmere (1971) Terpenoid amines as insect juvenile hormones. *Nature, Lond.* **233**, 488–9.

Downing F.S., N. Punja & C.N.E. Ruscoe (1970) Insect hormone mimics and their use in insect control. *Rep. Progr. appl. Chem.* **55**, 446–57.

Dyte C.E. (1969) Evolutionary aspects of insecticide selectivity. *Proc. IVth Br. Insectic. Fungic. Conf.* (1969), 393–7.

Earle N.W., E.N. Lambremont, M.L. Burks, B.H. Slatten & A.F. Bennett (1967) Conversion of β-sitosterol to chloresterol in the boll weevil and the inhibition of larval development by two aza sterols. *J. econ. Ent.* **60**, 291–3.

Earle N.W., I. Padovani, M.J. Thompson & W.E. Robbins (1970) Inhibition of larval development and egg production in the boll weevil following ingestion of ecdysone analogues. *J. econ. Ent.* **63**, 1064–9.

Ellis P., E.D. Morgan & A.P. Woodbridge (1970) Is there new hope for hormone mimics as pesticides? *Pestic. Abstr.* **16**, 434–46.

Fieser L.F. & M. Fieser (1959) *Steroids*. Chapman & Hall, London.

Gill J.S. & C.T. Lewis (1971) Systemic action of an insect feeding deterrent. *Nature, Lond.* **232**, 402–3.

Grayson J.M. & D.G. Cochran (1968) The phenomenon of cross resistance in insects: empirical, theoretical and genetical considerations. *Wld Rev. Pest Control* **7**, 172–5.

Ilan J., J. Ilan & S. Ricklis (1969) Inhibition by juvenile hormone of growth of *Crithidia fasciculata* in culture. *Nature, Lond.* **224**, 179–80.

Kaplanis J.N., W.E. Robbins, M.J. Thompson & A.H. Baumhover (1969) Ecdysone analog: conversion to alpha ecdysone and 20-hydroxy-ecdysone by an insect. *Science* **166**, 1540–1.

Kaplanis J.N., M.J. Thompson & W.E. Robbins (1970) The effects of ecdysones and analogues on ovarian development and reproduction in the housefly *Musca domestica* (L.). *Proc. XIIIth Int. Congr. Entomol.* (1968) 393.

Karlson P. (1956) Biochemical studies on insect hormones. *Vitamins and Hormones* **14**, 227–66.

Krishnakumaran A. & H.A. Schneiderman (1965) Prothoracicotropic activity of compounds that mimic juvenile hormone. *J. Insect Physiol.* **11**, 1517–32.

Krishnakumaran A. & H.A. Schneiderman (1969) Induction of moulting in Crustacea by an insect moulting hormone. *Gen. and Compar. Endocr.* **12**, 515–8.

Law J.H., C. Yuan & C.M. Williams (1966) Synthesis of a material with high juvenile hormone activity. *Proc. natn. Acad. Sci. U.S.A.* 55, 576–8.

Lupien P.J., C. Hinse & K.D. Chaudhary (1969) Ecdysone as a hypochloresterolemic agent. *Archs int. Physiol. Biochem.* 77, 206–12.

Maddrell S.H.P., D.E.M. Pilcher & B.O.C. Gardiner (1971) Pharmacology of the Malpighian tubules of *Rhodnius* and *Caurausius*: the structure-activity relationship of tryptamine analogues and the role of cyclic AMP. *J. exp. Biol.* 54, 535–73.

Masner P., K. Slama, J. Zdarek & V. Landa (1970) Natural and synthetic materials with insect hormone activity. X. A method of sexually spread insect sterility. *J. econ. Ent.* 63, 706–10.

McGovern T.P., R.E. Redfern & M. Beroza (1971) Juvenile hormone activity of acetals applied topically and as a vapour to the yellow mealworm. *J. econ. Ent.* **64**, 238–41.

Mordue W. (1969) Insect hormones and analogues and their potential contribution to insect control. *Proc. Vth Br. Insectic. Fungic. Conf.* (1969) **2**, 386–91.

Mors W.B., H.J. Montiero, B. Gilbert & J. Pellegrino (1967) Chemoprophylactic agent in schistosomiasis: 14,15-Epoxygeraniol. *Science* **157**, 950–1.

Muftic M. (1969) Metamorphosis of miracidia into cercariae of *Schistosoma mansoni* in vitro. *Parasitology* 59, 365–71.

Pallos F.M., J.J. Menn, P.E. Letchworth & J.B. Miaullis (1971) Synthetic mimics of insect juvenile hormone. *Nature, Lond.* **232**, 486–7.

Riddiford L.M. (1970) Prevention of metamorphosis by exposure of insect eggs to juvenile hormone analogues. *Science* **167**, 287–8.

Riddiford L.M. & C.M. Williams (1967) The effects of juvenile hormone analogues on the embryonic development of silkworms. *Proc. natn. Acad. Sci. U.S.A.* 57, 595–601.

Robbins W.E., J.N. Kaplanis, M.J. Thompson, T.J. Shortino, C.F. Cohen & S.C. Joyner (1968) Ecdysones and analogues: effects on development and reproduction of insects. *Science* **161**, 1158–60.

Robbins W.E., J.N. Kaplanis, M.J. Thompson, T.J. Shortino & S.C. Joyner (1970) Ecdysones and synthetic analogues: moulting hormone activity and inhibitive effects on insect growth, metamorphosis and reproduction. *Steroids* **16**, 105–25.

Roller H. (1968) In C.M. Williams and W.E. Robbins. I.B.P. Conference on insect-plant interactions. *Bioscience* 18, 791–9.

Roller H. & J.S. Bjerke (1965) Purification and isolation of juvenile hormone and its action in lepidopteran larvae. *Life Sciences* 4, 1617–24.

Romanuk M. (1970) On the selectivity of the effect of some juvenile hormone analogues. *Proc. VIIth Int. Cong. Plant Prot.* 1970.

Ruscoe C.N.E. (1972) Growth disruption effects of an insect antifeedant. *Nature, New Biology* **236**, 159–60.

Sacher R.M. (1971) A hormone-like mosquito larvicide with favourable environmental properties. *Proc. VIth Br. Insectic. Fungic. Conf.* (1971) **2**, 611–16.

Sato Y., M. Sakai, S. Imai & S. Fujioka (1968) Ecdysone activity of plant-originated moulting hormones applied on the body surface of lepidopterous larvae. *Appl. Ent. Zool.* 3, 49–51.

Schneiderman H.A., A. Krishnakumaran, V.G. Kulkarni & L. Friedman (1965) Juvenile hormone activity of structurally unrelated compounds. *J. Insect. Physiol.* **11**, 1641–9.

SCHNEIDERMAN H.A., A. KRISHNAKUMARAN, P.J. BRYANT & F. SEHNAL (1970) Endocrinological strategies in insect control. *Agric. Sci. Rev.* 1970, 13–25.

SCHWARZ, M., N. WAKABAYASHI, P.E. SONNET & R.E. REDFERN (1970) Compounds related to juvenile hormone. VII. Activity of selected nitrogen-containing terpenoid compounds on the yellow mealworm. *J. econ. Ent.* 63, 1858–60.

SHANTA C.S. & E. MEEROVITCH (1970) Specific inhibition of morphogenesis in *Trichinella spiralis* by insect juvenile hormone mimics. *Can. J. Zool.* 48, 617–20.

SLAMA K. (1969) Plants as a source of materials with insect hormone activity. *Entomologia exp. appl.* **12**, 721–8.

SLAMA K., M. SUCHY & F. SORM (1968) Natural and synthetic materials with insect hormone activity. 3. Juvenile hormone activity of derivatives of p (1,5-dimethylhexyl) benzoic acid. *Biol. Bull. mar. biol. Lab., Woods Hole* **134**, 154–9.

SLAMA K., M. ROMANUK & F. SORM (1969) Natural and synthetic materials with insect hormone activity. 2. Juvenile hormone activity of some derivatives of farnesinic acid. *Biol. Bull. mar. biol. Lab., Woods Hole* **136**, 91–5.

SLAMA K. & C.M. WILLIAMS (1966) The juvenile hormone. V. The sensitivity of the bug *Pyrrhocoris apterus* to a hormonally active factor in American paper pulp. *Biol. Bull. mar. biol. Lab., Woods Hole* **130**, 235–46.

SPIELMAN A. & V. SKAFF (1967) Inhibition of metamorphosis and ecdysis in mosquitoes. *J. Insect Physiol.* **13**, 1087–95.

SPIELMAN A. & C.M. WILLIAMS (1966) Lethal effects of synthetic juvenile hormone on larvae of the yellow fever mosquito, *Aedes aegypti. Science* **154**, 1043–4.

SVOBODA J.A. & W.E. ROBBINS (1967) Conversion of beta sitosterol to chloresterol blocked in an insect by hypocholesterolemic agents. *Science* **156**, 1637–8.

THOMAS P.J. & P.L. BHATNAGAR-THOMAS (1968) Use of a juvenile hormone analogue as insecticide for pests of stored grain. *Nature, Lond.* **219**, 949.

VINSON J.W. & C.M. WILLIAMS (1967) Lethal effects of synthetic juvenile hormone on the human body louse. *Proc. natn. Acad. Sci. U.S.A.* 58, 294–7.

WALKER W.F. & W.S. BOWERS (1970) Synthetic juvenile hormones as potential coleopteran ovicides. *J. econ. Ent.* 63, 1231–3.

WALTON P.D. (1968) Future prospects for commercial use of hormone insecticides. *Internat. Pest Control.* Sept.–Oct. 1968, 13–14.

WHITE D.F. (1968) Postnatal treatment of the cabbage aphid with a synthetic juvenile hormone. *J. Insect Physiol.* **14**, 901–12.

WHITE D.F. & K.P. LAMB (1968) Effect of a synthetic juvenile hormone on adult cabbage aphids and their progeny. *J. Insect Physiol.* **14**, 395–402.

WIER S.B. (1970) Control of moulting in an insect. *Nature, Lond.* **228**, 500–58.

WIGGLESWORTH V.B. (1969) Chemical structure and juvenile hormone activity: comparative tests on *Rhodnius prolixus. J. Insect Physiol.* **15**, 73–94.

WIGGLESWORTH V.B. (1970) *Insect hormones.* Oliver & Boyd, Edinburgh.

WILLIAMS C.M. (1967) Third-generation pesticides. *Scient. Am.* **217**, 13–17.

WILLIG A., H.H. REES & T.W. GOODWIN (1971) Biosynthesis of insect moulting hormones in isolated ring glands and whole larvae of *Calliphora. J. Insect Physiol.* **17**, 2317–26.

WRIGHT J.E., W.F. CHAMBERLAIN & C.C. BARRETT (1971) Ovarian maturation in stable flies: inhibition by 20-hydroxyecdysone. *Science* **172**, 1247–8.

WRIGHT J.E. & J.N. KAPLANIS (1970) Ecdysones and ecdysone-analogues: effects on fecundity of the stable fly, *Stomoxys calcitrans. Ann. ent. Soc. Amer.* 63 622–3.

ZAORAL, M. & K. SLAMA (1970) Peptides with juvenile hormone activity. *Science* **170**, 92–3.

Genetic control of insects

G. DAVIDSON *Ross Institute of Tropical Hygiene, London School of Hygiene and Tropical Medicine, Keppel Street, Gower Street, London*

It seems generally agreed that present-day methods of insect pest control by the use of chemicals are not the ideal, both from the point of view of efficiency and because they can lead to contamination of the environment with substances detrimental to other living organisms including man himself. Alternative methods must be sought, and genetical methods aimed at producing reproductive failure or at replacing harmful insect populations by harmless ones seem to offer perhaps the greatest potential. Here an attempt is made to review progress in those methods developed to the point of field application. Literature on laboratory aspects is already voluminous; some of it has been reviewed by Proverbs (1969).

The sterile insect technique

The most tested technique to date involves the swamping of natural populations with insects (usually males) sterilized by irradiation or by exposure to chemosterilants. One of the effects of such treatments is to induce dominant lethal mutations in the gamete chromosomes which, when they combine with those of the wild individual, produce lethal effects in developing embryos.

The success of such a technique depends on adequate mating between introduced and wild insect. The competitiveness of the introduced insects may be affected by the mass-rearing procedure necessary to produce the required numbers and the sterilizing treatment given. Mass-rearing procedures usually involve artificial methods, allowing the survival of far greater proportions of individuals than would survive in the field. A proportion of these might be expected to be less competitive than natural populations subjected to natural selection pressures. Sterilization by irradiation seems to affect subsequent survival, dispersal and mating activity more than does treatment with chemosterilants; but present efforts aim at careful control of irradiation dosage to achieve a high degree of sterility (not necessarily complete) with the least effect on competitiveness. The stage of development of the insect gonads at the time of sterilization seems all-important, with the optimum effect being at the time nearest to gamete maturity. For practical reasons the most convenient stage is usually the pupa, and deferment of treatment until near to the time of eclosion seems the ideal.

Where sterile males are released, the greatest effect on reproduction will be in those insect populations where females mate once only. The wild female mated by the sterile male may not then be subsequently mated by a fertile male and will lay sterile eggs for the rest of her life. Bryan (1968), using a forced mating technique, has shown this to be the case with *Anopheles gambiae* species B mated with a sterile hybrid male produced by crossing species B with *Anopheles melas*. She also showed that the reverse was true, namely that fertile females subsequently force-mated with sterile males continued to lay fertile eggs. Craig (1967) has shown the cause of single mating in *Aedes aegypti* to be a mating inhibitor (which he has called 'Matrone') in the accessory gland fluid of the male which is passed into the female with the seminal fluid.

The choice of the male insect for sterilization and release has largely been dictated in the medical and veterinary fields by the less important rôle this sex usually plays as a transmitter of human or animal disease (tsetse flies are a notable exception in that both sexes suck blood and transmit trypanosomiases). In the agricultural field the male may be chosen because it disperses more and mates more than once. A reliable sex-separation technique is therefore a usual requirement for large-scale sterile-insect release schemes. Such was not available in the case of the screw-worm and so both sexes were sterilized and released. Good separation techniques based on distinct size differences in the pupae are available for sexing *Culex pipiens fatigans* and *Aedes aegypti* and Whitten (1969) has devised a technique for the sexing of the Australian sheep blowfly, *Lucilia cuprina*, by selecting for a sex-limited pupal-colour mutant which can be detected automatically by exposing pupae to transmitted light. A simple machine sorting 7,000 pupae per hour has been made.

Concerned about the fact that, where the release of only one sex is being made, half the factory production may be wasted, Whitten & Taylor (1970) consider possible roles for sterile females. Where male polygamy and female monogamy prevail, as they do in many insects, the efficiency of sterile female releases will depend on the upper limit of male mating capacity—an average of 10 matings per male in *L. cuprina*, for example, and perhaps 4 in *A. gambiae* species A (Cuéllar *et al.* 1970). Under these conditions their sterilizing efficiency will be considerably less than that of males. However, if the female is polygamous (and Whitten & Taylor 1970, suggests that a selection for this condition can be made with *L. cuprina*) her sterilizing effect is enhanced. One would hesitate, however, to release large numbers of blood-sucking, disease-carrying female insects such as mosquitoes.

Knipling (1965), who was involved in the screw-worm eradication campaign from its inception, has since expounded the dynamic principles upon which the control of insects by insecticides and by the release of sterile insects depends. Table 1 is a modification of one of his tables and shows what

Table 1. Relative trends of hypothetical insect populations subjected to control by insecticide and by sterile-insect releases

Generation	Uncontrolled population	Insecticide treatment (90% kill in each generation)	Sterile insect releases (9,000,000 in each generation)	Ratio of sterile: normal insects
Parent	1,000,000	1,000,000	1,000,000	9:1
1	5,000,000	500,000	500,000	18:1
2	25,000,000	250,000	131,625	68:1
3	125,000,000	125,000	9,540	954:1
4	125,000,000	62,500	50	180,000:1
5	125,000,000	31,250	0	
6	125,000,000	15,625		
7	125,000,000	7,812		
8	125,000,000	3,906		
Total requirements for theoretical elimination of populations		Treatment for 18 generations	45,000,000 sterile insects	

Assumed uncontrolled population increase is 5-fold with a maximum supportable density of 125,000,000.

After Knipling (1965) *Proc. XII Int. Congr. Ent.*, 251.

happens to a hypothetical population of one million insects, whose rate of growth is 5-fold per generation, and whose environment can support a maximum population of 125,000,000, when it is subjected to insecticidal treatment inflicting a 90% kill in each generation and when 9,000,000 sterile insects are released into it in each generation. It is assumed that the sterile insects are equally competitive and that the female insect mates once only. Initially effects are the same, namely that the population is reduced to one-tenth, which with its 5-fold rate of increase results in a population size of 500,000. This in the case of the insecticide treatment is again subjected to a further

90% kill but in the case of the sterile male release it is subjected to a 95% sterilizing influence because the proportion of sterile to normal insects increases from 9:1 to 18:1. In subsequent generations this proportion increases rapidly. The net result is that while the insecticide treatment requires 18 generations to achieve eradication, the sterile male release requires only 5 generations or the release of 45,000,000 insects.

While serving to illustrate the general principle behind the sterile insect technique, this is a gross over-simplification of natural happenings and can certainly not be used for prediction purposes in practical control schemes. Geier (1969) points out that no insect population grows at a fixed rate and that the substitution of what he considers to be a more realistic rate of increase into Knipling's 'model' does not result in eradication. Both Cuéllar (1969a, b) and Conway (1970) have now produced detailed mathematical models applicable to mosquito populations and with the aid of a computer have simulated sterile male releases producing surprisingly similar results (see the discussion by Conway after the paper by Davidson & Kitzmiller 1970). Cuéllar's model and computer predictions served as a foundation for a field trial of sterile hybrid males for the control of a single species of the *Anopheles gambiae* complex (Davidson *et al.* 1970) and proved extremely useful, though since that time it has been realized that no allowance had been made for such things as a declining population with decreased availability of breeding sites, an accumulation of sterile males and a possible harassment effect of these large numbers of males on a few remaining females.

Sterilization by irradiation

The mass-release of insects sterilized by irradiation requires an irradiation source of such a size as to permit exposures of large numbers of insects for the shortest possible time. Gamma-irradiators are almost universally used, and nowadays compact, easily-transported sources are available which, though expensive, last a long time (^{60}Co has a half-life of 6 years and ^{137}Cs one of 30 years) and involve little or no risk in their use. In most insects it appears that the minimum irradiation dosage to sterilize males will certainly sterilize females and usually affects the longevity and activity of the latter. This can be an advantage where sex-separation is difficult.

Beyond a doubt the most impressive practical application of the sterile-insect method has been the eradication of the screw-worm (*Cochliomyia hominivorax*) from the island of Curaçao and from Florida and the south-western United States, by the release of vast numbers of factory-reared insects of both sexes sterilized by exposure of the pupae to 7,500 rad of gamma radiation from a ^{60}Co source. Bushland (1971), another of the original participants in this work, has recently reviewed the history and progress of this enormous undertaking to remove a pest which was causing damage to livestock estimated at 120 million dollars a year. The cost so far has been of the order of 60 million dollars, including a recurring annual cost of 6 million dollars to maintain a barrier-release of sterilized flies along the 1,500-mile-long border between the United States and Mexico to prevent invasion by flies from the latter country. In the Florida project, 400 males and 400 females per square mile per week were released from aircraft, in flight lanes 12 miles apart. This required a factory capacity for rearing 60 million flies per week. Later, for the southwestern states project, a factory was built at Mission, Texas, capable of producing 150 million flies per week. This quantity requires 20,000 pounds of dried blood and 200,000 pounds of meat to feed the maggots during their larval cycle of 5 to 6 days. The pupae are irradiated in batches of 35,000 some two days before eclosion and releases are made after the adults emerge.

The screw-worm eradication campaign remains the one big success in the use of irradiated insects, but in recent years a number of small-scale successes have been achieved with this technique against various fruit fly species in different parts of the world, e.g. the melon fly (*Dacus cucurbitae*) in Rota, Mariana Islands (Steiner *et al.* 1965), the oriental fruit fly (*Dacus dorsalis*) also in the Mariana Islands (Steiner *et al.* 1970, though here a combination of the release of sterilized flies with male-annihilation through attraction to a lure-insecticide was used), the Mexican fruit fly (*Anastrepha ludens*) on the California–Mexico border (quoted by Proverbs 1969), the Queensland fruit fly (*Dacus tryoni*) in New South Wales (Andrewartha *et al.* 1967), the Mediterranean fruit fly (*Ceratitis capitata*) on the Spanish island of Tenerife (Mellado 1971) and the Italian island of Capri (Nadel & Guerrieri 1969), and against the codling moth (*Carpocapsa pomonella*) in Canada (Proverbs *et al.* 1966).

Among the insects of public health importance, some attempts have been made to control the housefly (*Musca domestica*) and the blowfly (*Lucilia sericata*) by the release of irradiated males. Rivosecchi (1962) attempted to control houseflies in a coastal area of Latina province, Italy, by releasing 45,500 males sterilized by exposure to 2,000 rad between April and July. Sterile females began to show themselves one month after releases started, and in the following two months the female flies were markedly reduced. However, from August onwards there was again an increase in fertile flies. It was concluded that the area was insufficiently isolated to achieve eradication and that the sterile males were under-competitive. MacLeod and Donnelly (1961) tried without success to eradicate the blowfly from a small island off the north-east coast of England (Holy Island) during the years 1956 to 1958. The sterilization dosage was 6,000 to 7,000 rad which reduced longevity by 40 per cent and hence competitiveness. Donnelly (1965) estimates that normal males successfully copulate 6 and sometimes 12 times while the sterile male is for the most part only capable of copulating once and this does not necessarily prevent the female from subsequently accepting a normal male, nor from producing viable offspring.

The results of three attempts to control mosquitoes by the release of irradiated males have been published. The first concerned *Anopheles quadrimaculatus* in Florida and involved the release of 433,600 irradiated males (irradiated at 12,000 rad) over a period of 14 months (Weidhaas *et al.* 1962). No reduction in total numbers of mosquitoes nor increase in sterility of wild females resulted, and it was concluded that the laboratory strain used for mass-production was no longer competitive because it had been laboratory-bred for some 200 generations over the previous 25 years. Perhaps the irradiation dosage used was too high also. An attempt to control *Culex pipiens fatigans* in India, by the release over a period of 35 days of 24,000 males irradiated at 7,700 rad, resulted in only 6% reduction in the hatching of egg rafts and was discontinued because of objections from the people in the village where the trial was made (Krishnamurthy *et al.* 1962). Finally a field release of 4,777,000 male *Aedes aegypti* exposed to 11,000–18,000 rad over a period of 43 weeks near the town of Pensacola, Florida, produced little effect on the wild population (Morlan *et al.* 1962). Failure has been attributed to too high an irradiation dosage affecting male competitiveness and perhaps inadequate dispersal of the release points, for such a poor flier as *A. aegypti*.

Sterilization by chemosterilants

Chemosterilants are relatively cheap sterilizing agents but present certain toxicity hazards to man and his environment (they can be both mutagenic and carcinogenic). Being general toxicants for all living organisms they cannot be used for large-scale field applications, and handling precautions (though not over-complicated ones) are needed in their use. Their ideal method of use is in combination with an efficient specific attractant. Such an auto-sterilization mechanism placed in the insect's environment can do away with the necessity of having a factory source of insects and special sterilization and release procedures. Of the compounds most used so far, viz. metepa, tepa, thiotepa, hempa and apholate, thiotepa seems to be favoured in producing the least deleterious effects on treated insects. Unlike radiation, chemosterilants seem to have less sterilizing effect on the female than on the male.

Adequate sterilization, with no after-effects on male performance, from pupal immersion in chemosterilant solution, was first demonstrated for mosquitoes by White (1966) using *Aedes aegypti* and thiotepa. This method was adopted by Patterson *et al.* (1971) for large-scale experiments with *C.p. fatigans*, where a 4-hour exposure in 0·7% thiotepa was the routine. There have been two recent attempts to control this last mosquito with chemosterilants; one took place on the island of Seahorse Key off the coast of Florida in 1969 (Patterson *et al.* 1970). Male sterilization was achieved by forcing emerging adults to walk over surfaces treated with tepa. During 8 weeks 141,000 sterile males were released, an average of 2,500 per day. The ratio of sterile to normal males was calculated to vary from 2:1 to 5·7:1 and an 85% sterility was achieved before unfavourable weather conditions led to abandonment of the experiment. Another field trial in Kenya was unsuccessful (Bransby-Williams 1971). Here 2-day-old males were fed on 0·015% apholate in sugar solution to sterilize them and 32,350 were released over a period of 10 weeks from 3 points in a village where the wild density was of the order of 8 to 12 females per house in the first 5 of these weeks and fell to one per house in the tenth week. During most of this period, however, 77 to 100% of collected egg-rafts hatched.

Considerable success has been achieved using a combination of chemosterilant and sweetened bait against houseflies. Weekly, or preferably biweekly, applications of tepa, apholate, metepa or hempa to areas where the flies congregate, e.g. refuse dumps (LaBrecque *et al.* 1962, Gouck *et al.* 1963), poultry droppings (LaBrecque *et al.* 1963, Meifert & LaBrecque 1971) and privies (Meifert *et al.* 1967), all gave good control in the United States and the West Indies. Sacca & Stella (1964) similarly controlled flies in Italy by spraying garbage dumps with 0·0625 to 0·2% concentrations of tepa in syrup with 1% malt extract added. Magaudda *et al.* (1969) on the other hand, released male flies sterilized by feeding on 0·05% tepa in sugar syrup into an isolated population on the island of Vulcano near Sicily and achieved eradication, albeit temporarily.

Attempts have also been made to eradicate tsetse-flies (*Glossina morsitans*) from islands in Lake Kariba, Rhodesia (Dame & Schmidt 1970). Three experiments were made. In the first, males emerged from field-collected pupae were exposed to 10 mg tepa per square metre for 30–60 minutes and then released. The survival of these males was shown to be only 17% of that of wild males and the highest ratio of sterile to normal males achieved was only 12:100. No control resulted. In a second trial 26,000 sterile males were released over 20 months into an island population of some 600–1200 *G. morsitans*. The last female was seen three months before the last release but some of the population decline was attributed to a decrease in host animals. At the time of the third trial the fly density was estimated at 120 per square kilometre and was reduced by two aerial applications of dieldrin before sterilized flies were introduced. Sterilization was achieved by immersing puparia in 5% tepa and allowing the sterilized males to emerge 'on site';

98% control was achieved in nine months though never complete eradication.

Among agricultural pests, a measure of success was achieved in Alabama by dipping adults of the cotton boll weevil (*Anthonomus grandis*) in apholate solution (Davich 1969), though the conclusion was reached that the treated insects were not fully sterile nor were they equally competitive. Before irradiated flies were released in 1966, good control of the Mexican fruit fly (*Anastrepha ludens*) had been achieved by the release of tepa-sterilized flies. No infested fruit was found in either 1964 or 1965 (Shaw *et al.* 1966).

Hybrid sterility

Hybrid sterility resulting from the crossing of two closely related species is a well-known biological phenomenon. Such hybrids, in addition to being sterile, may show heterosis or hybrid vigour in such characters as enhanced size, longevity and even sexual activity. The strength, stubbornness and stamina of the mule are well known. What is perhaps less well known is the sexual aggressiveness of this animal which leads breeders to isolate it from the parent stocks of horses and donkeys.

Such a combination of male hybrid sterility and vigour has now been found to result from crosses between the 5 sibling species of the *Anopheles gambiae* complex, some of which form the main malaria vectors of Africa (Davidson *et al.* 1967). Moreover, some of the 20 possible crosses, notably between the freshwater species A and B males and saltwater species *A. melas* and *A. merus* females, result in an F_1 generation predominantly male—a ready-made sex-separation technique. Numerous laboratory-cage competition experiments (Davidson 1964, Davidson *et al.* 1967, Davidson 1969a, b) have shown these males to be competitive with normal males. Field releases of males derived from a cross between species B males and *A. melas* females were made in 1968 in an isolated village in Upper Volta where a species A population was on the decline following the wet season (Davidson *et al.* 1970). In all some 300,000 were released over a period of about 2 months, though it is extremely doubtful whether the numbers were adequate in the first month. Over most of this period 75% of the males captured in houses and in outside resting places proved to be the released sterile ones but, in spite of this, only 3·36% of the local wild females laid sterile eggs (as compared with 1·35% in control villages). Female densities declined to almost zero and the possibility exists of an enhanced female mortality resulting from sheer sterile male pressure or harassment without actual copulation (Cuéllar, in press).

A number of factors could have contributed to this lack of significant matings between introduced sterile males and natural females, such as an adverse climatic change at the time of peak release, the relatively artificial conditions in which the sterile males were reared up to the pupal stage, and inadequate releases over too short a period of time. However, it is generally concluded that the mating behaviour of the sterile males, though highly competitive in the small confines of a laboratory cage against single-species populations conditioned to life in such cages, is not so under natural conditions allowing the natural mating behaviour of wild populations. Additionally, the use of a cross between two species to control a third may have been a further barrier to the mating of sterile males with natural females. It is hoped to mount a repeat trial involving the release of greater numbers and using a hybrid involving the population to be controlled.

The possibility of genetic control by sterile hybrids is also being investigated in the *A. punctulatus* complex from New Guinea and adjacent regions, again important malaria vectors. The previously recognized *A. farauti* of this complex is now known to consist of two species, and crosses between them produce not only sterile males but also sterile females (Bryan 1970).

Cytoplasmic incompatibility

Within some insect species, e.g. *Culex pipiens* and *Aedes scutellarius*, crosses between various populations are sterile. Sterility is considered to be due to a cytoplasmic factor transmitted through the egg, which kills the sperm of the incompatible male after its entry into the egg. Between some populations sterility results in both directions; between others sterility results from one cross but not the reciprocal. Thus a potential exists for control by the mass rearing and release of males of one population into an area populated by incompatible crossing types.

More than 20 different crossing types are known in the *Culex pipiens* complex (Laven 1967, 1969a, b) and it is possible to specify at least one, sometimes several, strains which are incompatible with a certain population anywhere in the world. Desirable genetic traits can be introduced into an incompatible strain without changing the incompatibility, and so strains from temperate regions can be adapted to tropical environments. Such a strain was developed for use in Burma by incorporating the genome of a Fresno, California, strain into the cytoplasm of a strain from Paris. This was shown to be incompatible with an isolated field population near Rangoon and reared there in large numbers. Males were separated and released over a period of three months, for the last two of which releases reached 5,000 per day. By the 12th week nothing but sterile rafts could be found (Laven 1971).

Translocations

One common result of the irradiation of cells is chromosome breakage. If this occurs in two non-homologous chromsomes in the same cell, and the broken pieces become attached to the wrong partners, reciprocal translocations result. Meiosis in individuals heterozygous for translocations results in the production of a proportion of unbalanced gametes without a complete genetic complement, and when these fertilize normal gametes inviable embryos are produced. For a single translocation such semi-sterility is usually inherited by one half of the progeny of a heterozygous-normal mating. If an individual inherits the same translocation from both parents it becomes a translocation homozygote and may be normally fertile and viable (though most will probably not be so). Thus the possibility arises of isolating such a homozygote, breeding it on a large scale and releasing it into a wild population. Every mating between it and a wild individual, while normally fertile, will result in only semi-sterile translocation heterozygotes which at meiosis will produce a proportion of inviable gametes. When this individual mates, only a proportion of its offspring will mature. The net result will be a population reduction limited by the degree of inherited sterility but continuing for some time after releases are stopped. It is highly probable that fewer releases would need to be made to achieve this level of reduction than if conventional sterile males were used. Some doubt remains over the final outcome of translocation releases made with a view to population reduction, especially in those insects with a high biotic potential and whose populations are strongly buffered by density-dependent factors (Curtis 1968a). This is not the only potential use of translocations, however. If sufficient translocation individuals are released it is possible that the wild population may eventually be replaced by the translocation population. If the introduced translocation population carried genes for insecticide susceptibility or refractoriness to disease, then a wild, insecticide-resistant or disease-carrying insect population could be replaced by an insecticide-susceptible or harmless one (Curtis 1968b, Whitten 1971a). The simple genetic nature of most insecticide resistances is well-known, and the incorporation of susceptible alleles into translocation populations should present no difficulties. The mode of inheritance of the ability to transmit disease is less well-known, though the few cases worked out in detail show indications of simple mechanisms (Ward 1963, Macdonald 1967, Kilama & Craig 1969). Additionally there are conditional lethal genes which might be used, genes which allow survival in the artificial conditions of the laboratory or factory, but which will be disadvantageous to the insect under natural conditions. High- and low-temperature lethals are known as well as genes preventing diapause (Smith 1971).

Three main types of simple translocation can occur:

1 between an autosome and the Y-chromosome (or chromosome I bearing the dominant male genetic factor in *Culex* and *Aedes*);
2 between an autosome and the X-chromosome (or chromsome I bearing the recessive sex allele in *Culex* and *Aedes*);
3 between two autosomes.

In the first, semi-sterility is inherited only through the male so long as no crossing-over occurs. In the second, it can be inherited through both male and female but only the female can become homozygous, while in the third it is inherited equally through both sexes and can be homozygous in both. Of course combinations of these translocation types can exist in the same individual and the more there are the greater the degree of sterility. In fact Whitten (1971b) is of the opinion that eradication by translocation releases can only be achieved in those insects with relatively high chromosome numbers carrying several translocations, and is unlikely to be achieved in mosquitoes which have only three chromosomes.

Translocations have already been isolated in *Culex pipiens* (Laven *et al.* 1971), *Culex tritaeniorhynchus* (Laven *et al.* 1971, and Sakai *et al.* 1971), *Aedes albopictus* (Laven *et al.* 1971), *Aedes aegypti* (Rai & McDonald 1971), *Anopheles gambiae* species A and B (Krafsur 1972), *Musca domestica* (Wagoner *et al.* 1969) and *Glossina austeni* (Curtis 1971) among insects of medical importance. Most of these isolations are in the heterozygous condition and only a handful are viable translocation homozygotes.

The only published field release using translocations is that of Wagoner (1969), who released both sexes of a population of *Musca domestica* heterozygous for translocations between chromosomes II, III and V at a 9:1 ratio to the wild-type. The wild population was reduced to 3·2% of the control level in one generation and to 0·25% in the second generation.

A special type of translocation, a compound chromosome, is known in *Drosophila melanogaster* and involves the exchange of whole chromsome arms between a homologous autosome pair. Both left arms attach to one centromere and both right arms to the other. Other insects, and especially mosquitoes with their few metacentric chromosomes, are considered good potential candidates for the isolation of such chromosomal exchanges (Foster *et al.* 1972, Whitten 1971b). Compound chromosome strains breed true and are fully viable though their fertility is only 25 to 50% of the wild-type. When they mate with the wild-type, however, no viable offspring are produced. The hybrids die in the embryonic stage. Thus, theoretically, if the compound strain is released into a wild population in excess of 4 times the number of the latter, it should replace it within a small number of generations and form an ideal transport mechanism for introducing conditional lethals (temperature-sensitive or insecticide-susceptible genes) or other genes advantageous to man.

Conclusion

Genetic control methods possess the obvious advantage of being species specific, though some of them could produce an upset of the balance of nature where they lead to eradication and the vacation of an ecological niche. For this reason those methods involving population replacement are to be preferred.

By no means all the suggested genetic methods have been dealt with here. Such mechanisms as meiotic drive, sex distorters and species competition exist, but remain to be convincingly demonstrated as practical methods. There is certainly no shortage of ideas. Man's ingenuity has been challenged and the challenge is being met. Whether nature will produce some resistance mechanism to such novel developments remains to be seen.

The efficient application of these methods involves considerable knowledge of the basic biology of the insect to be controlled. Information, on population size, rates of increase, seasonal fluctuation, dispersal, mating behaviour and longevity will be of paramount importance. There are some who say we should wait until we have this knowledge in detail before we attempt any field applications, and would embark on long-term laboratory research programmes to gain it. There are others like Lindquist (1969) who believe that if we wait until all the information is to hand, no field trials will ever be made and interest and enthusiasm will wane. Much can be learned from small-scale pilot schemes, even though they may be only partially successful. The trials themselves expose many problems which may not be thought of in the laboratory and should be looked upon as an essential step in the development of the technique.

References

ANDREWARTHA H.G., J. MONRO & N.L. RICHARDSON (1967) The use of sterile males to control populations of Queensland fruit fly *Dacus tryoni* (Frogg) (Diptera, Tephritidae): II. Field experiments in New South Wales. *Aust. J. Zool.* **15**, 475–99.

BRANSBY-WILLIAMS W.R. (1971) A field release of male *Culex pipiens fatigans* sterilized by apholate. *E. Afr. med. J.* 48, 68–75.

BRYAN J.H. (1968) Results of consecutive matings of female *Anopheles gambiae* species B with fertile and sterile males. *Nature, Lond.* **218**, 489.

BRYAN J.H. (1970) A new species of the *Anopheles punctulatus* complex. *Trans. R. Soc. trop. Med. Hyg.* 64, 28.

BUSHLAND R.C. (1971) Sterility principle for insect control. Historical development and recent innovations. In *Sterility principle for insect control or eradication* (Proc. Symposium, Athens, 1970), 3–14. I.A.E.A., Vienna.

CONWAY G.R. (1970) Computer simulation as an aid to developing strategies for anopheline control. *Misc. Publ., ent. Soc. Amer.* 7, 181–93.

CRAIG G.B. (1967) Mosquitoes: female monogamy induced by male accessory gland substance. *Science* 156, 1499–501.

CUÉLLAR C.B. (1969a) A theoretical model of the dynamics of an *Anopheles gambiae* population under challenge with eggs giving rise to sterile males. *Bull. Wld Hlth Org.* 40, 205–12.

CUÉLLAR C.B. (1969b) The critical level of interference in species eradication of mosquitoes. *Bull. Wld Hlth Org.* 40, 213–19.

CUÉLLAR C.B. (in press) Dynamic aspects of the Pala sterile hybrid field experiment. Proc. Panel, Vienna, 1971. FAO/IAEA.

CUÉLLAR C.B., B. SAWYER & G. DAVIDSON (1970) Upper limit in male multiple mating in *Anopheles gambiae* species A. *Trans. R. Soc. trop. Med. Hyg.* 64, 476.

CURTIS C.F. (1968a) A possible genetic method for the control of insect pests with special reference to tsetse flies (*Glossina* spp.). *Bull. ent. Res.* 57, 509–23.

CURTIS C.F. (1968b) Possible use of translocations to fix desirable genes in insect populations. *Nature, Lond.* 218, 368–9.

CURTIS C.F. (1971) Experiments on breeding translocation homozygotes in tsetse flies. In *Sterility principle for insect control or eradication* (Proc. Symposium, Athens, 1970), 425–35. I.A.E.A., Vienna.

DAME D.A. & C.H. SCHMIDT (1970) The sterile-male technique against tsetse flies, *Glossina* spp. *Bull. ent. Soc. Amer.* 16, 24–30.

DAVICH T.B. (1969) Sterile-male technique for control or eradication of the boll weevil, *Anthonomus grandis Boh.* In *Sterile-male technique for eradication or control of harmful insects* (Proc. Panel, Vienna, 1968), 65–72. I.A.E.A., Vienna.

DAVIDSON G. (1964) The five mating types in the *Anopheles gambiae* complex. *Riv. Malariol.* 43, 167–83.

DAVIDSON G. (1969a) The potential use of sterile hybrid males for the eradication of member species of the *Anopheles gambiae* complex. *Bull. Wld Hlth Org.* 40, 221–228.

DAVIDSON G. (1969b) Genetical control of *Anopheles gambiae*. *Cah. O.R.S.T.O.M. sér. Ent. méd. et Parasitol.* 7, 151–4.

DAVIDSON G. & J.B. KITZMILLER (1970) Application of new procedures to control. Genetic control of anophelines. *Misc. Publ. ent. Soc. Amer.* 7, 118–29.

DAVIDSON G., J.A. ODETOYINBO, B. COLUSSA & J. COZ (1970) A field attempt to assess the mating competitiveness of sterile males produced by crossing two member species of the *Anopheles gambiae* complex. *Bull. Wld Hlth Org.* 42, 55–67.

DAVIDSON G., H.E. PATERSON, M. COLUZZI, G.F. MASON & D.W. MICKS (1967) The *Anopheles gambiae* complex, Chapter 6 in *Genetics of Insect Vectors of Disease*, eds. J.W. Wright and R. Pal, 211–50. Elsevier, Amsterdam.

DONNELLY J. (1965) Possible causes of failure in a field test of the 'sterile males' method of control. *Proc. XII int. Congr. Ent.* (1964), 253–4. Roy. ent. Soc., London.

FOSTER G.G., M.J. WHITTEN, T. PROUT & R. GILL (1972) Chromosome rearrangements for the control of mosquitoes and other insect pests. *Science* 176, 875–80.

GEIER P.W. (1969) Demographic models of population response to sterile-release procedures for pest control. In *Insect ecology and the sterile-male technique* (Proc. Panel, Vienna, 1967), 33–44. I.A.E.A., Vienna.

GOUCK H.K., D.W. MEIFERT & J.B. GAHAN (1963) A field experiment with apholate as a chemosterilant for the control of houseflies. *J. econ. Ent.* 56, 445–6.

KILAMA W.L. & G.B. CRAIG (1969) Monofactorial inheritance of susceptibility to *Plasmodium gallinaceum* in *Aedes aegypti. Ann. trop. Med. Parasit.* 63, 419–32.

KNIPLING E.F. (1965) The sterility principle of insect population control. *Proc. XII int. Congr. Ent.* (1964), 251–2. Roy. ent. Soc., London.

KRAFSUR E.S. (1972) Production of reciprocal translocations in *Anopheles gambiae* species A. *Trans. R. Soc. trop. Med. Hyg.* 66, 22–3.

KRISHNAMURTHY B.S., S.N. RAY & G.C. JOSHI (1962) A note on preliminary field studies on the use of irradiated males for reduction of *C. fatigans* Wied. populations. *Indian J. Malariol.* 16, 365–73.

LABRECQUE G.C., D.W. MEIFERT & R.L. FYE (1963) A field study on the control of the housefly with chemosterilant techniques. *J. econ. Ent.* 56, 150–2.

LABRECQUE G.C., C.N. SMITH & D.W. MEIFERT (1962) A field experiment in the control of houseflies with chemosterilant baits. *J. econ. Ent.* 55, 449–51.

LAVEN H. (1967) Speciation and evolution in *Culex pipiens.* Chapter 7 in *Genetics of Insect Vectors of Disease*, eds. J.W. Wright and R. Pal, 251–75. Elsevier, Amsterdam.

LAVEN H. (1969a) Incompatibility tests in the *Culex pipiens* complex. I. African strains. *Mosquito News* 29, 70–4.

LAVEN H. (1969b) Incompatibility tests in the *Culex pipiens* complex. II. Egyptian strains. *Mosquito News* 29, 74–83.

LAVEN H. (1971) Une expérience de lutte génétique contre *Culex pipiens fatigans* Wied., 1828. *Ann. Parasitol.* 46, 117–48.

LAVEN H., E. JOST, H. MEYER & R. SELINGER (1971) Semisterility for insect control. In *Sterility principle for insect control or eradication.* (Proc. Symposium, Athens, 1970), 415–24. I.A.E.A., Vienna.

LINDQUIST A.W. (1969) Biological information needed in the sterile-male method of insect control. In *Sterile-male technique for eradication or control of harmful insects.* (Proc. Panel, Vienna, 1968), 33–7. I.A.E.A., Vienna.

MACDONALD W.W. (1967) The influence of genetic and other factors on vector susceptibility to parasites. Chapter 19 in *Genetics of Insect Vectors of Disease*, eds. J.W. Wright and R. Pal, 567–84. Elsevier, Amsterdam.

MACLEOD J. & J. DONNELLY (1961) Failure to reduce an isolated blowfly population by the sterile males method. *Entomologia exp. appl.* 4, 101–18.

MAGAUDDA P.L., G. SACCA & D. GUARNIERA (1969) Sterile male method integrated by insecticides for the control of *Musca domestica* in the island of Vulcano, Italy. *Ann. Inst. Sup. San.* 5, 29–38.

MEIFERT D.W. & G.C. LABRECQUE (1971) Integrated control for the suppression of a population of houseflies, *Musca domestica* L. *J. med. Ent.* 8, 43–5.

MEIFERT D.W., G.C. LABRECQUE, C.N. SMITH & P.B. MORGAN (1967) Control of houseflies on some West Indies islands with metepa, apholate and trichlorfon baits. *J. econ. Ent.* 60, 480–5.

MELLADO L. (1971) La tecnica de machos esteriles en el control de la mosca del Mediterraneo. Programas realizados en España. In *Sterility principle for insect control or eradication* (Proc. Symposium, Athens, 1970), 49–54. I.A.E.A., Vienna.

MORLAN H.B., E.M. MCCRAY & J.W. KILPATRICK (1962) Field tests with sexually sterile males for control of *Aedes aegypti*. *Mosquito News* **22**, 295–300.

NADEL D.J. & G. GUERRIERI (1969) Experiments on Mediterranean fruit fly control with the sterile-male technique. In *Sterile-male technique for eradication or control of harmful insects* (Proc. Panel, Vienna, 1968), 97–105. I.A.E.A., Vienna.

PATTERSON R.S., M.D. BOSTON, H.R. FORD & C.S. LOFGREN (1971) Techniques for sterilizing large numbers of mosquitoes. *Mosquito News* **31**, 85–90.

PATTERSON R.S., H.R. FORD, C.S. LOFGREN & D.E. WEIDHAAS (1970) Sterile males: their effect on an island population of mosquitoes. *Mosquito News* **30**, 23–7.

PROVERBS M.D. (1969) Induced sterilization and control of insects. *A. Rev. Ent.* **14**, 81–102.

PROVERBS M.D., J.R. NEWTON & D.M. LOGAN (1966) Orchard assessment of the sterile male technique for control of the codling moth, *Carpocapsa pomonella* (L.) (Lepidoptera: Olethreutidae). *Can. Ent.* **98**, 90–5.

RAI K.S. & P.T. MCDONALD (1971) Chromosomal translocations and genetic control of *Aedes aegypti*. In *Sterility principle for insect control or eradication* (Proc. Symposium, Athens, 1970), 437–52. I.A.E.A., Vienna.

RIVOSECCHI L. (1962) Un esperimento sul campo con maschi irradiati de *Musca domestica* in una zona rurale della provincia di Latina. *Riv. Parassitol.* **23**, 71–4.

SACCA G. & E. STELLA (1964) Una prova di campo per il controllo di *Musca domestica* L. mediante esche liquide a base del chemosterilante Tepa (= Aphoxide). *Riv. Parassitol.* **25**, 279–94.

SAKAI R.K., R.H. BAKER & A. MIAN (1971) Linkage group-chromosome correlation in a mosquito. Translocations in *Culex tritaeniorhynchus*. *J. Hered.* **62**, 90–100.

SHAW J.G., W.P. PATTON, L.M. SANCHEZ-RIVIELLO, L.M. SPISHAKOFF & B.C. REED (1966) Mexican fruit fly control. *Calif. Citrograph* **51**, 209/14.

SMITH R.H. (1971) Induced conditional lethal mutations for the control of insect populations. In *Sterility principle for insect control or eradication* (Proc. Symposium, Athens, 1970), 453–65. I.A.E.A., Vienna.

STEINER L.F., E.J. HARRIS, W.C. MITCHELL, M.S. FUJIMOTO & L.D. CHRISTENSEN (1965) Melon fly eradication by overflooding with sterile flies. *J. econ. Ent.* **58**, 519–22.

STEINER L.F., W.G. HART, E.J. HARRIS, R.T. CUNNINGHAM, K. OHINATA & D.C. KAMAKAHI (1970) Eradication of the oriental fruit fly from the Mariana Islands by the methods of male annihilation and sterile insect release. *J. econ. Ent.* **63**, 131–5.

WAGONER D.E. (1969) Suppression of insect populations by the introduction of heterozygous reciprocal chromsome translocations. *Bull. ent. Soc. Am.* **15**, 220.

WAGONER D.E., C.A. NICKEL & O.A. JOHNSON (1969) Chromosomal translocation heterozygotes in the housefly, *Musca domestica* L.: egg hatch, sex ratios and transmission to progeny of 193 translocations. *J. Hered.* **60**, 301–4.

WARD R.A. (1963) Genetic aspects of the susceptibility of mosquitoes to malarial infection. *Expl Parasit.* **13**, 328–41.

WEIDHAAS D.E., C.H. SCHMIDT & E.L. SEABROOK (1962). Field studies on the release of sterile males for the control of *Anopheles quadrimaculatus*. *Mosquito News* **22**, 283–91.

WHITE G.B. (1966) Chemosterilization of *Aedes aegypti* by pupal treatment. *Nature, Lond.* **210**, 1372.

WHITTEN M.J. (1969) Automated sexing of pupae and its usefulness in control by sterile insects. *J. econ. Ent.* **62**, 272–3.

WHITTEN M.J. (1971a) Insect control by genetic manipulation of natural populations. *Science* 171, 682–4.

WHITTEN M.J. (1971b) Use of chromosome rearrangements for mosquito control. In *Sterility principle for insect control or eradication* (Proc. Symposium, Athens, 1970), 399–410. I.A.E.A., Vienna.

WHITTEN M.J. & W.C. TAYLOR (1970) A role for sterile females in insect control. *J. econ. Ent.* 63, 269–72.

Selectivity in pesticides: significance and enhancement

D. PRICE JONES *Formerly Plant Protection Ltd., Jealott's Hill Research Station, Bracknell, Berkshire; now at 82 Shinfield Road, Reading, Berkshire*

Basic to the concept of selective action by pesticides is an understanding of what constitutes a pest. A working definition might be: an organism that, at some relevant point in time, is interacting with man in a way inimical to what man believes to be his interests. If viruses are to be included, then, as a matter of convenience, 'organism' must be sufficiently elastic to embrace particles capable of reproducing by proxy. It follows that a pest is something so designated by man, and that it has no existence except in relation to man. As pest status is a socio-economic concept, selectivity in pesticides is *ipso facto* a quality to be judged ultimately against a socio-economic background.

The present account is directed towards a biological analysis of the problems involved in developing selective action in the field. Many of these problems are also examined from the commercial point of view. It has been found convenient to build the review around crop pests, particularly insects, and to refer to other pests or their associated problems only for purposes of comparison or for extending the scope of a generalization.

The significance of selectivity

Pest species too numerous to count

The division between a pest and a non-pest is often so indistinct—and often so unstable—that an acceptable estimate of the total number of noxious organisms is impossible to achieve. In the U.K. alone, pests of crops (including pathogens and weeds) exceed 1,000 (Martin 1969, Fryer & Evans 1968), with perhaps less than half of these receiving some control treatment each year. Elsewhere in the Temperate Zone, the pest fauna is many times richer in species.

Economically important weeds of tropical crops number many more than the 110 or so selected for mention by Kasasian (1971). The number of important nematode diseases is said to be less than 100 (Mai 1971); on the other hand potential nematode problems are numerous, e.g. Williams (1969) records about 100 species of nematodes potentially parasitic on sugar cane. Plant viruses and mycoplasmas have by this time exceeded the 300 estimated by Bawden (1964); those transmitted by leaf hoppers alone amount to 70 (Ishihara 1968).

The pest insect potential on individual crops can be highlighted by reference to the 800 species recorded as feeding on coffee (Le Pelley 1969) and the 820 on rice (Grist & Lever 1969). A rough indication of the total number of insect pests is given by the Commonwealth Institute of Entomology cumulative card index currently containing about 70,000 species (R. G. Fennah, pers. comm.).

These estimates lead to two important conclusions in relation to selectivity. First, species-specificity is an unrealistic target for research on selectivity except for a very few important species. Second, the insects include by far the greatest number of pest species and, partly for this reason and partly because insecticides appear to be environmentally more troublesome than other pesticides, environmental problems related to selectivity, or lack of selectivity, are likely to centre mainly around the insects for some time to come.

Organisms, good, bad and indifferent

The distribution of pests and non-pests throughout the plant and animal kingdoms is a subject germane to any analysis of selectivity.

Among the plants, bacteria cause disease in plants and animals, blue-green algae may cause ecological disruption and green algae can disfigure swimming pools or choke irrigation channels or equipment. Mosses can make paving dangerous or disfigure lawns. Horsetails can be pernicious weeds. The flowering plants provide the worst weeds and many are also poisonous to mammals; some are parasitic on crops.

Among the animals, hardly a phylum is without its quota of harmful, dangerous or unpleasant creatures. Some, such as the Protozoa, Nematoda, Platyhelminthes and Arthropoda, contain important parasites of man and other animals. The Nematoda, Arthropoda and Mollusca contain undesirable plant feeders of considerable significance in agriculture. The Chordata, by virtue of their larger size, present somewhat different problems but they include amongst their numbers some important predators of man and animals, a few ectoparasites, living reservoirs of disease and a large number of herbivorous creatures readily attracted to cultivated crops.

Of greater significance is the extent to which noxious species are associated with harmless or even beneficial species in the same taxa. It is immediately apparent that this association is very close indeed. Even the bacteria include species that play an essential role in organic and mineral cycles and there are many beneficial fungi, some of them edible and indeed commercially produced. This association can be traced down through the taxonomic hierarchy to species level. Thus, in the plants, numerous cultivated species as well as pernicious weeds are found among the Angiosperms. The Gramineae contain a range of important crop plants closely linked with the development of civilization and also some of the most intractable weeds in present-day farming. *Hordeum sativum* is highly prized as cultivated barley but *Hordeum murinum* is just a minor weed.

Likewise among animals, the same association can be traced throughout the hierarchy. Of particular significance is that found in the insecta. Thus most of the important pests occur in the Orthoptera, Hemiptera, Lepidoptera, Diptera, Hymenoptera and Coleoptera; yet all of these contain beneficial insects, including important groups of insect predators and parasites that help to keep the pests in check. In the Lepidoptera, the Noctuidae include a large number of some of the most important pests of crops but, somewhat unexpectedly, the larvae of *Eublemma* are important predators of coccids. More perversely, the Coleopterous family, Coccinellidae, commonly regarded as a collection of valuable predators of aphids and other small insects, also contains a group of phytophagous insects injurious to cultivated plants.

This brief survey underlines the fact that pests and non-pests are inextricably distributed throughout the plant and animal kingdom. There is no convenient separation in the broad taxonomic scheme which would suggest that pests would have clear-cut morphological, physiological or biochemical characteristics compared with non-pests, thus leading to differential reactions to applied pesticides. It follows, therefore, that there can be no single solution to the problem of achieving pesticide selectivity. Equally, a solution requiring a tailor-made pesticide for each pest is unrealistic in view of the vast numbers of pests and potential pests. Some compromise is therefore necessary. One of the purposes of this discussion is to seek guide lines for such a compromise.

Structure of pest associations

Typically more than one pest is associated with any one crop, however restricted that crop may be: in fact, most crops have their own particular pest complexes which vary to some extent in both time and place. These pest complexes are structured in various ways, some of which can be exploited in the solution of problems of pesticide selectivity. Thus some components live in the soil, some on the surface; others affect aerial parts of crops in a variety of ways and may or may not be exposed at the time of inflicting their damage. For present purposes, however, it is proposed to examine the taxonomic structure a little more closely for possible clues to the operation of selectivity in the field.

Plant feeding insects and mites provide material for illustration. Table 1 classifies in major taxa all the pest species reported on apple in Britain (Massee 1954) and also those considered of major importance. Table 2 does

Table 1. Insect and acarine pests on apple (U.K.): frequency distribution in major taxa

Taxon	Individual species	Important pests
Lepidoptera	52	4
Coleoptera	21	4
Dermaptera	1	1
Hemiptera	21	5
Aphidoidea	6	4
Coccoidea	3	0
Others	12	1
Hymenoptera	5	3
Diptera	3	1
Acarina	2	1
Total	105	19

Compiled from information in Massee (1954).

the same for major field and fruit crops in the U.S.A., with the difference that, in some cases, groups of pests (e.g. two or three species of one genus) are considered as one pest if they are covered by the same control recommendations (U.S. Department of Agriculture 1964).

These two tables bring out the overriding importance of Lepidoptera (caterpillars), Coleoptera (beetles) and Hemiptera (bugs) in the pest structure of important crops. In the past, caterpillars (particularly) and beetles have, with rare and arguable exceptions, been controlled primarily by broad-spectrum, persistent insecticides and, as a consequence, many of these pest situations have become aggravated by the spray programmes adopted. The target pest has often become more difficult to control and in many cases minor pests have assumed major importance (Smith & van den Bosch 1967). In general, if caterpillars could be removed from pest complexes by the use of caterpillar-selective insecticides, the control of the remaining pests could be made easier and ecologically much safer. In recent years the botanical

Table 2. Insect and acarine pests of field and fruit crops (U.S.A.): frequency distribution in major taxa.

Taxon	Cotton	Corn	Tobacco	Potato	Sugar beet	Sorghum	Apple	Pear	Peach	Plum/Prune	Citrus	Grape
Lepidoptera	15	7	3	5	4	4	14	7	9	6	0	5
Coleoptera	5	4	4	10	4	1	5	2	4	1	0	5
Orthoptera	2	2	2	2	1	1	1	1	1	0	0	0
Dermaptera	0	0	0	0	0	0	0	0	1	0	0	0
Thysanoptera	1	0	0	1	0	0	0	1	0	0	1	0
Hemiptera	6	1	2	6	4	0	12	6	6	3	9	5
Aphidoidea	1	0	1	1	1	0	5	1	1	1	1	2
Coccoidea	0	0	0	0	0	0	2	2	2	2	7	2
Others	5	1	1	5	3	0	5	3	3	0	1	1
Hymenoptera	0	0	0	0	0	0	0	1	0	0	0	0
Diptera	0	1	0	0	1	0	0	0	0	0	0	0
Acarina	1	1	0	0	1	0	6	5	1	1	8	1
Total	30	16	11	24	15	6	38	23	22	11	18	16

Compiled from information in U.S. Department of Agriculture (1964); a unit entry normally indicates one species but occasionally may represent two or more species for which control recommendations are identical.

insecticide ryania has had some limited success against codling moth on apple, mainly where the moth is univoltine, but is unsatisfactory for general use. *Bacillus thuringiensis* looked extremely promising in the 1960s but, despite progress in strain selection and formulation, has not completely fulfilled that promise. It does, however, demonstrate that group-selectivity is feasible. The investigations of Beard (1952) with the highly specific venom of the parasite *Bracon hebetor* also show that caterpillar-specific materials are not impossible targets for exploratory research. In recent years, systemic insecticides with strong (though not selective) activity against caterpillars have been marketed, suggesting that another possible barrier to the development of caterpillar-specific insecticides has been removed. If all leads fail (which is unlikely) it would still be in the interest of the pesticide industry and of pest control generally to have caterpillars removed by other means, e.g. by the exploitation of sex pheromones.

Other insect groups that require investigating in this way are the Coleoptera and the Coccoidea. Their control in a selective manner would also contribute to the simplification of many crop pest situations. Aphids are already responding to control by increasingly selective materials (Proctor & Baranyovits 1969). Among insects important in the field of public health, the Diptera (flies) would figure prominently as subjects for group-specific selectivity.

Among fungi, certain mildews are now susceptible to control by systemic fungicides with, as far as is known at present (and information is accumulating rapidly), a minimal effect on other organisms, including other fungi.

Three families of plants, the Gramineae, Compositae and Cyperaceae, contribute disproportionately to the weed flora and would appear to be suitable targets for group-specific selectivity. In general, however, herbicides very rarely generate problems outside the crops to which they are applied, and the commercially important aspect of selectivity is discrimination between crop and weeds while achieving maximum weed control, preferably over a wide spectrum of weeds. Now that various hormone-type weedkillers have advanced broad-leaved weed control in cereals to a level undreamt of thirty years ago, the outstanding problem is the control of weed grasses in cereals. This too is yielding to intensive research.

Individual pests of major importance

It has been emphasized that there are far too many pests for selectivity to be developed on a species-specific basis. There are many reasons for this, but the completely uneconomic nature of such a quest is an overriding factor. This may not apply in the case of a few pests of major importance, each of which would offer a market in its own right. Such pests could, for example, include cotton boll weevil, American bollworm, tobacco budworm, and codling moth among crop pests; the mosquitoes, *Aedes aegypti* and *Anopheles gambiae*, and the snail vectors of *Bilharzia* among public health pests; the aquatic weed *Eichornia crassipes* and the subtropical crop weed *Cyperus esculentus* (nutgrass). While it is very unlikely that absolute species-specificity could be achieved with any such pest, sufficient selectivity could probably be achieved with some or many of them to ensure a high degree of freedom in the use of the pesticides concerned.

Even so, resistance would be an ever-present threat. In many situations a highly selective compound could be expected to retain its efficacy longer than a non-selective compound, but insufficient is known about this problem to provide estimates of probabilities. With such uncertainty, commercial enthusiasm would be low.

Pest associations in individual crops

Le Pelley (1969) has compiled a list of insects found feeding on coffee: these again are shown classified in major taxa (Table 3). About 800 species have

Table 3. Pest insects and acarina on coffee: frequency distribution in major taxa

Taxon	Spp in world list (1)	Important pest spp in Kenya (2)	Tanzania (3)
Lepidoptera	183	9	11
Coleoptera	276	7	5
Orthoptera	51	0	2
Isoptera	12	0	0
Thysanoptera	23	1	4
Hemiptera	246	8	21
Aphidoidea	1	0	1
Coccoidea	117	5	11
Others	128	3	9
Acarina	8	1	2

(1) Le Pelley (1969). (2) Coffee Board of Kenya (1961). (3) Tapley (1964).

been recorded in various parts of the world but not more than about 100 of these could possibly be regarded as important pests. In Kenya twenty-six are recorded as important, in Tanzania forty-five. The individual cultivator would, perhaps, find five species economically significant on his crop.

This suggests that relatively few highly selective insecticides would suffice to cope with these few important pests. Such a conclusion would be erroneous on two counts. First, it ignores the incidence of diseases and weeds. Second, it fails to recognize that the grower is not a producer of insecticides. He is only one small part of a market that must be large enough to warrant the production of a suitable product or products. Otherwise expressed, the pattern of available insecticides is determined by regional or world requirements rather than those of the individual grower.

The nature of selectivity

In toxicology it is usual to measure the toxicity of a compound in terms of the amount of toxicant required to produce a given effect, e.g. 50% mortality. The conditions under which the toxicity is measured are explicitly stated or just implied. It is, however, clearly understood that the toxicity as measured depends on the conditions under which the toxicant is applied and those under which the organism is held; and indeed upon the physiological condition of the organism at all relevant times. Quite clearly, the concept of toxicity does not apply exclusively to the toxicant, or to the toxicant and organism; it must also embrace the ecosystem to which the organism belongs. That most toxicological investigations are conducted under laboratory, and therefore artificial, conditions does not invalidate the concept; it merely limits the confidence with which information from such tests can be applied in the field.

If two organisms are compared in toxicological tests, it frequently happens that they give quantitatively different responses. The toxicant is then said to be acting selectively. Since this selective effect depends on two toxicities which are themselves expressions of the interaction toxicant/organism/ecosystem, it follows that selectivity must be regarded not just as a unique property of the pesticide but also as a function of the ecosystem into which that pesticide is injected.

Here the view of Winteringham (1969) may be quoted: 'In a given ecosystem, a certain mean concentration of insecticide will induce a defined response in populations of some animal or plant species. The difference in response between any two species is a measure of the selectivity of insecticidal action in relation to those species.' It is significant that Winteringham, despite his approach to the problem through the biochemical mechanisms of selective insecticidal action, was nevertheless fully aware of the role of the ecosystem.

These considerations lead me to suggest that, although the term 'intrinsic selectivity' is a convenient one, it is perhaps more accurate and certainly more fruitful to regard any one chemical as having a conspectus of chemical and physical properties that enables that chemical to be used in practice in a selective manner; or the reverse, a conspectus of properties that impedes its use in a selective manner. Such an approach appeals to me not merely because it leads to a clearer understanding of the nature of selectivity but because it offers hope of developing an effective policy for pesticides.

When it comes to enumerating those properties that should enable a candidate material to be developed into an effectively selective pesticide, some difficulties are inevitably encountered. These arise partly from incomplete knowledge of the essential properties, but even more so from the immense task of appraising the innumerable interactions involved. However, the following is an attempt to provide an initial list of desirable properties.

1. Differential toxicity in laboratory tests

An insecticide, for instance, should show a high toxicity towards the target insects, much lower toxicities towards other insects, especially beneficial ones, a very low mammalian toxicity and, if plants are involved, negligible phytotoxicity. Such a spectrum of toxicity, normally assessed during the screening of new compounds, is essential if a useful degree of selectivity is to be developed in the field.

2. Limited persistence

The persistence must be as short as is consistent with a satisfactory field performance; the shorter the better. This short persistence should apply to the environment as a whole, not merely to the target area. Fumigants may be non-persistent in the target area but may yet persist in other parts of the ecosystem. They have one important advantage: they disperse rapidly and become diluted to concentrations that are not immediately harmful, so gaining time for degradation to permanently harmless materials.

In setting criteria for the degree of persistence that could be tolerated, allowance can be made for other characteristics. Thus greater persistence could be tolerated in a material favoured with a narrow and acceptable spectrum of activity, but showing no signs of accumulating in food chains.

3. Ability to degrade to innocuous materials

The short persistence of the parent material should not be bought at the expense of producing harmful degradation products. Completely harmless end products of chemical or microbiological breakdown could be, for example, water or carbon dioxide. Sulphur dioxide, mineral phosphates, nitrates or sulphates would also probably be harmless in the minute amounts produced. Certain highly halogenated heterocyclic organic compounds and certain heavy metals would be less acceptable.

Ideally, degradation should take place under a range of conditions and be effected by more than one mechanism. In practice, however, some measure of stability is required for the handling, application and effective action of the material. Again compromise is essential.

4. Inability to undergo concentration in food chains

The ability to degrade quickly (as in (3)) in a variety of situations is perhaps the best safeguard against concentration in food chains. Concentrated storage in intermediate links is usually in fat, skeleton, scales or feathers; sometimes in kidney or liver. Lipoid solubility would appear to be undesirable in this context but is useful in formulating the material and may sometimes ensure better field performance. The emphasis, undoubtedly, is on degradability.

5. Miscellaneous properties

The above characteristics are all desirable in almost any pesticidal material. There are many other characteristics that may or may not be desirable. Thus, in a particular compound, high volatility could conceivably present a hazard in confined spaces in certain applications suggested by its spectrum of activity; in another material, high volatility could be harnessed to give a fumigant effect in a restricted space, leaving no residual hazards. A high degree of adsorption on soil may be undesirable in a pesticide required to exert its activity in the soil, but in a material intended for above-ground activity (e.g. the herbicide paraquat) the rapid inactivation in soil can be exploited in the interests of environmental safety.

There is therefore a miscellany of properties that can be manipulated, in the context of the particular problem concerned, to enhance the degree of pest control while minimizing the challenge to the environment.

The exploitation of selectivity

While there is no prospect whatsoever of arriving at a mechanism or system of selectivity that would apply to all pests, there are undoubtedly many general principles that are appropriate to a very large number of pest situations.

Controlled dosage rate

Table 4 relates the concentration of the aphicide menazon to the percentage mortality of various test insects in a variety of laboratory screening tests. The useful degree of selectivity experienced in practice was foreshadowed in tests of this kind. Such selectivity was observed at the lowest dosage rates (expressed here as concentration in the spray) but diminished as the dosage rate increased. From this and many similar observations on other pesticides, the general conclusion could be drawn that the interests of selectivity in the field would be best served by keeping application rates as low as possible.

Table 4. Menazon selectivity related to spray concentration

	% kill at concentration (ai) indicated				
Pest	0·3%	0·1%	0·05%	0·025%	0·005%
Aphis fabae	100	100	100	100	100
Acyrthosiphon pisi	100	100	100	100	100
Tetranychus telarius					
Walking stages	–	100	–	95	20
Eggs	0	0	0	0	0
Dysdercus fasciatus	92	25	35	35	13
Phaedon cochleareae	40	0	5	0	0
Sitophilus granarius	0	0	0	0	0
Plutella maculipennis					
Caterpillars	0	0	0	0	0
Musca domestica					
Adults	30	30	0	0	0
Larvae	100	70	–	25	–

After Price Jones (Jones 1961).

This is excellent advice but difficult to apply to its full extent in the field. In practice, application rates are determined by a variety of factors. Approval authorities require proof of biological effectiveness when the product is used as recommended, and the standards they set reflect what are considered to be the best interests of the growers and, in some cases, the food processors. The pesticide industry, for its part, is under considerable pressure to set application rates sufficiently high to reduce the frequency of complaints of failure to a tolerable level. The combined effect of these factors is probably to keep the application rate at a level slightly above that demanded by the requirements of selectivity, probably even above the average needs of the grower, if indeed these could be accurately assessed.

The socio-economic system that determines dosage levels is not immutable but it will take more than pious hopes to change it. It will need constructive thinking and concerted effort by all parties concerned.

Effective placement

There have been occasions in the past when pesticides have been applied over very large areas without reference to the local distribution of the target pest. This applies more particularly to insecticides, for example to DDT used in 'blanket' spraying in emergency anti-mosquito campaigns, or to the use of dieldrin in the attempted eradication of Japanese beetle. Applications of insecticides to forests in North America have occasionally taken on this character, when the spraying has continued over rivers, lakes and other open areas. Scrub control on rangeland, using herbicides, also belongs to this category.

Nevertheless, despite the publicity, such cases constitute a minute proportion of pesticide applications. Typically, a pesticide is, for economic reasons if for no other, confined to the area effectively occupied by the pest. Thus, in agriculture, spraying is restricted to the fields on which the target pest occurs in potentially damaging numbers or is likely to occur in such numbers, these fields being only a proportion of the total number devoted to that crop. In this way there is a selective application of the pesticide—an input of pesticide into the ecosystem which, in a crude way, maximizes the benefit to man while minimizing the effect on the ecosystem as a whole. In the U.K., large acreages of sugar beet are sprayed each year for the control of virus spread by *Myzus persicae*. The aphid, however, is by no means confined to sugar beet or to other crops subject to spraying; hence only a small part of the total population is exposed to aphicides. It is no great surprise therefore that field populations of *M. persicae*, unlike glasshouse populations, have not as yet shown clear indications of resistance to insecticides. On the other hand, the hop aphid *Phorodon humuli* is subject to heavy insecticide pressure, as its summer populations occur mainly on commercial hops, where spraying is intense; hence the occurrence of resistance could reasonably have been anticipated.

Within the crop, applications can presumably become more selective by restricting the distribution of pesticide in various ways. Thus in a row crop, band spraying along the row may reduce the amount of material applied. Likewise in soil treatment, the abandonment of soil application and the adoption of row placement, seed treatment or, in a planted crop, spot treatment can all lead to reduced amounts of pesticide being applied. There is a great deal of information indicating effective control from reduced amounts of material applied in this way. For instance, Table 5 records some comparisons culled from experience on wheat in Britain. However, what is by no means clear is whether restrictive placement is selective in relation to beneficial organisms. The ecosystem is structured in relation to the crop and may

Table 5. Influence of placement method on effective application rate of insecticide (g/ha)

		Method of application			
Pest	Insecticide	Broadcast	Combine-drilled	Seed treatment	Reference
Wireworms, *Agriotes* spp.	γ-BHC	840	280–420	56–84	Plant Protection Ltd (1971)
Wheat bulb fly, *Leptohylemyia coarctata* (Fall.)	Aldrin	–	1120–2240	140*	Maskell & Gair (1961)

* Seed treatment rate also applies to dieldrin, heptachlor and γ-BHC.

be just as much (or more) affected by placement techniques. Such evidence as is available is mostly circumstantial but points towards increased selectivity. An exception is the case of bird poisoning by insecticidal seed treatments in Britain in the 1950s, where the effect was clearly related to the particular placement method employed.

Ultra-low-volume (ULV) spraying

The pioneer investigations of Sayer (1959) and others in East Africa demonstrated that certain pesticides (mainly dieldrin in this case) could be applied in a concentrated form in very low volume (ULV) with considerable reduction in the amount of toxicant applied per acre. During the 1960s ULV techniques were developed with a range of insecticides, fungicides and weedkillers, with malathion as the most investigated material. These applications involved neat or highly concentrated formulations with toxicological characteristics rather different from those of more normal low-volume applications. While such techniques have been successful in their prime task in a large number of situations, only a very limited amount of work has been done on the effects on non-target organisms (Linn 1968, Washino *et al.* 1968). These preliminary results, while favourable, are inadequate for an appraisal of ULV applications as a whole.

Systemic applications

Systemic activity by pesticides has been known for a very long time. The phenoxyacetic herbicides in the 1940s were the first to undergo large-scale commercial development, and undoubtedly marked an important advance in herbicide selectivity. Although this selectivity was partly related to the systemicity, there was little deliberate exploitation of the systemicity as such. Not until the organophosphates were developed in the 1950s and early 1960s was the possibility of using systemic applications to enhance selective activity fully appreciated. Ripper *et al.* (1951) drew attention to the significance of systemic applications but it was the work of Stern *et al.* (1959) that clearly established the role of systemic materials in selective pest control. In this investigation the substitution of the broad-spectrum but systemic demeton for the broad-spectrum but non-systemic malathion and parathion favoured the predators and eventually led to a higher level of control of the aphid.

This potential has since been extensively explored with organophosphates and carbamates, the selective toxicity being directed mainly against aphids and plant-feeding mites, but also embracing certain other pests, e.g. leaf miners. The susceptibility of aphids can be attributed in part to the large throughput of insecticide-laden sap; that of mites possibly to concentration of toxic material in epidermal cells or to local fumigant action. On the whole, large gross-feeding insects such as caterpillars tend to be immune to most compounds acting strictly as systemics.

In recent years systemic fungicides have become available and are now receiving extensive commercial development. They appear to meet the current stringent requirements for environmental care, including safety to wildlife, from which we can assume a reasonable measure of selectivity has been achieved. Nevertheless, most ecologists will prefer to wait and see.

Controlled release of pesticides

The demand for short-persistence chemicals is in conflict with certain agronomic requirements for the control of pests. Thus, convenience or pest control efficiency may require granules to be drilled along with the seed, with the biological effect to follow some weeks or months later. Used in such situations, short-lived materials require some method of immobilization until their intervention is required. Various methods of coating granules have been tried and some measure of success obtained, but such developments are still in the early stages. One of the major difficulties is the almost inevitable reliance on water as a releasing or release-triggering mechanism and the imperfect connection between moisture change and the pest problem concerned.

Another device is the use of encapsulation, which can isolate the toxicant from the environment or from organisms other than those that consume them along with food. This too awaits successful commercial exploitation.

The chemist, the biochemist and the biologist

This commentary on selectivity has been mainly concerned with the nature of the problem, the type of selectivity that can be reasonably sought and the various ways of achieving it. Little has been said about biochemical mechanisms of selectivity, as they fall outside the main contribution of the biologist. However, an understanding of the roles of chemist, biochemist and biologist is essential to any appraisal of the effort involved in improving the selective action of pesticides.

The chemist is primarily concerned with the provision of new compounds for biological screening. In his selection of compounds for synthesis he is guided initially by some quite general principles of chemical reactivity, by a general knowledge of biologically active structures and by a tentative application of structure-activity relationships in one group of compounds to some other not too distantly related group. He does not—and cannot—at this stage aim at selective action. Once biological activity has been established within a new group of chemicals, the group is explored by further synthesis and screening until some understanding of the relationship between structure and activity has emerged. At this stage, the patterns of activity of several candidates have become apparent and, with mammalian toxicity data already available, initial decisions can be made. Not until this stage can the requirement for selective action be usefully considered, but thereafter it becomes an essential factor in all relevant decisions. Commercial considerations must inevitably play a major part in the selection of material for development, but a minimal measure of selectivity is always imposed by the prospect of social constraints, particularly by official organizations responsible for the registration or approval of new products.

The role of the biochemist in the search for selectivity is also too frequently misinterpreted. It is true that, in medical and veterinary pharmacology, biochemical mechanisms are universally exploited for selective activity within the subject animal, but the resultant drugs are very largely confined to that animal by their method of administration; with a few exceptions, e.g. hormone or antibiotic contamination of meat, and infectious resistance in bacteria, the impact on the environment is inevitably extremely limited in extent and significance. With pesticides, however, the methods of application necessarily involve a more significant involvement of the materials with the environment; hence the need for a much wider concept of selectivity. In this context the task of the biochemist becomes formidable. As already indicated, the number of different kinds of pests is very large indeed, making biochemical subtlety at species level a highly impractical approach. Even for the individual pest of major importance, the biochemical approach is potentially valuable mainly in finding a chemical which is sufficiently active for commercial use and the main activity of which is directed at a biochemical system believed to occur only in the target pest. This is a considerable step forward but the method can be employed for only a limited number of major pests (better still for certain groups of pests) and must still be accompanied by all the toxicological, metabolic and ecological studies required by a material emerging from random screening. Such biochemical subtlety was used in developing selectivity between clover and broadleaved weeds in applications of phenoxybutyric herbicides to cereals undersown with grass/clover mixtures (Wain 1955). This, however, is a very restricted kind of selectivity and is more significant in an agronomic than an environmental sense.

This is not to belittle the work of the biochemist whose studies on metabolism and mode of action are essential in establishing the safety of a new material. A new pesticide could take many years to reveal its full effect on the environment; biochemical investigations, on the other hand, can explore the possibilities in a relatively short period and eliminate at least the main hazards.

The vast majority of pesticide selectivity problems are likely to be solved empirically, that is, by laboratory and field testing, selection of the most suitable candidates and the progressive incorporation of these into enlightened spray programmes or, better still, into integrated control systems. This philosophy is foreshadowed by Winteringham (1969) who calls for instability in the pesticide molecule rather than selectivity dependent on biochemical subtlety.

It follows that the role of the biologist in the development of selectivity is firstly to establish the biological aims and ensure that these are compatible with economic and social requirements; secondly, to examine candidate materials and to select for development those that, by reason of their chemical, physical and biological properties, offer the least challenge to the environment; thirdly, to integrate these into effective pest management systems. The responsibility of the biologist does not, however, end there. It remains his duty to monitor the commercial applications of pesticides to ensure that the intervention of some completely unexpected factor does not result in harmful effects on the environment. Most of the major ecosystems with which we are concerned tolerate a considerable measure of distortion before change becomes irreversible. Biologists must use this latitude wisely.

Summary and conclusions

Species classifiable as pests or potential pests number many thousands and are inextricably associated with harmless or beneficial species in all the main plant and animal taxa. Consequently, no simple segregation of pests and non-pests is possible as a prelude to the development of a few selective pesticides; nor is it meaningful to contemplate the development of a very large number of species-specific pesticides.

The solution is therefore to seek materials with a restricted spectrum of activity but with a conspectus of properties that enables them to be used selectively. Such a conspectus would include short persistence, degradability to harmless products and inability to enter food chains; also physical, chemical and biological characteristics enabling them to be formulated and applied in a manner least challenging to the environment.

Some important pests merit erection as individual targets for species-specific pesticides. Also certain taxa or associations of pests can usefully be treated as group-selective targets. Thus lepidopterous pests (caterpillars) demand special attention because of their key role in the pest complexes of many crops and because of the adverse effects of imperfect control measures employed in the past. Likewise, grasses in cereal crops merit similar attention.

The development of a portfolio of more selective pesticides must clearly depend initially on the efforts of the chemist and biochemist, but these efforts will be largely wasted unless the biologist (ecologist or agronomist) can establish the basic philosophy and also frame the field recommendations to make maximum use of the progressively better materials as they appear.

References

BAWDEN F.C. (1964) *Plant Viruses and Virus Diseases.* 4th edition. 361 pp. Ronald Press, New York.

BEARD R.L. (1952) The toxicology of *Habrobracon* venom: a study of a natural insecticide. *Conn. Agr. Expt. Sta. Bull.* 562.

COFFEE BOARD OF KENYA (1961) *An Atlas of Coffee Pests and Diseases.* Coffee Board of Kenya, Nairobi.

FRYER J.D. & S.A. EVANS, eds. (1968) *British Crop Protection Council: Weed Control Handbook.* Blackwell, Oxford.

GRIST D.H. & R.J.A.W. LEVER (1969) *Pests of Rice.* Longman, London.

ISHIHARA T. (1968) Families and genera of leafhopper vectors. In *Viruses, Vectors and Vegetation*, ed. K. Marmarosch, 235–54. Interscience Publishers, New York & London.

JONES D. PRICE (1961) Menazon: development of a selective systemic aphicide. *Proc. 1st Br. Insectic. Fungic. Conf.* (1961) **2**, 433–40.

KASASIAN L. (1971) *Weed Control in the Tropics*. Leonard Hill, London.

LE PELLEY R.H. (1969) *Pests of Coffee*. Longman, London.

LINN J.D. (1968) Effects of low volume aerial spraying of Dursban and fenthion on fish. *Down to Earth* **24**, 28–30.

MAI W.F. (1971) Introduction. In *Plant Parasitic Nematodes*, eds. B.M. Zuckerman, W.F. Mai and R.A. Rohde, **1**, 2–3. Academic Press, New York & London.

MARTIN H., ed. (1969) *British Crop Protection Council: Insecticide and Fungicide Handbook for Crop Protection*. Blackwell, Oxford. 387 pp. (3rd edition).

MASKELL F.E. & R. GAIR (1961) Further field experiments on the control of wheat bulb fly, *Leptohylemyia coarctata* (Fall.). *Bull. ent. Res.* **52**, 683–93.

MASSEE A.M. (1954) *The Pests of Fruit and Hops*. Crosby Lockwood, London.

PLANT PROTECTION LTD. (1971) Information from Company archives (unpublished).

PROCTOR J.H. & F.L. BARANYOVITS (1969) Pirimicarb: a new specific aphicide for use in integrated control programmes. *Proc. 5th Br. Insectic. Fungic. Conf.* (1969) **3**, 546–9.

RIPPER W.E., R.M. GREENSLADE & G.S. HARTLEY (1951) Selective insecticides and biological control. *J. econ. Ent.* **44**, 448–58.

SAYER H.J. (1959) An ultra-low volume spraying technique for the control of the desert locust, *Schistocerca gregaria* (Forsk.). *Bull. ent. Res.* **56**, 371–86.

SMITH R.F. & R. VAN DEN BOSCH (1967) Integrated control. In *Pest Control: Biological, Physical and Selected Chemical Methods*, eds. W.W. Kilgore and R.L. Doutt, 295–340. Academic Press, New York & London.

STERN V.M., R.F. SMITH, R. VAN DEN BOSCH & K.S. HAGEN (1959) The integration of chemical and biological control of the spotted alfalfa aphid. *Hilgardia* **29**, 81–101.

TAPLEY R.G. (1964) Insect pests. In *A Handbook on Arabica Coffee in Tanganyika*, ed. J.B.D. Robinson, 142. Tanganyika Coffee Board.

U.S. DEPARTMENT OF AGRICULTURE (1964) *Insecticide recommendations of the Entomology Research Division for the control of insects affecting crops, livestock and households, for 1964. Agriculture Handbook No. 120*. U.S. Department of Agriculture, Washington. 207 pp.

WAIN R.L. (1955) A new approach to selective weed control. *Ann. appl. Biol.* **42**, 151–7.

WASHINO R.K., K.G. WHITESELL & D.J. WOMELDORF (1968) The effect of low volume application of Dursban on non-target organisms. *Down to Earth* **24**, 21–2.

WILLIAMS J.R. (1969) Nematodes attacking sugar cane. In *Nematodes of Tropical Crops*, ed. J. E. Peachey, 184–203. Technical Communication No. 40. Commonwealth Bureau of Helminthology, St. Albans, Herts.

WINTERINGHAM F.P.W. (1969) Mechanisms of selective insecticidal action. *Ann. Rev. Entom.* **14**, 409–42.

Integrated control in Britain

M. J. WAY *Imperial College Field Station, Silwood Park, Ascot, Berkshire*

Definitions

As long as the definitions of ecology, applied ecology and of many widely used ecological terms remain imprecise we cannot in fairness carp about varying interpretations of integrated control. Some see integrated control as nothing new, a craft probably practised since prehistoric times; others see it as now outmoded by what is known as pest management. Pest management, however, is a generalized, all-embracing term virtually synonymous with pest control, its particular appeal being its aura of scientific respectability!

I see integrated control as providing both a discipline and an objective. Thus the integrated control approach provides a valuable conceptual basis for present-day pest control research, whereas integrated control itself can be foreseen as probably the most widely applicable pest control component of any kind of human activity (not only agriculture) which is significantly interfered with by pests.

When the term integrated control was first coined by Californian workers (Stern *et al.* 1959) they gave full recognition to the indispensable role of insecticides in the control of many pests. They also highlighted the value of control by natural enemies. But the limitations of each method used without reference to the other led to their emphasis on the need to integrate the two methods in ecologically based control practices in which selectively acting insecticides were used to supplement rather than supplant natural enemies. This ideal, initiated by Ripper (1944) in Britain, was I think a milestone in control concepts and it provided a *rationale* for the earlier work in Britain and Canada (Massee 1958, Pickett *et al.* 1958) as well as a basis for subsequent elaboration of new and imaginative ideas on how to attain pesticide selectivity and how to enhance the action of native and introduced natural enemies. The process of elaboration has also led, however, to the acceptance of all other kinds of control measures as part of integrated control, so it is now defined, by F.A.O. for example, as 'a system . . . which utilises all suitable techniques in as compatible a manner as possible . . .' (Anon. 1967). Thus the outstanding work of Hull and his colleagues (Hull 1965) against aphid-transmitted virus yellows of sugar beet in Britain must be classified as an example of successful integrated control although it owes its success to a combination of insecticides, hygiene and cultural practices and not to subtleties inherent in integration with natural enemy components. Whilst the inclusion of all possible methods has provided even greater opportunities for imaginative new control possibilities (e.g. Davidson 1974, Cherrett & Lewis 1974, Ruscoe 1974), the emphasis on ecologically based integration has nevertheless become less challenging and, according to some people, certain *ad hoc* combinations of methods may represent integrated control.

Research strategy

The enlightened response of a team of applied entomologists to failures in purely chemical methods of insect pest control has for more than a decade provided the Californian driving force for success in the integrated control approach. In Britain our general lack of interest, with a few notable exceptions, is a reflection of freedom from comparable insecticide-created problems. It is significant, however, that, without exception, the research method leading towards successful integrated control is common to all those who have succeeded. It is the experimental method. A hypothesis is made about the value of a particular part or parts of a control practice and is tested by critical field experiments. An edifice of integrated control practices is then constructed by a succession of such experiments. In contrast, the accumulation of descriptive information, including, for example, life-table data, which it is hoped will provide the key to control, has so far proved unproductive at any rate in the solution of practical problems. Thus the monumental work of Morris and colleagues (Morris 1963) on the population dynamics of the spruce budworm has done little more than add ecological respectability to good control measures that had been, or could be, conceived much more simply on more or less empirical grounds. Pure ecologists are, of course, usually asking questions very different from those of the applied ecologist, so it is important to ask what kinds of pure research are likely to be most helpful to the applied worker, particularly in view of the increasing and very welcome interest that pure ecologists are now taking in practical problems. I foresee especially valuable practical dividends from experimental ecologists who, rather than observing overall phenomena, are devising techniques and doing experiments to test hypotheses about the relevance of particular mortality factors in populations of pest species. It is a good sign that pure research on pest species is becoming increasingly respectable in Britain!

Past research on integrated control in Britain

It is salutary to list some past British attempts to develop integrated control practices. Foremost is the classical work of Massee and his colleagues (Massee 1958) on control of apple pests. Despite much promise this was not developed to practical fruition. Was it just because total reliance on pesticides continued to provide a 'better buy'—to use Strickland's (1970) terminology and, if so, is this still true? Much of the work of Broadbent and his colleagues (e.g. Broadbent 1964, 1969) demonstrated combined insecticidal and cultural practices which decreased the spread of insect-transmitted virus diseases of various crops. Why, unlike the equivalent sugar beet work, did this remain largely unaccepted by the farmer? Collaborative work on wheat bulb fly in the 1950s, reviewed by Long (1960), extended our knowledge of the range of different control methods that could be used collectively against this pest, but, in view of the success of organochlorine seed dressings, the alterations in farm practice required by integrated control did not then seem justified. More recently, detailed research at the National Vegetable Research Station (Coaker & Finch 1971) has examined alternatives to sole reliance on organochlorine insecticides against cabbage root fly. The localized development of cabbage root fly resistance to the recommended insecticide dieldrin was attributed largely to the destruction by dieldrin of predators (Wheatley 1971) which otherwise killed 90–95% of eggs and larvae of the pest (Hughes 1959). It is unfortunate that the ideally selective chlorfenvinphos, which is perhaps the best present-day alternative, is less effective than were the organochlorines, despite the large natural enemy-induced mortality that the use of chlorfenvinphos permits (Mowat & Coaker 1967). At Imperial College we have similarly failed to obtain as good control of brassica aphids by integrated methods as by solely chemical control (Way *et al.* 1969). None of the above studies was made in conditions where it was imperative that alternatives to existing chemically-based procedures should be devised, although the integrated control of sugar beet yellows disease, referred to earlier, no doubt reflected inadequacies of the purely chemical approach. However, current British success against some glasshouse pests (Hussey & Bravenboer 1971, Wyatt 1974) strikingly demonstrates how enthusiastic official and grower support can be elicited even for sophisticated integrated control research and development when conventional methods begin to fail.

The future of the integrated control approach

In Britain the comparative lack of pesticide-created resistance and resurgence problems and now, the timely phasing out of polluting organochlorines (Anon. 1969) and their replacement by seemingly much safer substitutes, has, in general, militated against the adoption of the usually more sophisticated procedures required in integrated control. On a short-term basis no one can therefore criticize the farmer for adhering to the simple control procedures offered by the insecticide manufacturer; but, for the future this is no excuse for complacency by those concerned with research and development. An agricultural revolution, implications of which are discussed by Southwood (1972), is still in progress in Britain. The pace-setters for change are economic factors unconnected with pests other than weeds, so, in general, pest control measures, although an important ingredient, have been relatively unimportant in shaping the revolution. On the contrary, they have had to be continually modified to conform with the changes. Furthermore, practices like drilling to a stand, clean weeding and restricted rotations are mostly tending to accentuate pest problems and thereby putting increasing emphasis on the need for chemical control measures. At present we can only speculate on the effects of other changes, such as the widespread removal of hedgerows in parts of Britain, nor are we certain that organochlorine substitutes will retain a clean record in terms of health and conservation requirements.

Strickland (1970) has enunciated five principles which those concerned with pest control should adopt, and, in particular, he emphasizes that the complexities of crop protection require an integrated multi-disciplinary approach if future problems are to be effectively solved. This involves closer integration not only within a particular discipline, such as insect pest control, but between disciplines of pest, disease and weed control and between these and those disciplines in agronomy that are at present dictating agricultural progress; also with conservation interests. Although this is a formidable proposition we are doing little about it. Surely we can afford to devote more effort to some co-ordinated inter-disciplinary research, particularly since our relatively abundant population of applied biologists is by no means overwhelmed by daunting short-term pest problems. The present position of research and development in pest control seems however to be one where specialists in ADAS (Agricultural Development & Advisory Service) could become increasingly restricted to immediately practical problems by cost/benefit criteria (Anon. 1971). It is understandable that industry, like ADAS, should be primarily concerned with relatively short-term prospects, but, nevertheless, many firms are putting notable investment into futuristic possibilities. Research institute and university workers are concerned with a wide range of more fundamental problems, many of practical significance;

but they tend to be compartmentalized, often guarding their personal projects. It is of course inevitable and necessary that most research should be personal but this does not detract from the need to create an atmosphere in which every encouragement is given to appropriate collaborative pest control research and development, because it is only such work that will provide an opportunity to test the feasibility of truly integrated control practices.

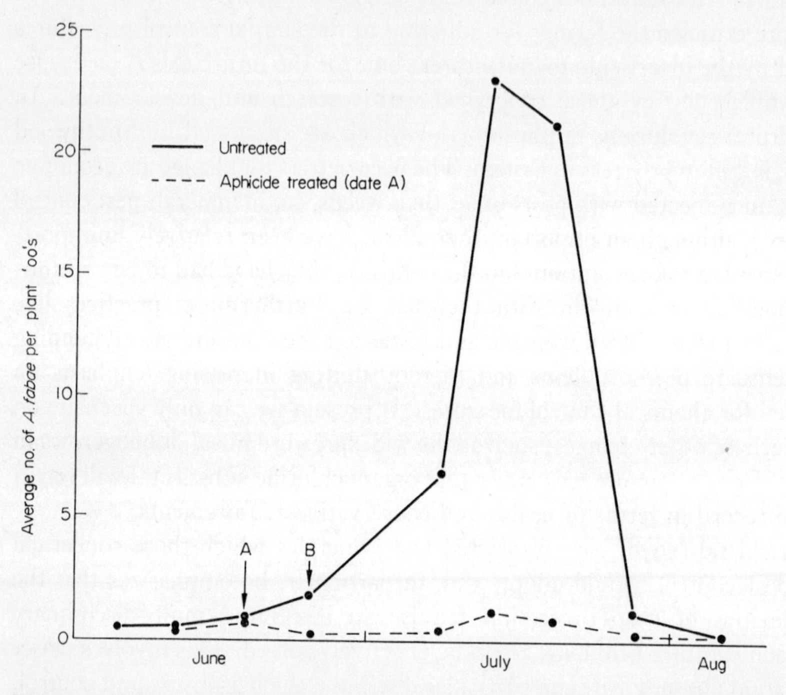

Figure 1. Effect of a single application of a suitable systemic aphicide on control of *A. fabae* on field beans. A selective insecticide applied at time A, immediately after aphid immigration has ceased, does not harm natural enemies, which are present only as adults. Delay in application until time B will cause death by starvation of immature stages of natural enemies.

The nature of the contribution that any one discipline can make to the inter-disciplinary approach will depend on the pest complex and the environment—e.g. the crop or cropping system. This can be illustrated by examples of the contribution of the applied entomologist. There are no doubt many British insect pest problems where there is no need to recommend more than the first step towards integrated control, as for example against the black bean aphid, *Aphis fabae*. A single spraying with a suitable systemic insecticide applied in early June directly after arrival of the pest on field beans will normally give excellent control (Fig. 1) (Way *et al.* 1954). From August to early June the aphid is mostly on unsprayed weeds or wild hosts where mortality is by non-chemical controlling factors. In these circumstances it may be confidently predicted that insecticide resistance will not be induced—nor is there evidence of resurgence problems. Where then lies the need for the integrated control approach? In two respects. First, it is needed to help preserve natural enemies in the agro-ecosystem as a whole. This requires early insecticide treatment with a selectively acting systemic aphicide—either a granular formulation which confers selectivity or a spray of an intrinsically selective chemical such as menazon. Adult predators already arrived on the crop will be unharmed and thus able to depart and act elsewhere (Fig. 1). Furthermore, early treatment will prevent wastage of predators through starvation of immature forms which would be present if the application was delayed until after the adults had begun to reproduce. Secondly, as with all good control practices, integrated or otherwise, an early warning system is needed to predict the need for application of an insecticide and to improve the timing of the application. It is with this aim in view that a group from ADAS, Research Institutes and Imperial College are working on the bean aphid at present.

The message from this simple example is that the integrated control approach should if possible be adopted in any control practice, irrespective of immediate value to the crop itself, since potentially useful natural enemies are being preserved to act against pests or potential pests elsewhere. In the bean aphid example, the requirements of the integrated control approach are the same as those for best protection of the particular bean crop, i.e. early spraying gives best kill of the aphid and least crop damage. In other pest situations, however, it could be that even simple refinements might be unacceptable to the farmer if they involved limitations that were not offset by immediate practical gains.

A more advanced stage of sophistication in integrated control is represented by a variety of within-and-between-crop procedures of the kind so well exemplified by Californian work (Stern *et al.* 1959; van den Bosch *et al.* 1971), by the work of Jones and colleagues (Jones *et al.* 1967, Whitehead 1972) on potato root eelworm control; by enlightened experiments such as those of ADAS in Cambridgeshire, where a group of farmers also collaborated in a wheat bulb fly control project by synchronizing sowing dates (Legowski & Maskell 1968), and by the work of Dempster & Coaker (1973). Potts (Potts & Vickerman 1974) has also obtained evidence of important within- and between-crop effects on pests related to herbicides and cropping systems, which should be followed up by team research.

The still evolving Californian programme of integrated control of pests of alfalfa and cotton is a classic demonstration of the required approach. The first development was to show that alfalfa fields should be 'strip cut' instead of clean cut. Strip cutting retained natural enemies which therefore continued to control the alfalfa aphid, *Therioaphis trifolii*. In contrast, fields that were clean cut were shown to be deserted by natural enemies, and so to be prone to severe aphid attacks on new growth formed after cutting. As a supplement, the systemic insecticide demeton could be used when natural enemy action proved inadequate. The critical feature of the demeton treatment was its use at one-sixth of the recommended rate. This still killed most of the aphids yet left a necessary residue to support and retain the natural enemies which were virtually unharmed by the selectively-acting low dosage (cf. Jones 1974). The later development of strains of alfalfa resistant to the aphid made strip-cropping practices largely redundant for alfalfa, but not in relation to cotton, because the strip-cropping of alfalfa was shown to retain *Lygus hesperus* that otherwise dispersed in destructive numbers from alfalfa to neighbouring cotton when the former was clean cut. Furthermore, insecticides used against *L. hesperus* on cotton upset the biological control of the American bollworm *Heliothis zea* by killing its natural enemies. The latest attack on this problem is to plant strips of alfalfa in cotton fields to retain *L. hesperus* and also provide an extra source of natural enemies of the bollworm (Stern 1969, van den Bosch *et al.* 1971). Other methods are also being used to encourage natural enemies, e.g. by supplementing their diet (Hagen *et al.* 1970). Complete success is not yet claimed; however, the startling feature of the control measures is that they confer useful biological and integrated control in an entirely artificial environment—a large area of irrigated desert comprising only about nine different crop species and virtually no wild plants. As DeBach (1964), Way (1966), and Kennedy (1968) have pointed out, this evidence, like that of the glasshouse crops work, demonstrates a form of stability in a very simple community, thereby questioning the assumption from ecological theory (e.g. Cole 1966, Elton 1958, MacArthur 1955, Pimentel 1970, Rudd, 1964, Watt 1971) that stability is attained only in complex communities. It seems therefore that we should be able to create very simple, entirely artificial, stable systems—especially in relatively isolated circumstances, as in a glasshouse, or irrigated desert, where outside interference is minimized. Furthermore, in these circumstances, tendencies to depart from the steady state can be corrected either by timed application of a suitable insecticide or by artificially assisting or supplementing the natural enemy populations. We should therefore welcome the increasing size of farms in parts of Britain because this may make it possible to create ecological islands free from undesirable elements of diversity. The ultimate in integrated control in our major arable areas may therefore include manipulation of both crop and non-crop environments involving destruction of elements which favour the pest, and preservation or establishment of elements favouring natural enemies or other mortality factors (Way 1966).

Feasibility of different methods of control

Some general conclusions about forms of control relevant to different crops and crop conditions are indicated in Table 1, which attempts to do for some British situations what Southwood & Way (1970) have already attempted on a more world-wide basis.

Table 1. Components of the environment which are likely to determine the success of the different methods of control that can be used against pests

	(1) Diversity in age and species of 'plants'	(2) Permanence of 'plant' population	(3) Size of unit or of group of adjoining units	(4) Stability of climate
Tropical rain forest	*****	*****	*****	*****
Mixed woodland (U.K.)	****	*****	***(**)	***
Conifer monoculture (U.K.)	*	****	***(**)	**
Top fruit (U.K.)	*	****	**	***
'Permanent' grassland (U.K.)	**	***(*)	****	***
Mixed arable monoculture (U.K.)	*(*)	*	**(**)	***
Single sp. arable monoculture (e.g. Canada)	*	*	*****	*
'Protected' crops (U.K.)	*(*)	**	*	*****
Stored products (short term) (U.K.)	*	*	*(**)	****

The number of stars indicates the magnitude of the component. Brackets indicate variation, e.g. **(**) = varying from ** to ****.

(a) Five-star combinations of components (1), (2) and (4), as in tropical rain forest, most favour automatic control by natural controlling mechanisms.

(b) Relatively small artificial decreases in magnitude of (1) and (2) and decreases in (4) can greatly diminish action of natural checks (e.g. as in U.K. mixed woodlands).

(c) Quantitatively, plant species diversity (1) is not necessarily important for biological control. Thus natural enemies can be a major cause of mortality in relatively very simple plant communities such as top fruit and protected crops. In these circumstances the addition of relatively little of the right quality of diversity may greatly assist biological control.

(d) Relative permanence of plant population (2) and stability of equable climate (4) are especially important for successful control by natural enemies.

(e) In extreme contrast to (a), existing or manipulated situations in which (2) or (4) have one-star status are unfavourable for natural enemies but can be used to exert valuable

The number of crosses represents the magnitude of a particular characteristic, and, as a standard for comparison, the tropical rain forest is included to show 5-star combinations of great diversity of plant species and age, great permanence, large size of ecological unit and generally an equitable all-the-year-round ecoclimate. It seems to be an ecological axiom that this is the ultimate in stability—a situation where no one species becomes sufficiently abundant to become a pest. In Britain the nearest approach to the tropical rain forest is probably the mixed deciduous woodland, and yet, it is a startling revelation by Varley & Gradwell (1962) that up to 60% loss of growth increment of oak trees may be caused by leaf-eating caterpillars in a typically diverse British woodland, to which must no doubt be added the damage from many other phytophagous species that are common on oak. Diversity as such is therefore not necessarily a desirable attribute—it is its quality, not quantity that matters. Maybe top fruit monocultures in particular would repay study in terms of manipulated diversity (Table 1). Irrespective of other factors, the somewhat inequable British climate must often disrupt biological balances and, in this context, there is no doubt that the stability of the 'climate' contributes notably to success of biological and integrated control of glasshouse pests.

Contrasting with tropical rain forest and man-created stable systems is the single species cereal monoculture. In an extreme form this is exemplified by parts of the Canadian prairies where the instability of the cropping system (one cereal crop for about five months every alternate year), the large areas of a single crop, and relative scarcity of non-arable land may put the pests

---

Table 1 (*contd.*)

control by directly harming the pest especially if (1) also has one-star status (e.g. in Canadian single-species monocultures or in short-term stored products). The limitations to natural enemy action set by one-star status of (2) can only be overcome by artificially introducing natural enemies, as in glasshouse crops procedures, or by providing adjacent sources of natural enemies (e.g. strips of perennial alfalfa in cotton fields).

(f) In large field units, relatively little pest damage may be done to annual crops because the abundant even-aged plants may 'swamp' available colonizing phytophagous species. Otherwise, size as such above a certain minimum is probably not important compared with isolation, e.g. ecological isolation of even small areas of crops such as top fruit or glasshouse crops may help biological control. Useful 'isolation' may also be attained within very large simply cropped areas where undesirable elements of crop or non-crop diversity are distant (e.g. in the entirely artificial conditions of parts of California or potentially, in large farm units in the U.K.). Some of the new methods of control (using sterilants, pheromones, etc.) are likely to be useful only in such relatively isolated situations.

(g) Conventional insecticides will remain a major component of most pest control measures. The degree of sophistication required to integrate insecticides with the other methods of control will, in general, vary in direct relation to proximity of conditions to (a) as distinct from (e).

themselves at a serious disadvantage. There is also the 'swamping' effect of large areas of the same crop. Extreme instability of the ecosystem may therefore be valuable in pest control!

Significance of stability

Sagar (1974) refers to the ecologically dispiriting nature of arable cropping systems dislocated by cultivation practices, but, on the contrary, this is ecologically meaningful. Thus Andrewartha & Birch (1961) refer to species requisites, including a 'place in which to live'; so, in terms of pest control, there is good justification on ecological grounds for artificial practices that create a place in which a pest cannot live! This is inherent in the evidence that only a few species with special qualities are able to succeed in the dislocated environments of arable systems (Southwood 1962). Ecosystem stability is therefore not necessarily a desirable aim in pest control. Furthermore, for a particular species, stability merely implies the maintenance of fairly consistent mean numbers irrespective of the degree of periodic displacement from the mean (Solomon 1971), or of damage caused. We generally assume, however, that a very stable population is one in which, most of the time, numbers not only oscillate relatively little but also remain too small to cause damage. The latter is not necessarily true for some species, notably 'low density' pests, and, also, there may be practical advantages in creating situations where the period when the crop is at risk is manipulated to coincide with troughs in the cycles of relatively unstable populations. Furthermore, relative instability may cause such violent oscillations that the species is locally eradicated. In these circumstances, the control of rare but predictably severe eruptions (e.g. Morris 1963) may be simpler and more rational than that of many present-day British insect pests which cause damage, so often borderline, that is inadequately predictable, therefore necessitating routine protective insecticide applications.

Conclusions

Conclusions from this survey of the situation in Britain are as follows:

1 The integrated control approach should provide the conceptual basis of all field research on pest control.

2 The adoption of even the simplest integrated control practices should be encouraged even in pest control situations where, at the time, the control of the target pest may not be immediately improved.

3 In the immediate future the more sophisticated forms of integrated pest control will no doubt have limited application but, if present changes in agriculture continue, they may well increasingly prove to be the best buy, especially in large agricultural units in certain parts of Britain.

4 The application of ecological principles to pest control research must be very critically examined in relation to each particular crop problem.

5 Some of our present resources should be devoted to a few major research projects on integrated control in which encouragement is given to collaborative work within and between pest control and other disciplines concerned with agricultural development.

References

ANDREWARTHA H.G. & L.C. BIRCH (1961) *The Distribution and Abundance of Animals*, 782 pp. University of Chicago Press, Illinois.

ANON. (1967) *Report of the First Session of the F.A.O. Panel of Experts on Integrated Pest Control.* F.A.O., Rome 1966.

ANON. (1969) *Further review of certain persistent organochlorine pesticides used in Great Britain*, 148 pp. H.M.S.O.

ANON. (1971) *The Agricultural Development and Advisory Service of the Ministry of Agriculture, Fisheries and Food: its advisory work*, 16 pp. Min. of Agric., London.

VAN DEN BOSCH R., T.F. LEIGH, L.A. FALCON, V.M. STERN, D. GONZALES & K.S. HAGEN (1971) The developing program of integrated control of cotton pests in California. In *Biological Control*, ed. C.B. Huffaker, 377–94. Plenum Press, New York & London.

BROADBENT L. (1964) Control of plant virus diseases. In *Plant Virology*, eds. M.K. Corbett and H.D. Sisler, 330–64. University of Florida Press.

BROADBENT L. (1969) In *Viruses, Vectors and Vegetation*, ed. C. Maramarosch, 593–630. Wiley Interscience.

CHERRETT J.M. & T. LEWIS (1974) Control of insects by exploiting their behaviour. In *Biology in Pest and Disease Control*, eds. D. Price Jones and M.E. Solomon, 130–46. Blackwell, Oxford.

COAKER T.H. & S. FINCH (1971) The Cabbage Root Fly *Erioischia brassicae* (Bouché). *Rep. natn. Veg. Res. Sta. for 1970*, 23–42.

COLE L.C. (1966) The complexity of pest control in the environment. In *Scientific Aspects of Pest Control*, 13–25. Nat. Acad. Sci.—Nat. Research Council, Washington D.C.

DAVIDSON G. (1974) Genetic control of insects. In *Biology in Pest and Disease Control*, 162–77, eds. D. Price Jones and M.E. Solomon. Blackwell, Oxford.

DEBACH P. (1964) (ed.) *Biological Control of Insect Pests and Weeds*, 844 pp. Chapman & Hall, London.

DEMPSTER J.P. & T.H. COAKER (1974) Diversification of crop ecosystems as a means of controlling pests. In *Biology in Pest and Disease Control*, eds. D. Price Jones and M.E. Solomon, 106–14. Blackwell, Oxford.

ELTON C.A. (1958) *The Ecology of Invasions by Animals and Plants.* 181 pp. Methuen, London.

HAGEN K.S., E.F. SAWALL & R.L. TASSAN (1970) The use of food sprays to increase effectiveness of entomophagous insects. *Proc. Tall Timbers Conf. Ecol. Anim. Control by Habitat Management* 2, 59–81. Tallahassee, Fla. (1970).

HUGHES R.D. (1959) The natural mortality of *Erioischia brassicae* (Bouché) (Dipt., Anthomyiidae), during the egg stage of the first generation. *J. anim. Ecol.* 28, 343–57.

HULL R. (1965) Control of sugar beet yellows. *Ann. appl. Biol.* 56, 345–7.

HUSSEY N.W. & L. BRAVENBOER (1971) Control of pests in glasshouse culture by the introduction of natural enemies. In *Biological Control*, ed. C.B. Huffaker, 195–216. Plenum Press, New York & London.

JONES D. PRICE (1974) Selectivity in pesticides: significance and enhancement. In *Biology in Pest and Disease Control*, eds. D. Price Jones and M.E. Solomon, 178–95. Blackwell Scientific Publications, Oxford.

JONES F.G.W., D.M. PARROTT & G.J.S. ROSS (1967) The population genetics of the potato cyst-nematode *Heterodera rostochiensis*: mathematical models to simulate the effects of growing eelworm-resistant potatoes bred from *Solanum tuberosum* ssp. *andigena*. *Ann. appl. Biol.* 60, 151–71.

KENNEDY J.S. (1968) The motivation of integrated control. *J. appl. Ecol.* 5, 492–99.

LEGOWSKI T.J. & F.E. MASKELL (1968) Control of wheat bulb fly by large area cropping restrictions. *Plant Path.* 17, 129–33.

LONG D.B. (1960) The wheat bulb fly, *Leptohylemyia coarctata* Fall.—a review of current knowledge of its biology. *Ann. Rep. Rothamsted Expt. Stat. for 1959*, 216–29.

MASSEE A.M. (1958) The effect on the balance of arthropod populations in orchards arising from the unrestricted use of chemicals. *Proc. 10th Int. Cong. Ent. Montreal* (1956) 3, 163–8.

MACARTHUR R. (1958) Fluctuations of animal populations and a measure of community stability. *Ecology* 36, 533–6.

MORRIS R.F. (1963) *The dynamics of epidemic spruce budworm populations.* Mem. Ent. Soc. Canad. No. 31, 332 pp.

MOWAT D.J. & T.H. COAKER (1967) The toxicity of some soil insecticides to carabid predators of the cabbage root fly *Erioischia brassicae* (Bouché). *Ann. appl. Biol.* 59, 349–54.

PICKETT A.D., W.L. PUTMAN & E.J. LE ROUX (1958) Progress in harmonising biological and chemical control of orchard pests in Eastern Canada. *Proc. 10th Int. Cong. Ent. Montreal* (1956) 3, 169–80.

PIMENTEL D. (1970) Population control in crop systems: monocultures and plant spatial patterns. In *Tall Timbers Conference on Ecological Animal Control by Habitat Management* 2, 209–21. Tallahassee, Fla. (1970).

POTTS G.R. & G.P. VICKERMAN (1973) Studies in arable ecosystems. *Adv. in Ecol. Res.* 8 (in press).

RIPPER W.E. (1944) Biological control as a supplement to chemical control of insect pests. *Nature, Lond.* 153, 448–52.

RUDD R.L. (1964) *Pesticides and the Living Landscape*. Faber & Faber, London.

RUSCOE C.N.E. (1974) Insect control by hormones. In *Biology in Pest and Disease Control*, eds. D. Price Jones and M.E. Solomon, 147–61. Blackwell Scientific Publications, Oxford.

SAGAR G.R. (1974) On the ecology of weed control. In *Biology in Pest and Disease Control*, eds. D. Price Jones and M.E. Solomon, 42–56. Blackwell Scientific Publications, Oxford.

SOLOMON M.E. (1971) In discussion of paper by M.J. Way in *Dynamics of Populations*, eds. P.J. den Boer and G.R. Gradwell, p. 240. Proc. Adv. Study Inst. Dynamics Numbers Popul. Oosterbeek 1970. Centre for Agric. Publ. & Docum., Wageningen.

SOUTHWOOD T.R.E. (1962) Migration of terrestrial arthropods in relation to habitat. *Biol. Rev.* 37, 171–214.

SOUTHWOOD T.R.E. (1972) Farm management in Britain and its effects on animal populations. In *Tall Timbers Conference on Ecological Animal Control by Habitat Management* 3, 29–51. Tallahassee, Fla. (1971).

SOUTHWOOD T.R.E. & M.J. WAY (1970) Ecological background to pest management. In *Concepts of Pest Management*, eds. R.L. Rabb and F.E. Guthrie, 6–29. University of North Carolina Press, Chapel Hill.

STERN V.M. (1969) Interplanting alfalfa in cotton to control *Lygus* bugs and other insect pests. In *Tall Timbers Conference on Ecological Animal Control by Habitat Management* 1, 55–69. Tallahassee, Fla. (1969).

STERN V.M., R.F. SMITH, R. VAN DEN BOSCH & K.S. HAGEN (1959) The integrated control concept. *Hilgardia* **29**, 81–101.

STRICKLAND A.H. (1970) Economic principles of pest management. In *Concepts of Pest Management*, eds. R.L. Rabb and F.E. Guthrie, 30–44. University of North Carolina Press, Chapel Hill.

VARLEY G.C. & G.R. GRADWELL (1962) The effect of partial defoliation by caterpillars on the timber production of oak trees in England. *Proc. 11th Int. Cong. Ent. Vienna 1960* **2**, 211–14.

WATT K.E.F. (1971) Dynamics of populations: a synthesis. In *Dynamics of Populations*, eds. P.J. den Boer and G.R. Gradwell, 568–80. Proc. Adv. Study Inst. Dynamics Numbers Popul. Oosterbeek 1970. Centre for Agric. Publ. & Docum., Wagenigen.

WAY M.J. (1966) The natural environment and integrated methods of pest control. In *Pesticides in the environment and their effects on wildlife*, ed. N.W. Moore. *J. appl. Ecol.* 3 (suppl.), 29–32.

WAY M.J., G. MURDIE & D.J. GALLEY (1969) Experiments on integration of chemical and biological control of aphids on brussels sprouts. *Ann. appl. Biol.* **63**, 459–75.

WAY M.J., P.M. SMITH & C. POTTER (1954) Studies on the bean aphid (*Aphis fabae* Scop.) and its control on field beans. *Ann. appl. Biol.* **41**, 117–31.

WHEATLEY G.A. (1971) Pest control in vegetables: some further limitations in insecticides for cabbage root fly and carrot fly control. *Proc. 6th Brit. Ins. Fung. Conf.* (1971) **2**, 386–95.

WHITEHEAD F.H. (1972) Nematicides for field crops. *Proc. 6th Brit. Ins. Fung. Conf.* (1971) 3, 662–72.

WYATT I.J. (1974) Progress towards biological control under glass. In *Biology in Pest and Disease Control*, eds. D. Price Jones and M.E. Solomon, 294–301. Blackwell, Oxford.

Section 3
Application of biological control

The use of biological methods in the control of vertebrate pests

R. K. MURTON *The Nature Conservancy, Monks Wood Experimental Station, Abbots Ripton, Huntingdon*

The classic concept of biological control is the limitation in numbers of noxious animals or plants by their natural enemies; some authors (e.g. Stern *et al.* 1959) extend the term to involve all predator-prey type interactions that effectively restrict prey density, irrespective of whether contrived by man or entirely natural. There is practically no applicability of biological control *sensu stricto* where vertebrates are concerned. Nor would such control be generally justified, for it has become increasingly evident that vertebrates become an economic nuisance only in restricted situations, and the position is not necessarily ameliorated by attempts to reduce population size below the 'natural' level. For example, several bird species can provide an economic hazard to man when they congregate on aerodrome runways, but at other times are subject to stringent protection. Bats in Britain are a valuable, and perhaps declining, component of our fauna which we should make every effort to safeguard, yet a bat in the belfry can sometimes cause a problem by fouling, and householders may demand remedial action. On the other hand, in a totally different context bats may provide a national problem, as illustrated by the vampire bats, particularly *Desmodus rotundus* in the New World tropics, whose sanguivorous habits make them important vectors of paralytic rabies (Greenhall 1968, Villa 1969).

These examples emphasize that every problem posed by a vertebrate has to be considered as a special case, that vertebrates are obvious and widely appreciated components of the ecosystem, so that a powerful conservation lobby may in any event complicate the attempt to deal with a problem in a completely rational and scientific manner. Biological methods of control—using the term to embrace the use of biological scaring mechanisms such as recorded distress calls, pheromones and repellents and such procedures as gene manipulation—are generally to be preferred. Essentially we must accept agents of control which have a biological basis but do not necessarily lead to reductions in pest numbers. In those situations where population control is considered feasible and desirable, it should be remembered that ecologists have produced techniques, for example key-factor analysis (Varley & Gradwell 1968, Varley 1970), whereby the density-dependent factors regulating populations or the factors responsible for fluctuations in numbers can be identified and quantified, enabling artificial control to be applied at the most effective stage of the cycle.

The idea of biological control is by no means new, for man was practising these methods long before Christ. Presumably this should be expected, since man is essentially a biological predator and natural enemy of many animals. As an example of the classical approach to biological control, Sweetman (1958) instances the use of predatory ants for the control of citrus pests as antedating written records; the earliest written records are dated A.D. 900–1200 (Liu 1939). A tomb painting from Thebes dated to the 18th Egyptian dynasty shows bird-hunters in the papyrus swamps handling a cormorant as if to evoke a distress call and other individuals are evidently being attracted to within missile range. There is reason to suspect that our ancestors had empirically learned many of the subtleties of the biological approach to pest control.

Part 1 of this contribution is a brief review which attempts to outline the potential of biological control methods for vertebrates, and Part 2 reports a specific research project.

Part 1. Biological methods for vertebrate control

(a) *Predator-prey interactions*

Ecologists often emphasize that climax plant communities and natural ecosystems exhibit a degree of diversity in their constituent plants and animals that encourages ecological stability (Elton 1958). The numerous predator-prey interactions operating in such complex systems tend to damp down the fluctuations in the numbers of any one species so that the chances of a pest outbreak are reduced. However, with certain forest insects, increase in the number of host trees, as distinct from number of competitors, increased instability, so the processes involved are clearly complex (Watt 1964). Outbreaks of forest pest insects are more frequent in pure stands of timber than in mixed woodland (Voûte 1946). This all suggests that agriculturalists should introduce diversity into crop monocultures to encourage predator-prey systems. Unfortunately, though laudable, this approach can seldom be realistic in extensive agricultural monocultures where maximum productivity at minimum cost must be the objective; often, when conditions are contrived to suit the needs of a predator, crop yields suffer disproportionately, as in the case of the small white butterfly *Pieris rapae* and its predators (Dempster 1969). The agricultural ecosystem is organized on a basis of instability leading to marked fluctuations in the standing crop so that the peaks in primary production can be harvested; ideally there should be as many peaks in production as the operating conditions allow. Only a few species can adapt to such fluctuating habitats; most predatory species require greater stability. Vertebrate predators are rarely obligate, and they usually do not attack the healthy component of their prey, while the maintenance of an alternative food supply would rarely be conducive to sensible agriculture. But exceptions do occur, and Gibb *et al.* (1969) have produced evidence in New Zealand that feral cats may be responsible for keeping rabbits (*Oryctolagus cuniculus*) at low density without help from man. Predators such as cats may be influential at low but not at high densities of their prey (e.g. Elton 1953).

Another reason why predators are usually of doubtful value for the control of vertebrates is that they have too low a reproductive rate relative to their hosts (contrast insects) and succeed only in cropping an expendable surplus of their prey. Moreover this surplus usually involves sick or ailing or otherwise unfit segments of the population, so that the main effect of the predator is to maintain a degree of biological fitness in the prey. Thus golden eagles (*Aquila chrysaëtos*) and hen harriers (*Circus cyaneus*) mostly capture those grouse (*Lagopus lagopus*) which are surplus to and are displaced from the social system (Jenkins *et al.* 1964). Van Dobben (1952) showed that 6·5% of the roach in the Ijesselmeer, Holland, were infected with an intestinal parasite *Ligula intestinalis*, but of those eaten by cormorants (*Phalacrocorax carbo*) 30% were infected. Goshawks (*Accipiter gentilis*) mostly eat wood-pigeon (*Columba palumbus*) when there is a population peak comprising many young birds just after the breeding season (Murton 1971a). Numbers of pocket gophers (*Thomomys bottae navus*) were not correlated with the presence or absence of coyotes (Robinson & Harris 1960, quoted by Howard 1967). Of the moles (*Talpa europaea*) eaten by buzzards (*Buteo buteo*) in Poland, 86% were juveniles, and these were mostly caught in late spring, a season when they penetrate the soil surface in search of new areas in which to establish themselves (Stoczeń 1962).

When man has artificially increased predator numbers, trouble has often followed. According to Van Wijngaarden & Bruijns (1961, quoted by Howard 1967), weasels (*Mustela nivalis*) and ermines (*M. erminea*) were introduced to the Friesian Island of Terschelling in 1931 to safeguard new forest plantations from water voles (*Arvicola terrestris*), and this they did successfully. But after three years the weasels became extinct, the ermine turned to other prey, and themselves had to be controlled. The introduction of cats to numerous oceanic islands to control the rats introduced by settlers has frequently led to the cats becoming a serious pest of the indigenous animals, once rats became scarce or were exterminated; feral cats do considerable damage to the sea-birds attempting to nest on Ascension Island (Stonehouse 1960). The mongoose *Herpestes auropunctatus* introduced to Jamaica is another example of man's folly (Pimentel 1955).

Vertebrates have sometimes been used to control invertebrates. For example, Indian myna birds (*Acridotheres tristis*) were imported into Mauritius in 1762 and have successfully controlled the red locust (*Nomadacris septemfasciata*) (Moutia & Mamet 1946). The poeciliid fish *Gambusia affinis* originates in the West Indies, but occurs throughout the warmer parts of the world where it has been introduced to control mosquito larvae. Although in many places the results of introduction have been beneficial, the swarming habits have in some places disrupted the balance of the native fauna (Krumholz 1948). The fish has become an important constituent of the diet of several heron species, particularly the little egret *Egretta garzetta* (Valverde 1958). Since most introductions have been ecologically harmful, we should have reservations about introducing to Britain the grass carp *Ctenopharyngodon idella* (cf. Stott 1974, this volume) to control water weed since these species feed by churning up the bottom mud and destroying plants, and can denude a water of cover for other species. Such drastic habitat disturbance can have many ramifications.

Certain forest birds have been held to be beneficial in reducing the likelihood of an insect pest outbreak, and in some circumstances the intensity of predation may be sufficient to damp oscillations in prey numbers, thereby reducing the risk of an epidemic. Tinbergen & Klomp (1960) calculated that, given density-dependent predation, the birds must take more than a quarter of the insects at risk for this to apply. But once an insect achieves plague proportions, or even its usual summer numbers, bird predation is usually ineffective (refs. in Murton 1971a). In many cases birds selectively remove parasitized pupae, for example, coal tits (*Parus ater*) were feeding on caterpillars of *Thera* spp. which had been parasitized by the chalcid *Litomastix* spp. (Gibb & Betts 1963).

(b) *Habitat control*

There are special situations, such as the environs of airfields, where modification of the habitat is feasible and probably justifiable (Wright 1968). It is obviously senseless to site garbage dumps which attract gulls near the approaches to airport runways, as has been done. Habitat modification might be considered a biological method if the precise requirements of a problem species could be defined and the habitat manipulated appropriately. But too frequently the approach involves such drastic spoliation of the countryside as scrub clearance, tree-felling and hedgerow removal, often aided with herbicides such as paraquat and 2,4,5-T. Such methods are hardly biological and are deplored by conservationists, since whole biotic communities are affected.

(c) *Disease*

Indigenous species have usually evolved immunity to endemic disease organisms, so these are unlikely to become sufficiently pathogenic to achieve the sustained scale of kill necessary for population control. But the same organisms may sometimes be highly lethal in man or his domestic animals. Recently the poultry industry has suffered considerable losses from fowl pest (Newcastle disease); strains of the same virus are endemic in wild birds which may be a reservoir for the disease (cf. rabies mentioned above). Moreover, it is becoming increasingly evident, with benefit of research and opportunities for full diagnoses of relatively minor ailments, that wild animals harbour many diseases which can cause illness in man. Thus while the epidemiologies of serious zoonoses such as leptospirosis are well enough known, this is not the case with minor and short-term infections; many cases of ornithosis are probably diagnosed as virus pneumonia, and similarly such ailments as histoplasmosis must frequently occur through infection from wild animals (see Murton 1971a for references). These facts make it risky to consider using diseases, particularly virus diseases, for the control of vertebrates, for epizootics might become epidemics.

The rabbit disease myxomatosis is usually quoted as the classic example of a vertebrate controlled by a pathogenic organism. The causative virus is closely related to fowl pox, and rabbits of the New World genus *Sylvilagus* are the natural hosts and have naturally acquired immunity. The disease was established in Australian populations of rabbits (*Oryctolagus cuniculus*) in 1950, and spread rapidly, producing a kill of over 99%. Subsequently, less virulent strains of the virus appeared, the rabbits acquired active immunity and also genetic resistance appeared. Rabbit populations subsequently recovered, but, owing to myxomatosis and the more efficient use of conventional control techniques, have not reached the pre-myxomatosis level, nor do they pose the same problem (Fenner & Ratcliffe 1965). The disease has also become enzootic in rabbits in Britain following the initial outbreak in Kent in 1953 (Armour & Thompson 1955). Mutations of the highly virulent strain of virus which originally reached Britain, and produced nearly 100% kill, have since occurred, to give various attenuated strains which produce kills in the order of 99% to under 50% of susceptible animals (modal percentage case-mortality rate about 70–95%); those rabbits that recover from the disease have acquired immunity and give to their young a measure of passive immunity (Mead-Briggs 1966). In addition there is evidence for the selection of rabbits with genetic resistance.

Yet rabbit numbers have not increased as rapidly as would be expected in the circumstances (see for example, Lloyd & Walton 1969). Even after 90% kill, full recovery of a population could theoretically occur in one season, while the rapid recovery of numbers following the attention of trappers is well appreciated (Sheail 1971). Neither predation nor the disease alone satisfactorily accounts for the slow recovery, and it is conceivable that we are witnessing an example of efficient integrated control, with the rabbit 'trapped by its own conventions'. The rabbit has evolved as a burrow-living herbivore, which habit gives it protection from predators. A complex social hierarchy ensures that the fittest and dominant animals have the best feeding prospects around the burrow system, to which they can rapidly retreat if danger threatens. Subordinate animals must feed furthest away from the burrow and those unable to establish themselves in the social system are displaced to die, or to initiate new colonies if the potential exists; they may even become top-living animals which hide away in temporary stops, in scrub or thick grass (Southern 1948, Mykytowycz 1959, Fenner & Ratcliffe 1965). Those genotypes which behave as typical rabbits, with a measure of dominance, are presumably at greater risk of contracting myxomatosis through the rabbit flea vector, whereas those which might escape the disease by leaving the burrow system must be at risk of predation. The supposed integration of mortality factors is represented schematically on the opposite page; the situation deserves more study.

It is still not yet clear to what extent the pattern of disease transmission and the evolution of the myxoma virus depend on the vectors involved; in Australia these are primarily mosquitoes, particularly *Anopheles atroparvus*, but in Europe the rabbit flea *Spilopsyllus cuniculi*. Fleas need leave their host only when it dies, so they should tend to transport the more virulent strains of myxoma; accordingly, Fenner & Marshall (1957) predicted that attenuated viruses should be selected against if fleas were the most important vectors. It is true that the more virulent strains of myxomatosis have persisted longer in Britain than was the case in Australia (Fenner & Chapple 1965).

Considerable scope existed in the case of myxomatosis for research into methods of artificially spreading the disease and of breeding new strains in order to maximize its efficiency as a means of reducing the rabbit problem. But an Advisory Committee of Myxomatosis appointed in 1953 acknowledged public sentiment for rabbits (two members resigned in protest) and recommended that the disease should not be purposely spread: following a fierce public debate, the Government banned the deliberate spreading of the disease under a clause in the 1954 Pest Act. It is unlikely that research into the field of pathogen manipulation for vertebrate control would be readily condoned by public opinion.

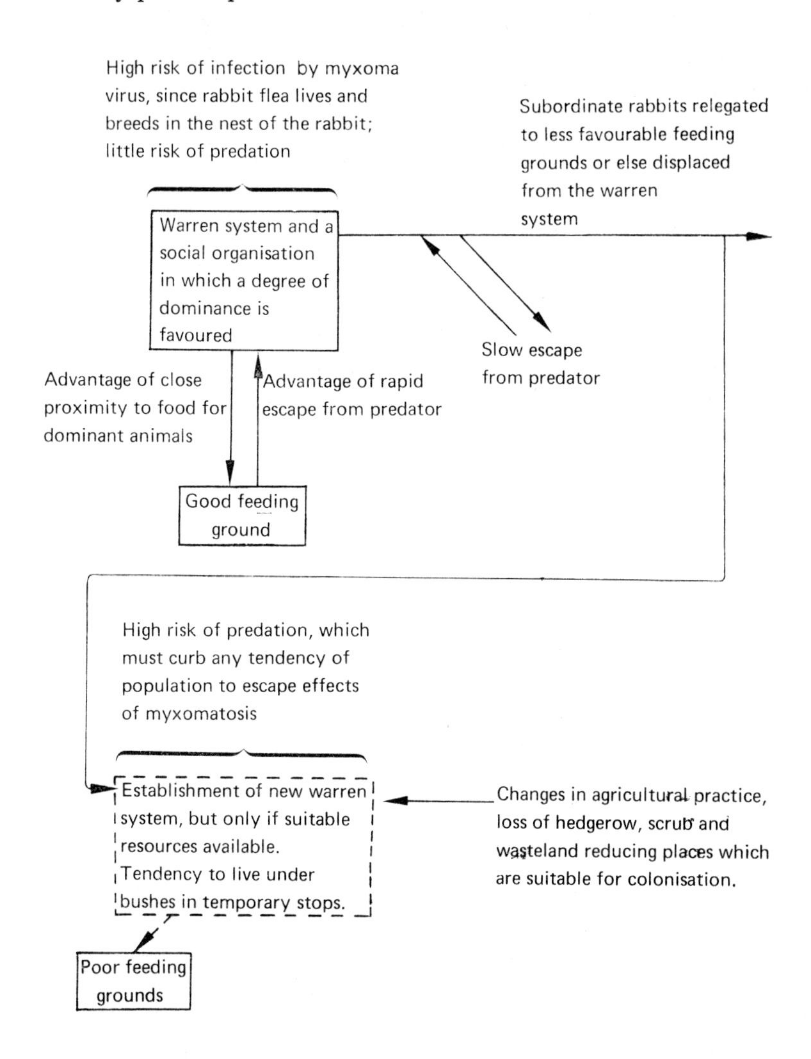

(d) *Reproduction inhibitors*

There is little scope for raising and releasing radiation-treated and sterile male vertebrates for the purposes of population control, and interest centres on the use of chemosterilants. The possibility of using anti-fertility agents has caused much excitement, and numerous synthetic oestrogens, progestins, anti-gonadotrophin substances, as well as inorganic chemicals have been shown to be effective in the laboratory, but very few experiments have produced unequivocal and worthwhile results under field conditions. Agents which are highly effective sterilants, especially those having long-term or permanent effects, are usually too dangerous to use outside the laboratory, cadmium chloride being but one example (Lofts & Murton 1967). Davis (1962) showed that triethylene-melamine (TEM) would inhibit gonadal growth in starlings (*Sturnus vulgaris*) and Elder (1964) tested this as well as various insecticides, fungicides and various synthetic oestrogens in a search for a suitable chemical.

Of the oral contraceptives tested by Elder, the one which showed most promise was the anti-cholesterol compound 22,25-diazocholesterol dihydrochloride, manufactured by G. D. Searle & Co. Ltd. (compound SC 12937). Laboratory work confirmed its potential usefulness (Lofts *et al.* 1968) and field trials were initiated with a study population of feral pigeons (*Columba livia* var.) at Salford, Manchester. Unfortunately, doses sufficient to sterilize in one operation are toxic, and, ideally, sub-lethal quantities need to be free-fed to the birds over a period. Using such techniques, various workers in the U.S.A. have claimed much success against town pigeons, but doubt attaches to the validity of the results (Murton *et al.* 1972); in some cases any reduction in pigeon numbers probably depended on direct mortality, while in others the methods of estimating population size before and after treatment were unreliable. Murton *et al.* (*loc. cit.*) discovered that only 30% of the feral pigeons in their study area actually bred, and so most of the drug was wasted on birds that contributed nothing to population replacement. Indeed, in this species the reproductive rate needs to be reduced by over 70% to prevent the replacement of adult losses. The effect of a sterilant is to increase the expressed rate of population increase (proportion of animals in one generation replaced by new individuals in the next), and this becomes most marked in species which realize a low reproductive rate in the wild but which have a high biotic potential for population increase (cf. the work on house-flies, *Musca domestica*, by Weidhaas & LaBrecque 1970).

The problems of using chemosterilants relate not so much to the difficulty of finding suitable agents, but to that of actually administering them to free-living subjects, especially as it is not feasible to treat all the individuals at once.

Given that only a proportion p of the population can be treated at random and independently during each control operation, and that n treatments are needed for the drug to be effective, the proportion of the flock effectively treated is p^n. Figure 1 illustrates the rapid fall-off in effectiveness of any treatment as p decreases and n increases. If one application of a drug is effective yet only a quarter of the population is treated, none of the drug is wasted. But if the total number of animals is M, the total number of animals treated on all occasions (some animals being treated and therefore counted more than once) is the number of animals treated per operation times the number of operations (n), i.e. Mpn. However, the number of animals treated

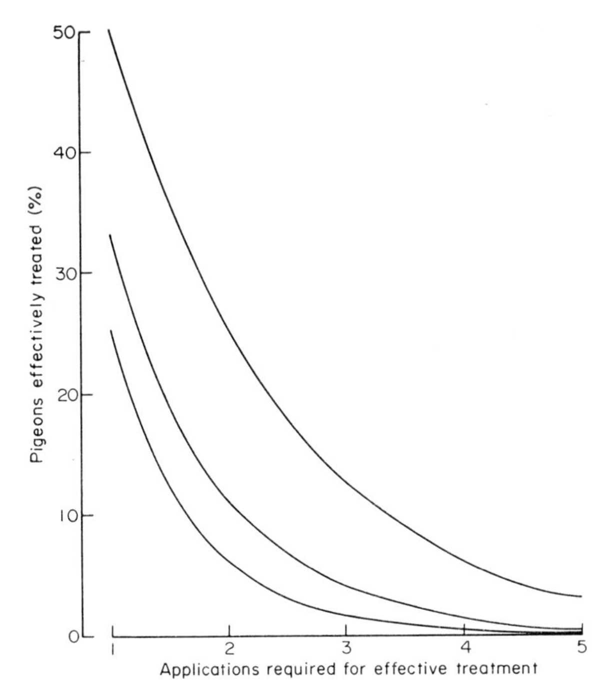

Figure 1. Percentage of animals effectively treated by a bait in relation to the proportion of the population at risk and the number of applications necessary for effective treatment. The graph assumes that animals are treated at random and that $\frac{1}{2}$ (top curve) $\frac{1}{3}$ (middle) or $\frac{1}{4}$ (bottom) of the population is treated at each occasion. See text for derivation.

effectively is Mp^n. Hence, the number of treatments applied effectively is the number of animals being treated effectively times the number of treatments received by each, or $Mp^n \times n$.

$$\text{Thus, the effective proportion is } \frac{Mp^n n}{Mpn} = p^{n-1}$$

$$\text{and the proportion wasted is } 1 - p^{n-1}$$

Thus with $p = \frac{1}{3}$ and $n = 1$, 33% of the population is effectively treated and none of the drug wasted, but if n increases to 2 only 11% of the population is effectively treated and 67% is wasted. The rapid increase in wastage rate shown graphically in Fig. 2 applies in many pest-control situations, where more than one treatment by an agent is necessary for results and animals are treated at random. It is unlikely to apply when baits with warfarin (3-(1-phenyl-2-acetyl)-4-hydroxycoumarin) are presented to rats (*Rattus*

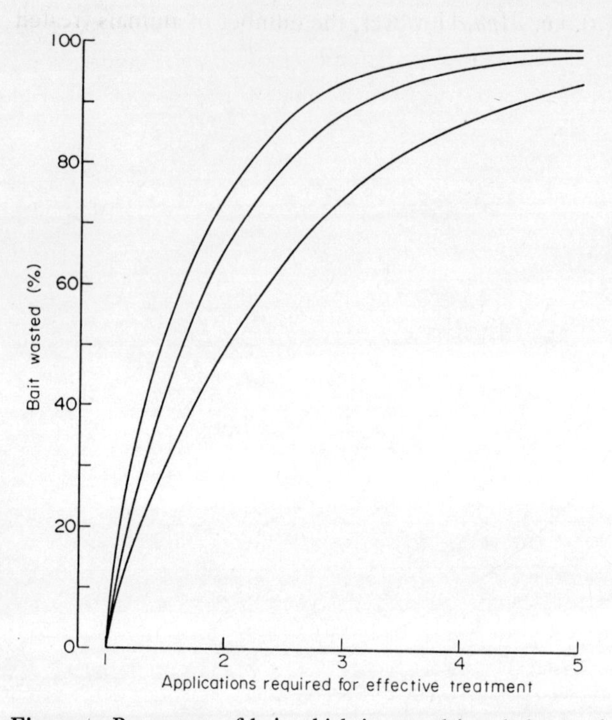

Figure 2. Percentage of bait which is wasted in relation to the number of applications necessary for successful treatment and the proportion of the population treated on each occasion ($\frac{1}{2}$ top curve, $\frac{1}{3}$ middle curve, $\frac{1}{4}$ bottom curve, assuming population treated at random). Thus if half the population can be treated and only one application of bait is needed, none is wasted, but if three applications are needed for an effective result then 87·5% is wasted.

norvegicus) in sewer systems and other urban sites, but it could apply to some extent in some rural situations.

(e) *Genetical manipulation*

The reason for non-breeding in the feral pigeon, mentioned above at (d) is significant for biological control. Town populations are polymorphic, being primarily composed of melanic varieties. Wild-type rock doves are photoperiodic and have a seasonal breeding cycle: their gonads regress in winter (Lofts *et al.* 1966, Murton & Clarke 1968). But in towns, food conditions make continuous breeding feasible, and certain melanic morphs, sensitive to day lengths too short to stimulate the wild-type, breed all through the year. The segregation of genotypes with the capacity for year-round breeding also necessitates the production of individuals with reduced fertility, and these are unable to establish themselves as breeders (Murton *et al.*, in preparation). There is scope for the manipulation of the gene pool of the population by selective release of trapped pigeons, and experiments to test the feasibility of this are beginning.

(f) *Behavioural methods*

Watt (1964) suggested that the effective biological control of vertebrates requires some self-accelerating method that forces populations down by eroding their homoeostatic capability. The concept of homoeostasis is of doubtful validity and, in those species studied in detail, there is evidence that apparent self-regulatory mechanisms are artifacts (Murton *et al.* 1971). Nevertheless, it is true that animals, especially vertebrates, have evolved a host of behavioural conventions which serve to increase their efficiency in combating environmental vicissitudes. It seems sensible to exploit behavioural conventions to manipulate problem animals. A good review of some possibilities given is by Frings & Frings (1967).

Birds that congregate on aerodrome runways pose an enormous potential hazard. In this and other situations they have been scared by the play-back of recorded distress calls (Frings & Jumber 1954, Busnel & Giban 1968, Boudreau 1968, Bremond *et al.* 1968). These calls are species-specific and very much more effective than non-biological noises. Thus both birds and mammals habituate rapidly to ordinary noises, even if loud, and ultrasonics bring no special benefits; rodents seem only to react under extreme laboratory conditions when sound levels of dangerous intensities (130 db) are employed (Sprock *et al.* 1967). Wright (in Busnel & Giban 1965) records that the use of gull distress calls kept an aerodrome clear over twelve months, whereas the birds habituated to acetylene bangers in only two days. Birds habituate to distress calls just as they do to model predators (Hinde 1954, Marler 1957) and research is now directed to techniques that can inhibit any waning in responsiveness or can reinforce response to a stimulus (Brough 1968).

There is no reason why substances with an obnoxious smell for man should deter other animals, and repellents based on this concept have proved ineffective (Duncan *et al.* 1960). But olfactory or gustatory chemicals which have biological meaning may yet be developed as attractants or repellents. The biological context has to be appreciated, for in many cases an animal will avoid a substance if it is presented in choice situations but not if, for instance, whole fields are treated. Factors involved in choice behaviour need to be appreciated (Tinbergen 1960, Allen & Clarke 1968, Murton 1971b). Selective attention theory is very relevant in this field, particularly the attention threshold model (Dawkins 1969a, b, Dawkins & Impekoven 1969). Mimicry depends on a balanced numerical relationship between mimic and model, and the extent to which situations can be contrived for control purposes needs exploration.

Certain chemicals, for instance drugs containing 4-aminopyridine, which are becoming available, disrupt various physiological and behavioural processes, resulting in atypical and inappropriate behaviour. Birds suffering from the injection of such agents may undergo convulsions, and emit distress calls, and they act as very strong repellents to others of their kind. These agents cause psychological stress in the subject and in consequence are disturbing to witness in action. For this reason they have been banned in England and no experimental work is contemplated.

Part 2. Experiments to devise a behavioural scarecrow for wood-pigeon control

There is survival value in the flocking behaviour of wood-pigeons (*Columba palumbus*) for it improves feeding efficiency (Murton 1971c). To facilitate flock co-ordination, the birds have white crescent-shaped markings on the wings. Flying pigeons are often attracted and attempt to join a feeding flock, a fact long appreciated by sportsmen, who employ artificial decoys to attract the birds within gunshot range.

Methods

Initially it was decided to examine the circumstances under which decoys would result in the largest shooting bags (Murton *et al.*, in preparation), but during the course of these studies it became evident that under certain circumstances decoys would scare pigeons.

Some artificial pigeon decoys of the kind commonly sold by gunsmiths were used, but because we wished to test large numbers the experiments reported here relied on dead wood-pigeons. These were partly mummified by injecting the body, wing muscles and viscera with 10% formalin using a hypodermic syringe, after which the carcasses were gradually air-dried. They remained adequately preserved under these conditions. Some birds were set-up with their wings outstretched as if flying, others were positioned with folded wings to resemble feeding birds. Groups of 5 to 200 'decoys' were positioned on fields known to be frequented by pigeons, a regular grid pattern always being adopted with the birds at 0·9 m (3 ft) centres; the birds either all had their wings closed or all had them open. Many separate trials were conducted spanning a period of about two years, and many different fields were utilized. Some of the variability in the recorded results might be explained by seasonal factors or by differences between fields in their attractiveness to pigeons, but this paper is not concerned with such matters and the records can be regarded as a wide and reasonably representative sample of typical conditions.

Once set up, the decoys were watched from a van at a distance of about 100–200 yards, and the behaviour of all pigeons that passed overhead and were judged to be near enough to see the decoys was recorded. Sometimes a single bird passed overhead, sometimes a flock of up to several hundred. Many records of single birds were obtained, and regarded as independent observations. But the number of occasions on which a large flock passed was limited, and the members of such a flock were obviously influenced by each other.

Results

For simplicity, Fig. 3 presents observations on single birds only; no point is based on less than 58 independent records. The graph depicts the percentage of birds that gave a positive response—i.e. they dipped, circled, tried to alight or did alight—towards decoys with their wings closed, according to

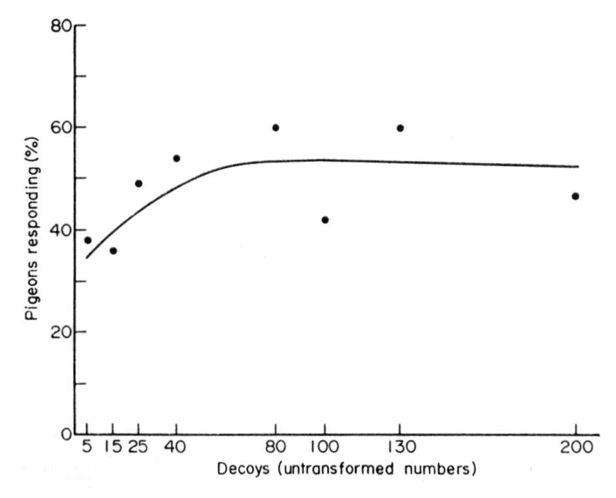

Figure 3. Percentage of single passing wood-pigeons which responded to decoys (dead wood-pigeons positioned on the ground with closed wings and at 0·9 m centres) in relation to the number of decoys.

the number of decoys present. Evidently, increasing the number of decoys up to 40–80 increased the proportionate response, but further increases did not improve the response. The curve bears considerable resemblance to the simple type of functional response curve of a predator depending on prey density (Murton 1968, Royama 1970).

Since Fig. 3 suggested a logarithmic relationship, all records involving from 1–10 birds passing overhead were combined (several hundred birds were now involved) and the percentage responding plotted against the $\log_{10}$ of the number of decoys (Fig. 4). It is apparent that the response rate was

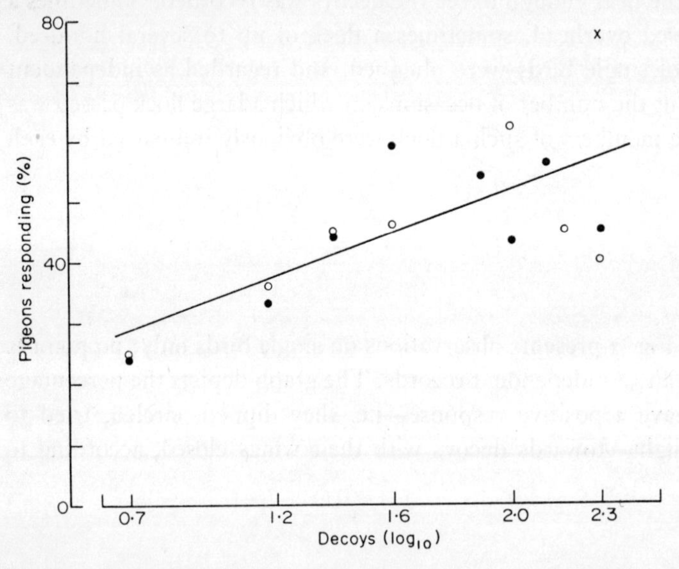

Figure 4. Percentage of passing wood-pigeons (flocks of 1–10 birds combined) responding to decoys set up with either closed ● or open ○ wings, in relation to the number of decoys. One experiment (*X*) involved pigeon wings without bodies. Note that $\log_{10}$ number of decoys has been plotted on the abscissa. See text for regression formula.

about the same irrespective of whether the decoys had open or closed wings. This is surprising, since pigeons with open wings expose two large white patches, so that 200 individuals with open wings are considerably more conspicuous to the human observer than a similar sized group with closed wings. Evidently the visual acuity of pigeons is so good that they see closed-winged birds as readily as birds with open wings. In one experiment, open wings detached from bodies were used, scattered about at odd angles, but again at 0·9 m centres. Passing pigeons responded as readily to these as to whole birds (Fig. 4). On the information available, there is no justification for distinguishing any differences in response rate to 200 detached wings,

200 bodies with open wings or 200 birds with closed wings. Accordingly, the variability in Fig. 4 is presumed to depend on sampling error (but see below) and all records have been used to fit the regression line $y = 16{\cdot}3 + 17{\cdot}6 \log_{10} x$, where y = % response and x = number of decoys. With $r_{14} = 0{\cdot}701$, $P = 0{\cdot}01 - 0{\cdot}001$, the coefficient of determination shows that 49% of the variability in the data is accounted for by the regression.

When flocks of 10 or more birds were combined in the same way, there was a tendency for the proportion of responding birds to increase with the number of decoys, but the data were variable and no significant correlation emerged ($r_{16} = 0{\cdot}441$; $P = 0{\cdot}10 - 0{\cdot}05$). The reason is apparent in Fig. 5,

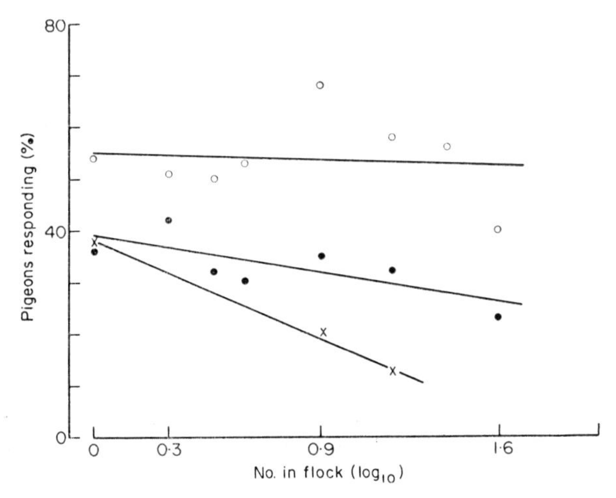

Figure 5. Percentage of wood-pigeons responding to decoys depending on whether they were passing alone or in flocks of various sizes. Single birds and groups of 2, 3, 4, 5 plotted separately as logs, and the means for flock sizes 6–10, 11–20, 21–30, 31–50, 51–100, 101–200, again as logs. Bottom regression line, fitted to X's, refers to 5 decoys with wings closed: $y = 38{\cdot}15 - 20{\cdot}9x$; $r_1 = -0{\cdot}999$. Middle regression line fitted to ●'s refers to 15 decoys with wings closed: $y = 38{\cdot}62 - 7{\cdot}98x$; $r_5 = -0{\cdot}745$. Top regression line, fitted to ○'s, refers to 40 decoys with wings closed: $y = 55{\cdot}26 - 1{\cdot}87x$; $r_6 = -0{\cdot}132$.

which considers the response rate of different-sized flocks when either 5, 15 or 40 decoys with wings closed were used, and Fig. 6, showing the response of different-sized flocks to 25 decoys with wings open. A single bird was not much less attracted to a flock of 15 than to one of 40 (a decrease in response from about 55% to 40%) but a flock of 15 was proportionately less influenced by small numbers of decoys (an approximately 53% response became reduced to 30%). Evidently single birds were readily attracted to any group of other pigeons but apparently an individual in a flock was tempted to leave only if it espied a larger gathering. It seems reasonable to regard the individual as being in a conflict situation so that it will tend to leave the birds it is with if it sees a larger flock. Obviously this process could not continue, otherwise all pigeons would end up in the same flock; so interest attaches to the possibility that when very large numbers of decoys were involved there was a fall-off in response (cf. Fig. 4). Of course, an individual must be variably motivated according to its state of hunger, or of involvement in reproductive activities, so that allowance ought, if possible, to be made for such factors.

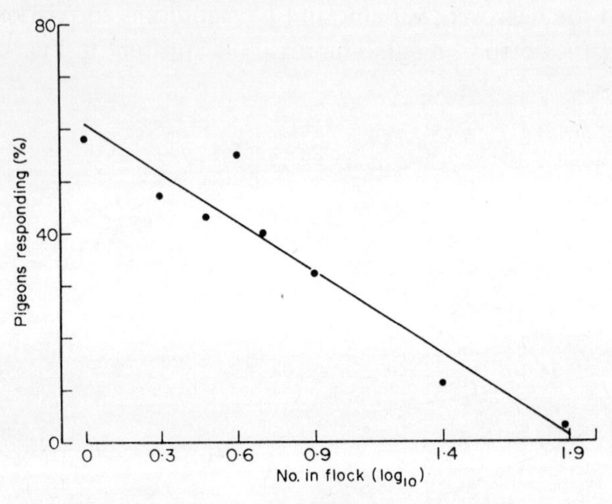

Figure 6. Percentage of wood-pigeons responding to 25 decoys positioned with open wings, according to the number of birds in the flock. Notes as for Fig. 5. $y = 60{\cdot}59 - 31{\cdot}3x$; $r_6 = -0{\cdot}955$; $P > 0{\cdot}001$.

Table 1 shows that, although the initial response of pigeons was about the same whether decoys had their wings opened or closed, their subsequent reactions were quite different. In experiments using decoys with wings closed, 6,668 birds passed overhead, some as singles and some in flocks of various sizes, and 34% responded, while, of 5,351 passing open-winged decoys, 39% responded ($\chi^2 = 34{\cdot}86$; $P > 0{\cdot}001$). But *in toto* 54% of responding birds would settle amongst closed-wing decoys whereas only 4% of those responding would settle amongst decoys with open wings (Table 1). Clearly dead pigeons mounted in a particular open-winged posture can provide a very strong deterrent to other pigeons. Moreover, of 763 pigeons which passed when only wings were laid out, 6% dipped, 89% attempted to settle, 4% settled and immediately left and 1% settled and stayed. In all cases those birds attempting to settle flew low over the decoys, hovering as if really wanting to settle, yet the sight of open-winged birds was sufficient to deter them. Since pigeons initially responded in the same way to open- or closed-winged decoys, their failure to settle amongst closed-winged birds

Table 1. Response of wood-pigeons to mounted dead wood-pigeons, in relation to size of flock and to number and arrangement of the decoys.

		Decoys with wings closed % of responding birds which:					Decoys with wings open % of responding birds which:				
No. birds in passing flock	Total no. birds responding	Dipped	Attempted to settle	Settled and stayed	Settled and left	Total no. birds responding	Dipped	Attempted to settle	Settled and stayed	Settled and left	χ^2	d.f.
5–30decoys												
1–10	105	60	0	38	2	65	77	0	17	6	9·782	2
10 +	13	0	0	100	0	1	(100)	0	0	0	–	
Total	118	53	0	45	2	66	77	0	17	6	10·243	1
40–100 decoys												
1–10	97	58	2	38	2	143	15	42	6	37	131·056	3
10 +	56	16	0	84	0	162	35	43	2	19	157·972	3
Total	153	42	1	55	1	305	26	43	4	27	232·872	3
130–200 decoys												
1–10	86	43	0	55	2	156	45	42	9	3	80·464	3
10 +	35	14	0	86	0	361	0	99	0	1	311·525	1
Total	121	35	0	64	1	517	14	82	3	1	375·387	3
Grand Total	392	43	1	54	2	888	23	63	4	10	659·660	3

Table 2. Reaction of wood-pigeons to open-wing decoys according to their spacing

Distance decoys positioned from each other (m)	Number of pigeons passing within range	% showing					
		No response	Dip in flight	Attempt to settle	Settle* and stay	Settle but leave	χ^2† between rows
0·9	97	21	7	18	52	2	
							74·40
1·8	115	20	14	23	40	3	
							23·93
2·7	106	23	30	33	9	5	
							6·28
3·7	113	39	34	22	3	2	

* Even though birds sometimes settled and stayed, they did not pitch amongst the decoys, but instead moved to the edge of the artificial flock.

† Chi-square values refer to differences between the adjacent rows, for those birds that responded. Comparing responses to decoys at 0·9 m with decoys at 3·7 m $\chi^2_{(2)} = 63{\cdot}26$. In all cases the 'settle and stay' and 'settle and leave' categories have been combined to give a 2 × 3 array with d.f. = 2. The χ^2 values indicate significant differences.

could depend on their recognition of the open-wing posture or specifically of the white signal mark. There is scope for experimentation to determine whether the white mark and its size and distribution relative to the wing could be exploited as a super-stimulus.

It might be argued that the experiments were successful because pigeons were not strongly motivated to alight on the fields in question. In a Cambridgeshire study area, at Carlton, there usually existed a very large manure dump (about 45 × 45 m in area) stored ready for application to nearby fields. This has always provided a very great attraction to pigeons, for in summer they come to find earthworm cocoons and insect food, and also to drink the liquid oozing from the stack; unless disturbed, birds are usually present throughout each day. For instance, on one sampling occasion in July, when no decoys were present, 84 pigeons passed within range of the dump and 81% responded (as birds continued to arrive, some of the responses were to individuals already present); 82% of the birds which responded settled and stayed. Table 2 indicates quite a different result when decoys were sited on the dump during the next few days, and also the effect of spacing decoys at different distances. Although, as shown in the table, some birds were seen to settle and stay, they would not alight amongst the decoys already present, and were instead relegated to the edges of the flock and in most cases to the sides of the manure tip where they could not feed effectively.

Artificially contrived pigeon flocks seem only to deter pigeons from settling in their immediate vicinity. But relative to the cost of much pigeon damage, the costs involved in covering whole fields with models would be low. At the test site an extended trial has suggested that pigeons do not habituate over periods of at least a month. However, many factors remain to be varied and tested, and development trials are in progress. There are grounds for suspecting that effective 'behavioural scarecrows' which have biological significance can be developed, and there is some prospect that these could be integrated with the use of distress calls and with other 'biological' methods to keep animals away from vulnerable sites. Our ancestors relied considerably on scarecrows, and perhaps they had empirically discovered techniques now forgotten or ineffectively applied.

References

ALLEN J.A. & B. CLARKE (1968) Evidence for apostatic selection by wild passerines. *Nature, Lond.* **220**, 501–2.

ARMOUR C.J. & H.V. THOMPSON (1955) Spread of myxomatosis in the first outbreak in Great Britain. *Ann. appl. Biol.* 43, 511–18.

BOUDREAU G.W. (1968) Alarm sounds and responses of birds and their application in controlling problem species. *The Living Bird* 7, 27–46.

Bremond J.-C., P.H. Gramet, T. Brough & E.N. Wright (1968) A comparison of some broadcasting equipments and recorded distress calls for scaring birds. *J. appl. Ecol.* 5, 521–9.

Brough T. (1968) Recent developments in bird scaring on airfields. In *The Problems of Birds as Pests*, eds. R.K. Murton and E.N. Wright, 29–38. Academic Press, New York & London.

Busnel R.-G. & J. Giban (1965) Colloque sur le problème des Oiseaux sur les Aérodromes. Nice 25–27 novembre 1963. *Instit. National de la Recherche Agronomique, Paris.*

Busnel R.-G. & J. Giban (1968) Prospective considerations concerning bio-acoustics in relation to bird-scaring techniques. In *The Problems of Birds as Pests*, eds. R.K. Murton and E.N. Wright, 17–28. Academic Press, New York & London.

Davis D.E. (1962) Gross effects of triethylenemelamine on gonads of starlings. *Anat. Rec.* **142**, 353–7.

Dawkins R. (1969a) A threshold model of choice behaviour. *Anim. Behav.* **17**, 120–133.

Dawkins R. (1969b) The attention threshold model. *Anim. Behav.* **17**, 134–41.

Dawkins R. & Monica Impekoven (1969) The 'peck/no-peck decision-maker' in the black-headed gull chick. *Anim. Behav.* 17, 243–51.

Dempster J.P. (1969) Some effects of weed control on the numbers of the small cabbage white (*Pieris rapae* L.) on brussels sprouts. *J. appl. Ecol.* 6, 339–45.

Duncan C.J., E.N. Wright & M.G. Ridpath (1960) A review of the search for bird-repellent substances in Great Britain. *Annals. des Epiphyties* 8, 205–12.

Elder W.H. (1964) Chemical inhibitors of ovulation in the pigeon. *J. Wildl. Mgmt.* **28**, 556–75.

Elton C.S. (1953) The use of cats in farm rat control. *Br. J. Anim. Behav.* **1**, 151–5.

Elton C.S. (1958) *The Ecology of Invasions by Animals and Plants.* Methuen, London.

Fenner F. & P.J. Chapple (1965) Evolutionary changes in myxoma virus in Britain. *J. Hyg. Camb.* 63, 175–85.

Fenner F. & I.D. Marshall (1957) A comparison of the virulence for European rabbits (*Oryctolagus cuniculus*) of strains of myxoma virus recovered in the field in Australia, Europe and America. *J. Hyg. Camb.* 55, 149–91.

Fenner F. & F.N. Ratcliffe (1965) *Myxomatosis.* Cambridge University Press.

Frings H. & M. Frings (1967) Behaviour manipulation (visual, mechanical, and acoustical). In *Pest Control, Biological, Physical and Selected Chemical Methods*, eds. W.W. Kilgore and R.L. Doutt, 387–454. Academic Press, New York & London.

Frings H. & J. Jumber (1954) Preliminary studies on the use of a specific sound to repel starlings (*Sturnus vulgaris*) from objectionable roosts. *Science N.Y.* **119**, 318–19.

Gibb J.A. & M.M. Betts (1963) Food and food supply of nestling tits (Paridae) in Breckland pine. *J. Anim. Ecol.* **32**, 489–533.

Gibb J.A., G.D. Ward & C. Patricia Ward (1969) An experiment in the control of a sparse population of wild rabbits (*Oryctolagus c. cuniculus* L.) in New Zealand. *N.Z. Jl. Sci.* **12**, 509–34.

Greenhall A.M. (1968) Bats, rabies and control problems. *Oryx* 9, 263–6.

Hinde R.A. (1954) Factors governing the changes in strength of a partially inborn response, as shown by the mobbing behaviour of the chaffinch (*Fringilla Coelebs*). *Proc. R. Soc.* B **142**, 306–58.

Howard W.E. (1967) Biochemical and chemosterilants. In *Pest Control, Biological, Physical and Selected Chemical Methods*, eds. W.W. Kilgore and R.L. Doutt, 343–86. Academic Press, New York & London.

JENKINS D., A. WATSON & G.R. MILLER (1964) Predation and red grouse populations. *J. appl. Ecol.* 1, 183–95.
KRUMHOLZ L.A. (1948) Reproduction in the western mosquito fish (*Gambusia affinis affinis*) (Baird & Girard), and its use in mosquito control. *Ecol. Monog.* 18, 1–43.
LIU G. (1939) Some extracts from the history of entomology in China. *Psyche* 46, 23–8.
LLOYD H.G. & K.C. WALTON (1969) Rabbit survey in West Wales (1961–67). *Agriculture*, January 1969, 32–6.
LOFTS B. & R.K. MURTON (1967) The effects of cadmium on the avian testis. *J. Reprod. Fert.* 13, 155–64.
LOFTS B., R.K. MURTON & R.J.P. THEARLE (1968). The effects of 22, 25-diazacholesterol dihydrochloride on the pigeon testis and on reproductive behaviour. *J. Reprod. Fert.* 15, 145–8.
LOFTS B., R.K. MURTON & N.J. WESTWOOD (1966) Gonadal cycles and the evolution of breeding seasons in British Columbidae. *J. Zool. Lond.* 150, 249–72.
MARLER P. (1957) Specific distinctiveness in the communication signals of birds. *Behaviour* 11, 13–39.
MEAD-BRIGGS A.R. (1966) Rabbits and myxomatosis. *Agriculture*, May 1966, 196–201.
MOUTIA L.A. & R. MAMET (1946) A review of 25 years of economic entomology in the Island of Mauritius. *Bull. ent. Res.* 36, 439–72.
MURTON R.K. (1968) Some predator-prey relationships in bird damage and population control. In *The Problems of Birds as Pests*, eds. R.K. Murton and E.N. Wright, 157–69. Academic Press, New York & London.
MURTON R.K. (1971a) *Man and Birds.* Collins, London.
MURTON R.K. (1971b) The significance of a specific search image in the feeding behaviour of the wood-pigeon. *Behaviour* 40, 10–42.
MURTON R.K. (1971c) Why do some bird species feed in flocks? *Ibis* 113, 534–6.
MURTON R.K. & S.P. CLARKE (1968) Breeding biology of rock doves. *Brit. Birds* 61, 429–448.
MURTON R.K., A.J. ISAACSON & N.J. WESTWOOD (1971) The significance of gregarious feeding behaviour and adrenal stress in a population of wood-pigeons *Columba palumbus*. *J. Zool. Lond.* 165, 53–84.
MURTON R.K., R.J.P. THEARLE & J. THOMPSON (1972) Ecological studies of the feral pigeon *Columba livia* var. 1. Population, breeding biology and methods of control. *J. appl. Ecol.* 9, 835–74.
MYKYTOWYCZ R. (1959) Social behaviour of an experimental colony of wild rabbits. *C.S.I.R.O. Wildl. Res.* 4, 1–13.
PIMENTEL D. (1955) Biology of the Indian mongoose in Puerto Rico. *Jl. Mammal.* 36, 62–68.
ROBINSON W.B. & V.T. HARRIS (1960) Of gophers and coyotes. *Amer. Cattle Producer* 42, 2 pp.
ROYAMA T. (1970) Factors governing the hunting behaviour and selection of food by the great tit (*Parus major* L.) *J. Anim. Ecol.* 39, 619–68.
SHEAIL J. (1971) *Rabbits and their History*. David & Charles, Newton Abbot.
SOUTHERN H.N. (1948) Sexual and aggressive behaviour in the wild rabbit. *Behaviour* 1, 173–94.
SPROCK C.M., W.E. HOWARD & F.C. JACOB (1967) Sound as a deterrent to rats and mice. *J. Wildl. Mgmt.* 31, 729–41.
STERN V.M., R.F. SMITH, R. VAN DEN BOSCH & K.S. HAGEN (1959) The integrated control concept. *Hilgardia* 29, 81–101.

STOCZEŃ S. (1962) Age structure of skulls of the mole *Talpa europaea* Linnaeus 1758 for the food of the buzzard (Buteo buteo L.). *Acta theriologica* 6, 1–9.
STONEHOUSE B. (1960) *Wideawake Island*. Hutchinson, London.
STOTT B. (1974) Biological control of waterweeds. In *Biology in Pest and Disease Control*, eds. D. Price Jones and M.E. Solomon. Blackwell, Oxford, 233–8.
SWEETMAN H.L. (1958) *The Principles of Biological Control*. W.C. Brown Co., Iowa.
TINBERGEN L. (1960) The natural control of insects in pinewoods. Part 1. Factors influencing the intensity of predation by songbirds. *Arch. Néerl. Zool.* 13, 265–336.
TINBERGEN L. & H. KLOMP (1960) The natural control of insects in pine woods. Part 2. Conditions for damping of Nicholson oscillations in parasite-host systems. *Arch. Néerl. Zool.* 13, 344–79.
VALVERDE J.A. (1958) An ecological sketch of the Coto Donana. *Brit. Birds* 51, 1–23.
VAN DOBBEN W.H. (1952) The food of the cormorant in the Netherlands. *Ardea* 40, 1–63.
VARLEY G.C. (1970) The need for life tables for parasites and predators. In *Concepts of Pest Management*, eds. R. L. Rabb and F. E. Guthrie. Proc. Conf. North Carolina State University, Raleigh, North Carolina, 59–68.
VARLEY G.C. & G.R. GRADWELL (1968) Population models for the winter moth. In *Insect Abundance: Symp. of the R. Ent. Soc., Lond.*, ed. T.R.E. Southwood. Vol. 4, pp. 132–142. Blackwell, Oxford.
VILLA R.B. (1969) *The Ecology and Biology of Vampire Bats and their Relationship to Paralytic Rabies*. FAO Rep. No. TA 2656, Rome 1969, iii + 16 pp.
VOÛTE A.D. (1946) Regulation of the density of the insect populations in virgin-forests and cultivated woods. *Arch. Néerl. Zool.* 7, 435–70.
WATT K.E.F. (1964) Animal population ecology and control fundamentals. *2nd Vertebrate Pest Control Conf.*, 24–8. Univ. California, Davis, Agr. Exten. Serv., 160 pp.
WATT K.E.F. (1968) *Ecology and Resource Management*. McGraw-Hill, New York.
WEIDHAAS D.E. & G. LABRECQUE (1970) Studies on the population dynamics of the housefly *Musca domestica* L. *Bull. Wld. Hlth. Org.* 43, 721–5.
WRIGHT E.N. (1968) Modification of the habitat as a means of bird control. In *The Problems of Birds as Pests*, eds. R.K. Murton and E.N. Wright, 97–105. Academic Press, New York & London.

Biological control of water weeds*

B. STOTT *Salmon and Freshwater Fisheries Laboratory, Ministry of Agriculture, Fisheries and Food, London*

The role of water plants in freshwater ecology has yet to be fully described, but it is already quite clear that it is important. Water weeds provide shelter and food for aquatic organisms (Witcomb 1968) as well as performing valuable mechanical functions in consolidating and protecting the banks and beds of watercourses. They also contribute to the aeration of streams, although, as Edwards (1968) points out, the interactions may be complex. However, water plants also impede the passage of water; thus they may increase the rate of silt deposition, leading to an increase in the substrate suitable for rooted aquatic vegetation. Their growth progressively reduces the transporting efficiency of watercourses, creating a greater liability to flooding, and causing higher pumping costs to be incurred in low-lying areas dependent on artificial drainage. Human leisure activities may also be affected; angling, sailing and swimming may all require the growth of water plants to be controlled.

For some or other of these reasons—but mainly for flood control—the river authorities and drainage boards have regular programmes of water weed control (Robson 1967). The main control method is cutting, which has traditionally been done by hand although the development of machines is proceeding and several types are now available (Robson 1968). Control by means of herbicides is also becoming increasingly feasible (Fryer & Makepeace 1970), but the number of chemicals approved for use in water in the United Kingdom is as yet limited (Makepeace 1971).

Current control methods are reasonably effective, but they have drawbacks: briefly, cut plants often regenerate, sometimes rapidly; and while the use of herbicides can be very effective, difficulties may arise in practice. Also, both cutting and the use of herbicides may necessitate the removal of the dying plant material under Acts designed to control water pollution. Clearly, without going into details which would be outside the scope of this paper, there is room for improving present methods, and the grass carp, or white amur (*Ctenopharyngodon idella*), may help to control water weeds simply by grazing on them.

The grass carp is a Cyprinid fish occurring in the rivers of Siberia and China flowing into the Pacific, but it has been introduced into many countries including Austria, Bulgaria, Czechoslovakia, France, Germany, Hungary, Japan, Poland, Rumania, Russia, Taiwan, the United States and, to a limited and controlled extent, into this country. The fish has been imported into most of these countries as an herbivorous food fish rather than a possible means of controlling aquatic weed, although this attribute is often listed among the good qualities of the fish in pond culture literature (Krupauer 1971).

Details of the biology of the fish are given by Lin (1935) and Hickling (1966, 1967a). It is an active, tough, fast-growing species, and, while it is somewhat thermophilic, it is quite able to flourish in our climate. One of the initial doubts on the suitability of the white amur for pond culture was due to the fact that it could not be bred, and supplies of fry had to be regularly imported from China or the River Amur. This disadvantage was eventually overcome by inducing spawning by injecting the ripening parent fish with a suspension of ground-up pituitary glands taken from common carp (*Cyprinus carpio*) and subsequently stripping the female fish and fertilizing the ova *in vitro*. This technique, the development of which has been reviewed by Hickling (1967b), is now used in Asia and most Eastern Europe countries, and also in Austria and the United States, to produce very large numbers of fish for commercial culture (Krupauer 1971).

Fertilized eggs are incubated in Zuger vessels in which a flow of water keeps the eggs in gentle motion. At the usual incubating temperatures, between 20 and 30°C, hatching occurs in 19–36 hours, and after a further period of about 4 days when the fry have absorbed their yolk sac, the young fish are transferred to shallow nursery ponds in which good populations of zooplankton have been induced by previous fertilization with inorganic, or sometimes organic, manures. After the young fish have grown to 3 or 4 cm in length they start showing an interest in macrophytes and, by the end of the summer growing period when (in Eastern Europe) they are about 10 cm long and weigh about 25 g, they are herbivorous. Nevertheless, fish of this size, and larger ones, are not exclusively herbivorous, for in the laboratory they will eat *Tubifex* sp., *Daphnia* sp. and pelleted trout feed. In the field the gut is invariably filled with plant material during the growing season, but it seems most probable that some animal matter is taken—if only accidentally. Table 1 gives the grass carp's food plant preferences taken from several authors, together with a 'nuisance rating' of the plants in the United

Kingdom (Robson 1967). Our own experience generally confirms the data on preferences. Gross food conversion ratios have been calculated from laboratory data and are given in Stott & Orr (1970) and Stott *et al.* (1971). For *Elodea canadensis*, the ratio was 117·6:1 (wet weight) for three-year-old fish weighing 56·6 g and fed to excess at a water temperature of 20°C ± 1·0. Under good conditions the fish can eat its own body weight of plant material in a day.

Table 1. Food preferences of grass carp

Plant	Order of preference found by various authors				Nuisance rating (Robson 1967)
	a	b	c	d	
Elodea canadensis	1	1	1	1	10
Ceratophyllum demersum	3	1	–	–	19
Potamogeton pectinatus	1	2	–	–	4
Lemna minor	2	–	1	2 }	13
L. trisulca	2	–	–	– }	
Hydrocharis morsus-ranae	4	2	1	1	–
Typha angustifolia	–	3	2	– }	3
T. latifolia	3	2	1	– }	
Phragmites communis	4	2	1	2	1
Myriophyllum spicatum	–	1	–	–	14
Polygonum amphibium	4	3	3	1	–
Carex pseudocyperus	–	3	–	–	7
Nasturtium officinale	–	–	1	–	12
Nuphar luteum	5	–	3	3	18
Juncus effusus	–	–	–	2	11
Age of fish (years)	1, 2	2	1	3, 4	
Temperature (°C)	16–26	20–22	–	16–29	

a. Zolotova (1966) Fish given choice of plants; 5 categories of preference: (1) plants absolutely selected, (2) plants positively selected, (3) plants for the most part selected, (4) plants for the most part avoided, (5) plants avoided.
b. Penzes & Tolg (1966) Fish offered 3% body weight; 4 categories of preference: (1) high consumption in 8 hours, (2) moderate consumption in 24 hours, (3) poor consumption in 48 hours, (4) not eaten.
c. Krupauer (1967) Fish offered a choice; 3 categories of preference: (1) whole plant eaten, (2) youngest parts eaten, (3) not eaten.
d. Krupauer (1968) Fish offered a choice; 3 categories of preference: (1) preferred species; all parts eaten, (2) most parts eaten, (3) plants avoided or feebly eaten.

Apart from a limited preliminary trial (Pentelow & Stott 1965), the Ministry's experimental work started in 1968 when 2,200 one- and two-year-old fish were obtained from a breeding station in Hungary. One of the first experiments was on the degree of weed control achieved in relation to the number of fish grazing; the results suggest (Fig. 1) that an initial stocking rate of 250 kg/ha would reduce the potential growth by about 50% (Stott & Robson 1970, Stott *et al.* 1971). Had the fish been introduced into the experimental ponds at the start of the plant growing season, rather than after growth had started, then considerably fewer fish would probably have been as effective, and this difference is being examined in current experiments.

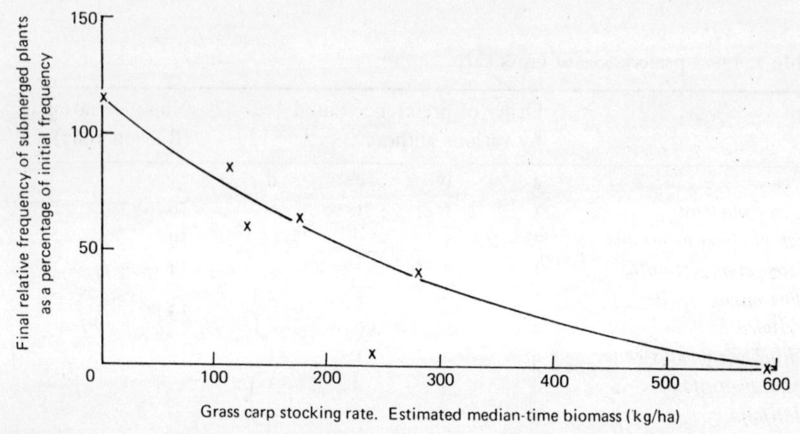

Figure 1.

Another line of experimental work being followed is designed to examine the effect which the grazing of grass carp might have on native fish. Doubtless a complete removal of water weed would be disadvantageous to most species —apart from its undesirability from every other point of view—but the degree to which plants can be removed without causing an appreciable effect is not known. So far only a limited experiment with common bream (*Abramis brama*) has been carried out, at the same time as the investigation into the

Table 2. Growth and survival of bream (*Abramis brama* L.) alone and with grass carp at two population densities

Pond	No. bream stocked 21.5.69	Initial biomass (kg)	Estimated no. present 22.9.69	Estimated biomass (kg)	Estimated median-time grass carp biomass (kg/ha)
1	30	1·0	26 (14–30)	1·89	0 (Control)
2	28	1·0	22 (14–28)	3·49	590
3	25	1·0	21 (8–25)	3·21	174

relation between stocking density and the degree of plant control. The results, shown in Table 2, indicate that for this species the presence of grass carp is distinctly beneficial, so far as growth rate is concerned. Of course, the response of other species is still unknown, and those requiring water weeds on which to deposit their spawn—as do bream—might be particularly sensitive to excessive weed removal over a long period.

In the context of inter-specific competition, it should be noted that polyculture, involving grass carp, silver carp (*Hypophthalmichthys molitrix*) and the common carp, is becoming a recognized practice in continental Europe; grass carp eat the macrophytes, silver carp the phytoplankton and common carp feed on the benthic fauna. Grass carp seem to have a low rate of food utilization. Stroganov (1963) and Hickling (1966) point out that about half the food eaten passes out of the fish to act as an organic fertilizer capable of supporting plankton growth, or even to be eaten by other fish (Hickling personal communication).

Whether or not the white amur will ever be used for controlling water weed in the United Kingdom is still in doubt, and at present no licences are being issued for their import, except for the comparatively few required for experimental purposes under the control of the Ministry. No adverse 'side effects' are yet apparent and the fact that the fish is unlikely to breed in this country may well encourage further exploration of its potential. The economics of the fish for weed control have been discussed (Stott *et al.* 1971) but better data are needed on some aspects of the problem, including the grass carp's mortality rate in a mixed population of fishes including predatory species. Currently, nine-month-old fish from Hungary delivered to London in batches of 10,000, cost about 22p each. At this price, with some of the cost apportioned to their probable sporting value, this biological method of controlling water weeds might prove reasonable in an enclosed water when compared with the cost of herbicidal treatment. From a biological viewpoint it would certainly be more attractive.

References

EDWARDS R.W. (1968) Plants as oxygenators in rivers. *Water Research* 2, 243–8.

FRYER J.D. & R.J. MAKEPEACE (1970) *Weed Control Handbook* II, 331 pp. Blackwell, Oxford.

HICKLING C.F. (1966) On the feeding processes of the white amur *Ctenopharyngodon idella* Val. *J. Zool.* 148, 408–19.

HICKLING C.F. (1967a) On the biology of a herbivorous fish, the white amur or grass carp, *Ctenopharyngodon idella* Val. *Proc. Roy. Soc. Edin.* B, LXX, 4, 62–81.

HICKLING C.F. (1967b) The artificial inducement of spawning in the grass carp, *Ctenopharyngodon idella* Val. *Proc. Indo-Pacific Fish Coun.* 12 (II), 236–43.

KRUPAUER V. (1967) Food selection of two-year-old grass carp. *Bulletin V.U.R. Vodnany* 3 (1), 7–17 (in Czech).

KRUPAUER V. (1968) The capacity for plant consumption of three- and four-year-old grass carp. *Zivocisna vyroba* 13 (XLI), 467–74 (in Czech).

KRUPAUER V. (1971) The use of herbivorous fishes for ameliorative purposes in Central & Eastern Europe. *Proc. Eur. Weed Res. Coun. 3rd int. Symp. Aquatic Weeds* 95–102.

LIN S.Y. (1935) Life history of Waan Ue, *Ctenopharyngodon idellus* (Cuv. & Val.). *Lingnan Sci. J.* 14, 129–35; 271–4.

MAKEPEACE R.J. (1971) The official clearance and approval of aquatic herbicides in the United Kingdom. *Proc. Eur. Weed Res. Coun. 3rd int. Symp. Aquatic Weeds* 305–14.

PENTELOW F.T.K. & B. STOTT (1965) Grass carp for weed control. *Prog. Fish Cult.* 27, 4, 210.

PENZES B. & I. TOLG (1966) Étude de la croissance et de l'alimentation de la 'Grass carp' (*Ctenopharyngodon idella* Val.) en Hongrie. *Bull. Français de Pisciculture* 39 (223), 70–6.

ROBSON T.O. (1967) A survey of the problem of aquatic weed control in England and Wales. *A.R.C. Weed Research Organisation Tech. Rep.* No. 5, 37 pp.

ROBSON T.O. (1968) The control of aquatic weeds. *Ministry of Agriculture, Fisheries and Food Bull.* No. 194. H.M.S.O. 54 pp.

STOTT B., D.G. CROSS, R.E. ISZARD & T.O. ROBSON (1971) Recent work on grass carp in the United Kingdom from the standpoint of its economics in controlling submerged aquatic plants. *Proc. Eur. Weed Res. Coun. 3rd int. Symp. Aquatic Weeds* 105–16.

STOTT B. & L.D. ORR (1970) Estimating the amount of aquatic weed consumed by grass carp. *Prog. Fish Cult.* 32, 1, 51–4.

STOTT B. & ROBSON T.O. (1970) Efficiency of grass carp (*Ctenopharyngodon idella* Val.) in controlling submerged water weeds. *Nature, Lond.* 226, 5248, 870.

STROGANOV N.S. (1963) The food selectivity of the Amur fishes. In Problems of the Fisheries Exploitation of Plant-eating Fishes in the Water Bodies of the USSR. *Akad. nauk. Turkm. SSR Ashkhabad* 181–91 (in Russian).

WITCOMB D.M. (1968) The fauna of Aquatic Plants. *Proc. 9th Brit. Weed Control Conf.* 1, 382–5.

ZOLOTOVA Z.K. (1966) Food preferences of grass carp. *Trudy vses. nauchnoissled. Inst. prud. ryb. Khoz.* 14, 39–50 (in Russian).

Biological control of plant diseases

J. RISHBETH *Botany School, University of Cambridge*

For the purposes of this account the rather wide definition of Garrett (1965) is accepted, namely that biological control involves any reduction in the incidence of a plant disease through the agency of one or more living organisms, other than the host itself or man. The first forms of biological control to be recognized were in field crops, and so it is appropriate first to consider some examples of agricultural interest, drawn mainly from soil-borne diseases. A landmark in this subject was the 1963 symposium at Berkeley, California, which dealt widely with the ecology of soil-borne pathogens as a prelude to biological control. In a foreword to the published proceedings (Baker & Snyder 1965), the point was made that, with intensification of agriculture, losses from root disease increased as the buffering action of natural biological controls diminished, but that these must still be generally effective for the incidence of such disease to be as low as it is. More recently Baker (1968) produced an excellent review of the mechanisms by which biological controls operate on soil-borne pathogens, and used mathematical models in attempting to predict the kinds of disease likely to be controlled by this means.

Diseases of field crops

Control measures may operate in several ways. A direct effect on the host may result from inoculation with a non-pathogenic or mildly virulent organism. For example, when wounded roots of tomato seedlings were exposed to *Cephalosporium* sp., vessels became obstructed and the ability of the wilt fungus *Fusarium oxysporum f. lycopersici* to spread in the tissues after subsequent inoculation was limited (Phillips *et al.* 1967). Again, the severity of wilt disease in tobacco, caused by the bacterium *Pseudomonas solonacearum*, was markedly reduced when a high proportion of avirulent cells was introduced at the same time as virulent ones, even though the dosage of the latter was sufficient to cause severe symptoms when inoculated alone (Averre & Kelman 1964). Such effects, though of great interest in furthering our understanding of resistance mechanisms, have not yet, so far as is known, been exploited for practical purposes.

Alternatively the pathogen may be affected by production of antibiotics. Thus seedling blight of maize, caused by *Fusarium roseum*, can be controlled by coating seed kernels with *Bacillus subtilis* or *Chaetomium globosum* (Chang & Kommedahl 1968). After tests in the glasshouse, this method was used in field trials where the results were as good as those obtained from treatment of kernels with captan or thiram, provided the soil temperature was below 20°. This procedure appears to be promising for development, by contrast with those involving direct inoculation of the soil: almost invariably in such cases, the population of inoculated organisms declines dramatically as a result of competition from the resident soil microflora.

More commonly, perhaps, a control procedure influences the pathogen through a reduction in nutrient supply, though the relative importance of this factor and antibiotic production is often difficult to determine. Major soil nutrients, sources of nitrogen as well as carbon, may be required by resting propagules of a fungal pathogen both for germination and for growth sufficient to initiate infection. For instance, *Fusarium solani f. phaseoli* is very sensitive to nutrient supply, and the root disease of beans that it causes can be reduced considerably by incorporating carbohydrate-rich material with the soil. In California this is achieved by turning in residues from a barley crop: soil microorganisms utilize much of the available soil nitrogen in decomposing the amendment. The critical importance of nitrogen is shown by the fact that control is nullified if this nutrient is added; also, amendments having a low carbon-nitrogen ratio, such as soybean and alfalfa residues (C/N 16), tend to make the disease worse (Snyder *et al.* 1959).

Root-rot of beans can also be reduced by applying chitin or laminarin to the soil, but here a different mechanism of control has been suggested (Mitchell 1963). *F. solani f. phaseoli*, like many other fungi, has chitinous hyphal walls, and laminarin or a closely related polysaccharide is also present. The soil amendments are thought to promote lysis of the fungal cell walls, which in the case of chitin-treated soil is associated with a marked increase in the population of actinomycetes. Interestingly enough the incidence of damping-off, caused by *Pythium debaryanum*, whose cell walls contain no chitin or laminarin, is not reduced by such amendments.

These are only two examples of a method commonly giving good results. Another is provided by *Verticillium* wilt of olives (Wilhelm & Taylor 1965) which, though a tree crop, is for convenience mentioned here. Californian growers are so confident of the soil-amendment method that they can actually use sawdust prepared from heavily infected trees! Also in California, potato scab is controlled by growing soybeans after harvest and turning them in whilst still green, a method similar to that used much earlier by Millard (1923) in Britain.

A leguminous crop is used in quite a different way to reduce the incidence of take-all, caused by *Ophiobolus graminis*, in successive spring-grown crops of wheat or barley. Whilst legumes are growing under good conditions of illumination, they take up more nitrogen from the soil than they excrete into it, and because little nitrogen is available the fungus dies out rapidly in the infected stubble (Garrett 1944). Under the Chamberlain system, which exploits this effect, spring-grown cereals are undersown with a suitable legume or legume-grass mixture, providing a catch-crop in autumn (Garrett 1950).

It has been known for many years that the first cereal crop planted after a ley is seldom badly affected by take-all. Scott (1970) frequently isolated from grasses a fungus, *Phialophora radicicola*, which had an infection habit very similar to that of *O. graminis* but was not virulent on wheat seedlings; in a glasshouse experiment it inhibited root infection by *O. graminis* to some extent. Balis (1970) further showed that inoculation of several grass species with *P. radicicola* reduced the extent to which *O. graminis* infected wheat following the grasses. Almost certainly *P. radicicola* exerts a significant degree of biological control over *O. graminis* under field conditions, though the mechanism is unknown. Whether such control can be increased by seed inoculation or any other means is a problem for further research.

Despite a great upsurge of interest amongst pathologists in biological control of soil-borne diseases during the last decade, Baker (1968) doubts whether there have been any really notable applications of new knowledge or even much modification of well-established practice over this period.

Diseases of trees

Recently several suggestions have been made about control of diseases involving die-back or canker formation that are often caused by wound-infecting organisms. Krstic & Hočevar (1959) found, for example, that natural infection by *Endothia parasitica*, which causes chestnut blight, could be reduced to about a fifth by inoculating certain fungal saprophytes at wound sites, the most effective competitor being *Penicillium rubrum*. Such inoculations had to be made in advance, for if *Endothia* spores were introduced simultaneously no control resulted. Similarly, infection by *Stereum purpureum*, which causes silver-leaf in plums and die-back in several other types of tree, has been prevented by introducing *Trichoderma viride* into wounds 2 days in advance of the pathogen (Grosclaude 1970). Protective treatments with organisms are thought to last longer than those involving fungicides, to be easier to apply and less likely to disturb the ecological balance of the orchard. A different situation exists in the western United States, where white pine blister rust, caused by the obligate parasite *Cronartium ribicola*, is controlled naturally to some extent by the fungus *Tuberculina maxima*. This fungus colonizes rust-infected tissue and may displace *Cronartium* in over 60% of cankers; indeed one survey showed that it had contributed to natural control where chemical treatments had failed (Leaphart & Wicker 1968). An extensive collection is being made of this and other canker-invading organisms with the aim of improving upon natural biological control. Study of antagonists and tissue defence mechanisms is an essential pre-requisite to any such development, and an investigation recently carried out on larch canker by Buczacki (1971) provides an excellent illustration of the detailed work that needs to be done. Perhaps the outstanding difficulty in exploiting such methods lies in protecting a sufficiently high proportion of potential infection sites. This would appear to be a huge task with forest trees, though one of more manageable proportions in orchards.

Fungal root diseases cause immense losses throughout the world in woody crops of many kinds. The main sources of infection are provided by the root systems of stumps or of trees already killed. In the last few decades, increased demand for timber has necessitated extensive planting, especially of conifers. Systematic management of these plantations has involved the creation of vast numbers of stumps, thus increasing the incidence of root diseases: improved methods of control are therefore urgently needed. An early method of biological control was developed by Leach (1939) in central Africa where the honey-fungus (*Armillaria mellea*) was particularly damaging to tea plantations. If large trees were felled in preparation for planting, they often acted as centres of infection. Leach suggested, and provided evidence to support the idea, that, if stems of such trees were ring-barked several months before felling, they would not be so dangerous. Provided ringing is carried out correctly, it prevents translocation of nutrients from the crown to the roots, and so facilitates rapid exhaustion of food reserves, especially starch. From its bases in adjacent roots, *Armillaria* is much less likely to become established in starved tissues than in healthy ones: Swift (1970) attributes this largely to its failure to compete effectively with other fungi under these conditions. Ring-barking has since been performed on indigenous trees before planting the oil-producing tung, losses of yield from this crop in treated areas only being one-tenth of those from the crop in untreated ones (Wiehe 1952). However, the method sometimes breaks down, and becomes ineffective, under temperate conditions.

The same fungus has been controlled in Californian citrus orchards by fumigating the soil with carbon disulphide. Bliss (1951) thought that this effect was entirely due to colonization of the treated soils by *Trichoderma viride* and its subsequent replacement of *Armillaria* in the infected roots, but Garrett (1958) showed that the fumigant to some extent also acts directly by killing the fungus. This method provides a good example of biological control mediated by chemical treatment.

Incidence of *Fomes lignosus* can be reduced in rubber plantations by a method combining chemical and biological actions in rather a different way (Fox 1965). Before replanting a diseased area, all trees of the previous crop are killed by chemical means whilst standing, or are felled and the stumps creosoted. A leguminous cover-crop is then established, commonly over about two-thirds of the area, but omitting the new planting rows which are kept bare to a width of 1·8 m. The cover-crop protects the soil from insolation, erosion and leaching, but, most importantly in the present context, provides conditions for vigorous growth and fructification of *F. lignosus*. However, because this growth stops short at the bare planting rows, food reserves of the fungus tend to become exhausted without a significant increase of root disease on the young trees.

Control of *Fomes annosus*

The application of biological control to this fungus, another important root parasite, will be considered in more detail. It is widespread in the northern hemisphere and characteristically invades living roots in contact with roots of infected stumps; thereafter it often progresses in the same manner from tree to tree. Particularly in dry areas, this form of attack may lead to extensive killing, for example of pines, but more generally it is followed by a progressive rot in stems of susceptible conifers such as larches and spruces. In East Anglian pine plantations there is commonly a progressive build-up of root disease where the soil is alkaline, following successive thinnings. The mean loss on such soils at the age of 30 years is about £11 per ha, but the maximum may be five times greater than this (Wallis 1960). In Sweden the incidence of butt-rot in Norway spruce is considerable and losses may be as high as £220 per ha at the age of 70 years. In 1958 the annual loss for this species alone was calculated at £4 million.

In addition to radial spread around existing infection centres, new foci appear chiefly through infection of freshly cut stumps by air-borne spores, produced by bracket-shaped structures. Stumps of pines are particularly likely to become infected with the fungus, which colonizes the surface and then grows down to the roots. It can be shown by trapping spores that there is a risk of such infection all over Britain, though it is greater in the vicinity of affected stands, where under appropriate conditions spores may be deposited at rates of $1/cm^2/h$.

It has been shown that the number of trees affected by root disease in a plantation is related to the proportion of stumps naturally infected by *F. annosus* at an earlier thinning (Rishbeth 1957). The same relationship was also demonstrated by the results of stump-treatment experiments. In one of them, for example, the proportion of pine stumps infected by the fungus after creosoting or painting the surface was 4% and 3% respectively, as compared with 36% for untreated controls. After 6 years the mean number of trees visibly affected by *F. annosus* in plots of 1 ha was five, five and thirty, respectively. Experiments in Denmark have given similar results (Yde-Andersen 1967). Nevertheless, early treatments such as creosoting had certain disadvantages; for instance stumps often became infected if the thinly impregnated surface was subsequently damaged. More seriously, no protection was given against infection of the stump from any roots containing the fungus at the time of felling.

Rather more satisfactory results were obtained from other chemical treatments, especially those that killed woody tissues rapidly and therefore destroyed their selectivity for *F. annosus*. Thus pine stumps treated with 40% ammonium sulphamate are frequently colonized by *Trichoderma viride*, competition from which is usually intense enough to prevent *F. annosus* becoming established. Moreover, another important competitor, *Peniophora gigantea*, often follows *T. viride* and helps to control any *F. annosus* entering the stump from infected roots. This again is an interesting example of biological control mediated by chemical treatment. Not all tissue-killing substances promote establishment of fungi sufficiently competitive with the parasite, however: thus pine stumps treated with 20% disodium octaborate are regularly colonized by the boron-tolerant *Botrytis cinerea*, but this does not prevent *F. annosus* entering from infected roots. For several years, some of the more effective chemical treatments have been used in Forestry Commission plantations, and for most conifer species they are still employed (Rishbeth 1967).

The results obtained with sulphamate-treated stumps, and the extent to which stump infection by *F. annosus* is often naturally controlled by *P. gigantea*, prompted investigation into the possibility of using this fungus for stump treatment. Experiments showed it to be remarkably effective, in contrast to *T. viride* and some 'blue-stain' fungi which gave poor or erratic results. In culture *P. gigantea* forms oidia (asexual spores). It was found that for pine stumps having a wood-surface area of some 0·02 m^2, a dosage of 1×10^4 oidia controlled infection from the largest number of *F. annosus* spores likely to be deposited under natural conditions (Rishbeth 1963). The fungus also tends to replace *F. annosus* in infected roots provided these are not too resinous. It rapidly breaks down cellulose and lignin, and has never been recorded as killing healthy pines. Within 6 to 12 months sporophores are commonly formed, but because these become desiccated rather rapidly during dry weather, by contrast with the more drought-resistant sporophores of *F. annosus*, natural control by *P. gigantea* is somewhat erratic. Moreover, in many young forests, *P. gigantea* is scarce or absent, and therefore its spores may not be available for colonizing stumps of the crucial first thinning. In order to give adequate protection, therefore, it is necessary to inoculate every stump, at least under British conditions.

Experiments demonstrating the effectiveness of stump inoculation were completed in East Anglian pine plantations by 1960. Two years later, trials set up by the Forestry Commission Research Branch confirmed these findings, and sufficient interest had developed in the method to justify attempts to prepare inoculum suitable for use on a large scale. Without this, the method could only be employed in very limited circumstances. Eventually, dehydrated tablets were devised which contained about 1×10^7 viable oidia and had a storage life of at least 2 months at 22°. For pines, stump inoculation first replaced chemical treatment in 1963, and since then its use has gradually been extended: by 1971 the method was employed in twenty-seven forests, of total area about 56,000 ha. Its cost varies from £1·25 to £3·75 per ha, which represents 1 to 3% of the value of thinnings.

Inoculum is now prepared in fluid form and distributed in sachets; the spores can be used for about 4 months. Sachets are tested in a Forestry Commission laboratory before a batch is released, and later during storage, by foresters. The contents of a sachet are mixed well with 5 l water and a non-toxic dye is added; the spore suspension is applied to stumps from a polythene container directly after felling. Complete coverage of a stump is indicated by a uniformly dyed surface. Spot checks are made on suspensions being used in the forest, and stump samples are collected periodically to determine the extent of colonization by *P. gigantea*: so far results have proved satisfactory. It seems very probable that, with continued application of this method, the ratio of *Peniophora* to *Fomes* spores in the air will increase, which in terms of tree protection would clearly be a desirable side-effect.

The preceding account has been concerned entirely with control of *F. annosus* attack arising from thinning, which is still a major hazard in British forests. However, serious problems also arise at the time of replanting, and it seems likely that in some circumstances stump inoculation will prove useful at this stage too. Certainly in the southern United States, where pine stands cover vast areas and *F. annosus* attacks are often severe, stump inoculation with *P. gigantea* will probably be combined with mechanical harvesting of trees in the near future. So far, stump inoculation has been used on a forest scale only in Britain.

In the future it is hoped to extend the method to stumps of Sitka spruce, another important timber species. Experiments with *P. gigantea* have generally proved disappointing, and various other fungi are being tested. Better results are often obtained if inoculation is combined with a chemical treatment. The possibility of adapting the method to help control *Armillaria mellea* is also being investigated. Since stumps of broad-leaved trees constitute the main sources of infection for this fungus, chemical treatment generally has to be combined with inoculation, to control regrowth and to promote rapid colonization by the competing fungus. For example the wood-rotting species *Polystictus versicolor* and *Polyporus adustus* have been successfully established in poplar stumps by inoculating them and then applying 40% ammonium sulphamate (Rishbeth 1971). With birch stumps, it was shown that fungi differ with respect to the chemical treatment most favouring their growth in stumps. Inoculation, with either spores or mycelial fragments, certainly seems feasible for stumps of broad-leaved trees, but as yet there is no evidence to show how well this restricts build-up of *A. mellea*. Again, various fungi are being screened for their ability to act as effective competitors.

Conclusions

It is apparent, from the examples selected, that biological control of plant diseases has reached very diverse stages of application. At one extreme, several recent experiments have given results that are of great theoretical interest but are probably incapable of development at present. At the other, various long-established cultural methods continue to give excellent results, but whereas in the past the reason for their success was obscure, modern work has elucidated at least some of the underlying mechanisms. Progress has been made, and is likely to continue, with the development of controls that already operate naturally to some extent, for instance in relation to infection of stumps or stem wounds on trees. Most probably the chance of success will be greater in situations where the microbial population is relatively simple, thus avoiding the complications of biological buffering. Possible examples that come to mind are, again, freshly exposed woody tissues, surfaces of seeds, fruits or tubers, and sterilized soil. There would appear to be ample scope for the development of further methods involving a combination of chemical and biological actions.

References

AVERRE C.W. & A. KELMAN (1964) Severity of bacterial wilt as influenced by ratio of virulent to avirulent cells of *Pseudomonas solonacearum* in inoculum. *Phytopathology* 54, 779–83.

BAKER K.F. & W.C. SNYDER (1965) Prelude to biological control. *Ecology of Soil-borne Plant Pathogens.* University of California Press, Berkeley.

BAKER R. (1968) Mechanisms of biological control of soil-borne pathogens. *A. Rev. Phytopath.* 6, 263–94.

BALIS C. (1970) A comparative study of *Phialophora radicicola*, an avirulent fungal root parasite of grasses and cereals. *Ann. appl. Biol.* 66, 59–73.

BLISS D.E. (1951) The destruction of *Armillaria mellea* in citrus soils. *Phytopathology* 41, 665–83.

BUCZACKI S.T. (1971) An investigation of larch canker and related problems. Ph.D. thesis. University of Oxford.

CHANG I. & KOMMEDAHL T. (1968) Biological control of seedling blight of corn by coating kernels with antagonistic organisms. *Phytopathology* 58, 1395–401.

FOX R.A. (1965) The role of biological eradication in root-disease control in replantings of *Hevea brasiliensis*. In *Ecology of Soil-borne Plant Pathogens*, eds. K.F. Baker and W. C. Snyder, 348–62. University of California Press, Berkeley.

GARRETT S.D. (1944) Soil conditions and the take-all disease of wheat. VIII. Further experiments on the survival of *Ophiobolus graminis* in infected wheat stubble. *Ann. appl. Biol.* 31, 186–91.

GARRETT S.D. (1950) The control of take-all under intensive cereal cultivation. *Agriculture, Lond.* 56, 514–16.

GARRETT S.D. (1958) Inoculum potential as a factor limiting lethal action by *Trichoderma viride* Fr. on *Armillaria mellea* (Fr.) Quél. *Trans. Br. mycol. Soc.* 41, 157–64.

GARRETT S.D. (1965) Toward biological control of soil-borne plant pathogens. In *Ecology of Soil-borne Plant Pathogens*, eds. K. F. Baker and W. C. Snyder, 4–17. University of California Press, Berkeley.

GROSCLAUDE C. (1970) Premiers essais de protection biologique des blessures de taille vis-à-vis du *Stereum purpureum* Pers. *Annls. Phytopath.* 2, 507–16.

KRSTIC M. & S. HOČEVAR (1959) Uticaj nekih antagonističikh mikroorganizama na infekcije pitomag Kestena od *Endothia parasitica* Anders. *Zasht. Bilja* 54, 41–52.

LEACH R. (1939) Biological control and ecology of *Armillaria mellea* (Vahl) Fr. *Trans. Br. mycol. Soc.* 23, 320–9.

LEAPHART C.D. & E.F. WICKER (1968) The ineffectiveness of cycloheximide and phytoactin as chemical controls of the blister rust disease. *Pl. Dis. Reptr.* 52, 6–10.

MILLARD W.A. (1923) Common scab of potatoes. *Ann. appl. Biol.* 10, 70–88.

MITCHELL R. (1963) Addition of fungal cell-wall components to soil for biological disease control. *Phytopathology* 53, 1068–71.

PHILLIPS D.V. *et al.* (1967) A mechanism for the reduction of *Fusarium* wilt by a *Cephalosporium* species. *Phytopathology* 57, 916–19.

RISHBETH J. (1957) Some further observations on *Fomes annosus* Fr. *Forestry* 30, 69–89.

RISHBETH J. (1963) Stump protection against *Fomes annosus.* III. Inoculation with *Peniophora gigantea. Ann. appl. Biol.* 52, 63–77.

RISHBETH J. (1967) Control measures against *Fomes annosus* in Great Britain. *Sect.* 24, *Congr. Int. Un. Forest Res. Org., Munich* 299–306.

RISHBETH J. (1971) Biological control for root diseases of trees. *Sect.* 24, *Congr. Int. Un. Forest Res. Org., Gainesville.*

SCOTT P.R. (1970) *Phialophora radicicola*, an avirulent parasite of wheat and grass roots. *Trans. Br. mycol. Soc.* 55, 163–6.

SNYDER W.C. *et al.* (1959) Effect of plant residues on root rot of bean. *Phytopathology* 49, 755–6.

SWIFT M.J. (1970) *Armillaria mellea* (Vahl ex Fries) Kummer in central Africa: studies on substrate colonisation relating to the mechanism of biological control by ring-barking. In *Root Diseases and Soil-borne Pathogens*, eds. T.A. Toussoun, R.V. Bega and P.E. Nelson, 150–2. University of California Press, Berkeley.

WALLIS G.W. (1960) Survey of *Fomes annosus* in East Anglian pine plantations. *Forestry* 32, 203–14.

WIEHE P.O. (1952) The spread of *Armillaria mellea* (Fr.) Quél. in Tung orchards. *E. Afr. agric. J.* 18, 67–72.

WILHELM J. & J.B. TAYLOR (1965) Control of *Verticillium* wilt of olive through natural recovery and resistance. *Phytopathology* 55, 310–16.

YDE-ANDERSEN A. (1967) Stump protection in conifer stands. *Sect.* 24, *Congr. Int. Un. Forest Res. Org., Munich* 314–20.

Control of nematode pests, background and outlook for biological control

F. G. W. JONES *Rothamsted Experimental Station, Harpenden, Hertfordshire*

Rather than give a catalogue of methods, I intend to review the circumstances under which control methods, chemical or biological, must operate. At present there are no examples of biological control of nematodes by classical methods, i.e. the introduction of new enemies into established populations; nor are there any certain means of encouraging old ones already present. Also, nematicides to match the efficiency and killing power of modern insecticides have still to be produced. Nematicides have never been used to blanket whole areas, but the methods used to control nematodes of pineapples in Hawaii (Christie 1959, Schmidt 1964) and root-knot nematodes in intensive tobacco-growing areas, e.g. in Rhodesia (Daulton 1965), are comparable.

Nematodes that harm crops live in a system that has four components: the soil, the plants, the nematodes and their enemies.

The soil environment

Although some nematodes inhabit inflorescences and other overground parts of plants, all harmful species occur in the soil at some stage in their life cycles, and most attack roots. Root ectoparasitic nematodes live entirely in the soil: endoparasitic species live mainly in roots, although some enlarge so much that they protrude into the soil. Intermediate forms are partly embedded in roots and partly in the soil. The habits of nematodes therefore decide the degree to which they escape the influence of unfavourable physical and biological factors in the soil.

Nematodes inhabit the spaces between soil particles and aggregates where the air is relatively stagnant and often contains much carbon dioxide. The relative humidity rarely drops below 98%, a level at which the forces tending to withdraw water from unwaterproofed microscopic animals are already large (Wallace 1963). The walls of the soil spaces are inert and restrictive, and movement of nematodes is determined by narrow channels (pore necks) that allow access between larger chambers: many of the necks are too narrow even for nematodes and may end blindly. Unlike roots, which are larger and more powerful, nematodes cannot force a passage through soil aggregates, although they may squeeze through pore necks with diameters a little less than their own and push aside fine particles in most soils (Jones *et al.* 1969, Townshend & Webber 1971).

The volume of the pore space in a soil can be estimated from its moisture characteristic (Fig. 1), and the fraction that can be penetrated by a nematode

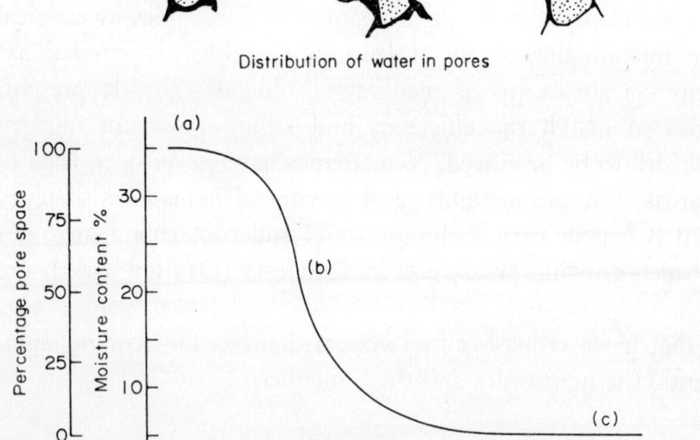

Figure 1. The relation between moisture percentage and suction in graded sand: (a) pores full, (b) pores emptying, (c) pores almost empty. This curve is the moisture characteristic on draining (Wallace 1963).

of given diameter calculated from the formula $h = 3000/d$, where h cm is the suction force of water and d μm the diameter of the nematode and of the pore necks just emptied of water at force h. Juveniles have relatively more space than adults, so for all stages to coexist the critical diameter is that of the largest individuals, i.e. the gravid females. Available space is clearly an important parameter determining the size of a population and its activity. Another, which so far defies assessment, is the average distance a nematode can travel before encountering blind ends. The amount of space that nematodes can occupy in a soil is sometimes only a small fraction of the soil volume. Figure 2 illustrates that the chalky boulder clay beneath deciduous

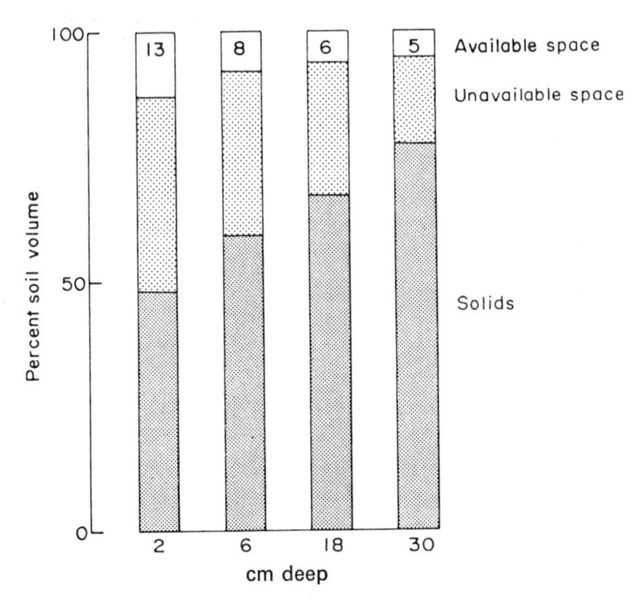

Figure 2. The space available to *Xiphinema diversicaudatum* in an untilled clay soil beneath deciduous woodland at Rothamsted (Jones, Larbey & Parrott 1969).

woodland provides little space for the dagger nematode, *Xiphinema diversicaudatum*, and that even this limited space decreases with depth. In such a soil, plant roots, nematodes and larger soil animals tend all to be confined to major pores, especially deep down where nematodes and their larger predators occupy the same spaces. Coarse, structureless sandy soils, provide more space for nematodes, and much remains even when the soil is partly compacted. These soils favour root ectoparasitic nematodes; also predators larger than the nematodes cannot follow them through pore necks. Figure 3 shows that the numbers of the migratory endoparasitic *Pratylenchus penetrans* in peach orchards in Canada also reflect the relative amount of space in different soils (Mountain & Boyce 1958).

The soil is buffered climatically and in other ways. North of the equator, during clear sunny weather in August, the temperature at the surface of an exposed soil has a pronounced diurnal rhythm which becomes much less below the surface and is barely perceptible below 10 cm (Fig. 4). Under a vegetation cover or when the soil is wet, fluctuations are smaller and descend less deeply. Soil temperatures differ less from place to place than do overground temperatures, and seasonal effects are moderated by the changes in sowing dates they dictate. Figure 5 shows accumulated temperatures in day degrees in a potato field at planting dates in different years. Even in 1960, when planting was delayed until May, the curves ran almost parallel from shortly after planting until June–July, the end of the first generation of the potato cyst-nematode, *Heterodera rostochiensis*.

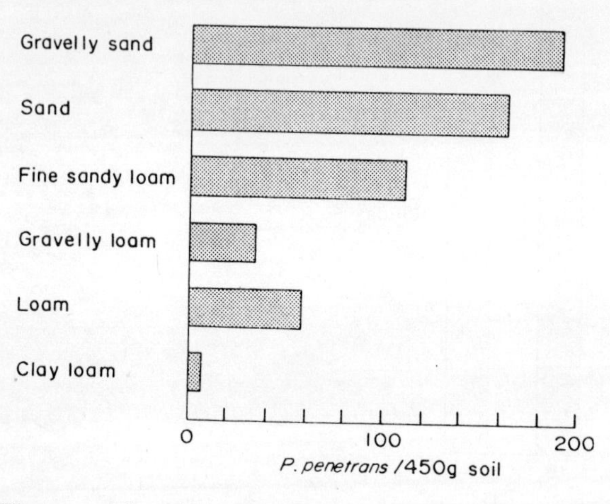

Figure 3. The relationship between soil type and numbers of *Pratylenchus penetrans* in peach nursery soils in Ontario, Canada (Mountain & Boyce 1958).

Although soil moisture is less well buffered than temperature, changes are slower than in overground habitats. In temperate regions, and especially in areas with oceanic climates, moisture is rarely limiting during the early part of the growing season, but may become so as the season advances and evapo-transpiration potentials increase. Under such circumstances many nematode populations show peaks during May–June, and again during September–October, when the moisture returns.

Soil nematodes are aquatic organisms, well adapted by their shape, movement and lack of appendages to life in soil. They are uninfluenced by the normal osmotic pressures of soil water, for they resemble the cells of plant roots in being able to maintain their turgor until the total suction forces approach the wilting point (Blake 1961). Structural moisture forces are important, however, and when they exceed 300 cm of water, surface tension thins the water films so much that the nematodes are clamped to soil particles and cannot move. Nematodes move best when suction forces are in the range 100 to 200 cm; i.e. when soil is draining after rainfall, when the larger pores contain air and the soil is well oxygenated. The same conditions probably also favour the motile zoospores of organisms endoparasitic in nematodes (Rapoport & Teschapek 1967).

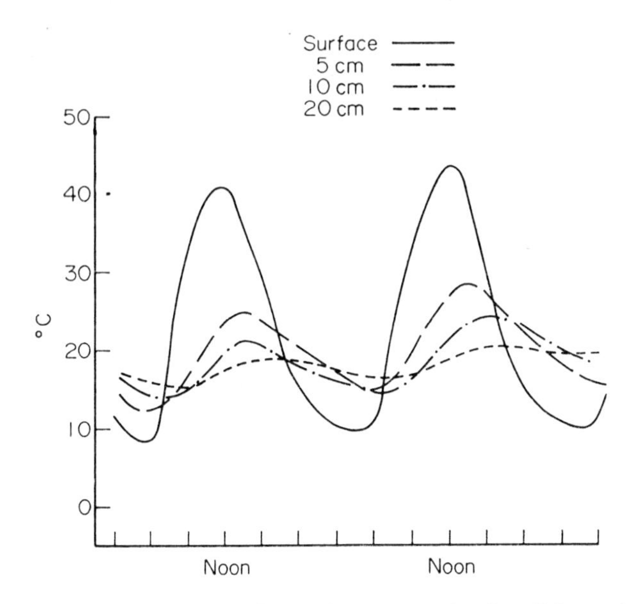

Figure 4. Temperature changes in an exposed soil at different depths during clear sunny weather in August (Russell 1950).

An outstanding feature of soil is its great bulk. There are 2,500 t/ha of top soil down to plough depth (20 cm), containing some 250 to 500 t of water, and nematodes often descend deeper. For example, the burrowing nematode, *Radopholus similis*, descends more than 2 m in the coral sand of Florida citrus orchards (Suit & Du Charme 1957). Except where nematicides are confined to seed rows to protect especially vulnerable seedlings, the quantity of chemical needed to kill 80 or 90% of the nematodes far exceeds that needed for insects exposed on plant surfaces, or inhabiting the top 5 cm of soil.

Because of the nature of the soil system, opportunities of modifying this environment to the detriment of nematodes are few. Occasionally flooding can be practised, but is ineffective unless it can be long continued, and irrigation benefits plants and nematodes. Small amounts of soil can be freed of nematodes by heat or by alternating electric currents at high potential. Adding large amounts of organic matter to encourage the microflora and fauna, in the hope that some will be antagonistic to nematodes, or will liberate compounds toxic to them while decaying, is a temporary palliative, perhaps owing much of its apparent benefit to the plant nutrients liberated. Noxious nematodes are sometimes reduced and yields increased by adding organic matter. Nematode-trapping fungi may be encouraged for a month or so but after a year the soil returns to its previous state. The texture of coarse sand soils can be changed by adding marl, so that it will perhaps provide less space for root ectoparasitic species such as the stubby root nematode, *Trichodorus* spp., and the needle nematode, *Longidorus attenuatus*,

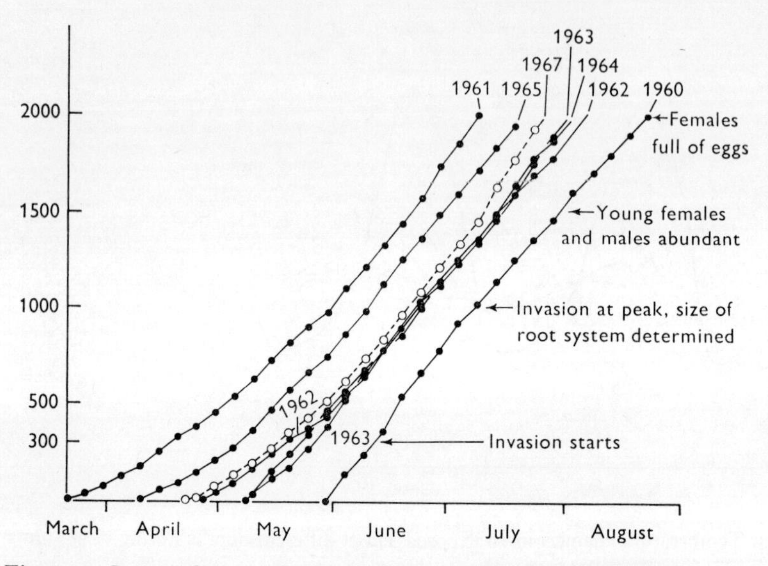

Figure 5. Curves of accumulated temperature in day degrees above 4·4°C (40°F) 10 cm (4 in) deep in the soil of a potato field at Woburn for different planting dates. Stages in the life cycle of *Heterodera rostochiensis* are shown (Jones & Parrott 1969).

that stunt sugar beet seedlings (Whitehead *et al.* 1971), but whether benefits would justify the cost is debatable. Fertilizers, especially nitrogen and potash, enable root systems to withstand attack. Without nitrogen, cereal roots are small, grow slowly, and absorb the same number of invading larvae of the cereal cyst-nematode, *H. avenae*, as the larger roots of plants receiving nitrogen. As a result they are branched, knotted and shallow, and so unable to obtain water during dry spells in June and July.

Nematode populations

The numbers of a single species of plant nematode may be as great as $2 \cdot 5 \times 10^{11}$/acre, which is equivalent to 25 million/m² or 100/g dry soil. Because nematode populations are immobile, each field or smaller cropping unit is effectively an island. Chance dispersal by wind, water currents or soil adhering to the feet of birds and animals is overshadowed by dispersal that results from farming and commerce. There are many types of nematode populations. Some species are slow breeders, taking longer than a year to complete one generation, for example dagger nematodes, *Xiphinema* spp., and needle nematodes, *Longidorus* spp., with generation times longer than that of the smallest British mammal, the pigmy shrew, *Sorex minutus*. At the other extreme are species such as the stem nematode, *Ditylenchus dipsaci*, and the bud and leaf nematode, *Aphelenchoides ritzemabosi*, which, like aphids, can multiply enormously in a few months. Between lie many species with one or two generations a year, often limited to one by the harvesting of the host crop. These can increase in numbers by factors of 30 to 150 in a year and so can recover by harvest in the year a nematicide is applied, whereas the slow breeders may take several years during which the effect of the nematicide remains apparent in succeeding crops (Cooke & Hull 1972).

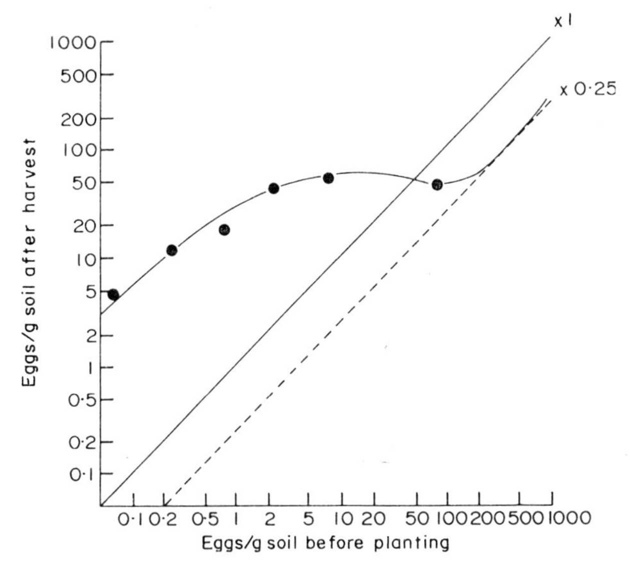

Figure 6. The curve relating preplanting to post-harvest numbers of *Heterodera rostochiensis*. Continuous line is the arithmetic sum of new and unhatched eggs. Broken line represents the unhatched fraction of the population that persists.

Of all nematodes, populations of the potato cyst-nematode, *H. rostochiensis*, have been studied most. When potatoes are not grown in infested fields, numbers of eggs decrease steadily by about a third every year regardless of population density. When a susceptible potato variety is grown, the relationship between preplanting and post-harvest numbers is as in Fig. 6. As the initial population density increases, the crude reproductive rate decreases from its maximum at small densities to less than one at large densities. The shape of the curve is peculiar because an appreciable fraction of the eggs remains unhatched, so the curve is the sum of unhatched eggs and those newly produced. When the reproductive rate approaches zero, the population after harvest consists almost entirely of unhatched eggs. Up to the curve's peak, population density after harvest is determined mainly by competition for root space. Larvae invading roots may become male or female,

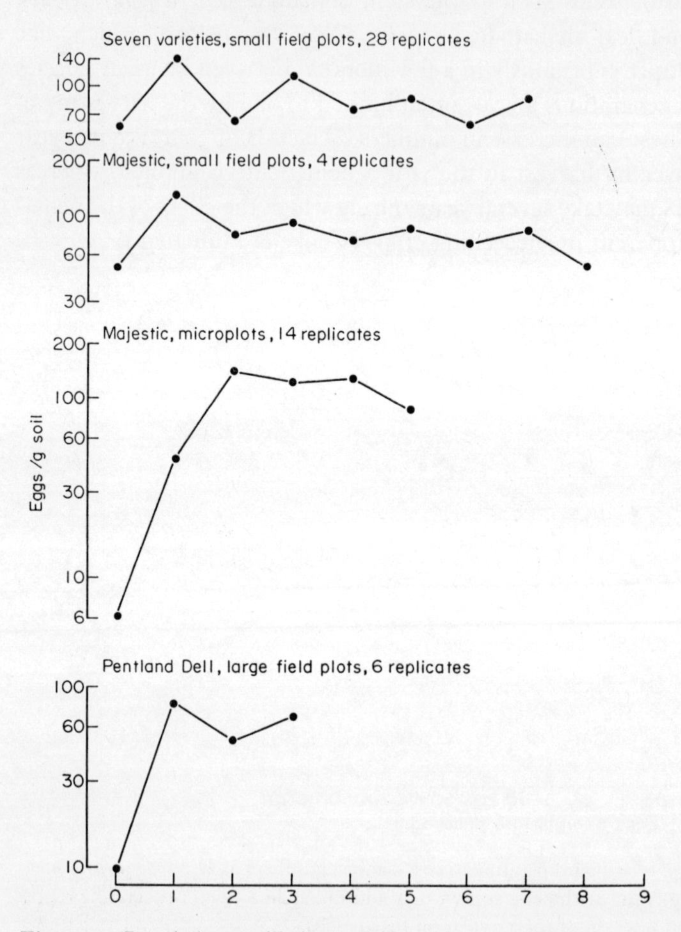

Figure 7. Population oscillations of *Heterodera rostochiensis* when susceptible potatoes are grown every year.

but only those that establish a large enough group of transfer cells (i.e. a feeding territory) become female; the remainder become males and soon leave the roots. Beyond the peak of the curve, the roots are shortened and stunted, so providing less food and space. The parasite-hostplant relationship is essentially the same as that between predator and prey, from which it can be predicted that nematode numbers will oscillate when potatoes are grown every year: in practice they do (Jones & Parrott 1969) (Fig. 7). Numbers would oscillate more, were it not for the fraction of the eggs within cysts that fails to hatch. This damps the oscillations and will also retard genetic changes when a population is selected by a resistant potato variety (Jones 1967).

Because the soil environment is stable, and about the same weight of seed potatoes is sown into soil dressed with about the same amount of fertilizer every year, and because for many years potato crops have been predominantly two genotypes, namely those of the varieties King Edward and Majestic, populations of the potato cyst-nematode have behaved with remarkable uniformity. This enables the relationship between the potato crop and nematode to be summarized as in Table 1. The penultimate column shows an

Table 1. Populations, yield lost and crop frequency: a summary of field experience with *H. rostochiensis*

Preplanting nos. Eggs/g soil	Post-harvest nos. Eggs/g soil	Crude multiplication rate	Crop loss	Years under other crops to: return to preplanting nos.	avoid crop loss
1	30	×30	None	9	1–2
10	100	×10	Little	6	4
50	200	×4	Quarter	4	6
100	300	×3	Half	3	7
200	250	×1·25	Most	1	6–7
300	210	×0·7	Total	0	6
Equilibrium numbers:					
230	230	×1·0	Most		6–7

estimate of the number of years needed without potato crops to maintain the infestation at the same level as the initial population, and the final column shows the number of years needed to avoid yield loss. When the initial populations are within reach of the equilibrium population density, to which all infested fields tend, 7 or 8 years between potato crops are necessary to avoid all loss, but the farmer often settles for a narrower rotation and accepts a small loss because it pays him to do so (Empson & James 1966, 1967).

The potato cyst-nematode is an alien that has followed the potato to Europe from the Andes, arriving some 250 years later (Jones 1970). The key factor determining the size of field populations is the supply of potato roots. Effective enemies seem not to occur in Europe, suggesting that the introduction of a specific enemy from the Andes or elsewhere might succeed in keeping populations smaller than those that injure potato roots seriously.

Other cyst-nematodes behave like the potato cyst-nematode but the cereal cyst-nematode, *H. avenae*, differs in its behaviour. In fields where cereals are grown frequently, populations are often limited to a few eggs/g soil, which cause little loss of yield. That an enemy or competitor is responsible for this was shown at Woburn and Rothamsted, where formalin was applied to soil to control root disease in experiments with fertilizers (Williams 1969). Although the formalin controlled the 'take-all' fungus and other fungi, and greatly reduced the number of *H. avenae* that invaded root systems, the number of successful females increased, and egg populations after harvest sometimes exceeded 100/g soil, which is damaging to a following spring cereal crop. The enemy or competitor seems to be part of the complex of organisms that develops in and around cereal roots when they are grown frequently. It is ineffective against other cyst-nematodes, which multiply greatly in the same soil on suitable host crops. The identity of the agent is unknown.

The population curves for other kinds of nematodes are simpler (Jones 1969a, Oostenbrink 1971, Seinhorst 1970). Seinhorst developed models relating preplanting to post-harvest numbers and to yield losses. Because much is known about the relationship between crop loss and preplanting numbers, advice on the cropping of infested fields in Gt. Britain, the Netherlands and elsewhere has often been based on estimates of nematode numbers obtained before planting. For cyst-nematodes especially, the safe levels have been worked out for areas with different soils and used successfully for many years.

Although some carbamoyl oximes, e.g. aldicarb, seem to destroy nematodes by interfering with their behaviour rather than by killing them directly, the opportunities of managing nematode populations by modifying behaviour is unpromising. Hatching factors are sugar-like and difficult to purify; their chemical structure still eludes us after 40 years of work. Potato roots contain some 20 compounds that stimulate hatching (Clarke 1972) and many artificial hatching agents are also known (Shepherd & Clarke 1971). Even when such compounds are produced throughout the soil by resistant potato plants, rarely do more than 80% of *H. rostochiensis* eggs hatch. Some nematodes moult only in the presence of host root-diffusates and, in those obligate bisexual species studied, females produce sex attractants to which males respond (Green 1971, Greet 1972). Plant roots produce specific stimuli that determine where nematodes feed, but there are also many non-specific stimuli, such as carbon dioxide. The possibility of using any of these substances in fields is slight, because their incorporation to plough depth is difficult in growing crops and they are soon destroyed by bacteria. Active analogues far more stable than the natural products would be essential, and these, like the natural products, would have to be water-soluble and perhaps slightly volatile. Many nematodes remain inactive when the stimuli from their hosts are lacking, and males are quiescent when there are no females within range. Inactivity under these circumstances conserves food reserves and has survival value. Artificial egg-hatching factors and other substances that arouse quiescent nematodes might be employed in the absence of a host crop, but none has been tried.

The host plant

For ectoparasites, the host plant provides food and many stimuli that lead to feeding: for endoparasites, the host plant provides both food and space, and a milieu protective and reasonably constant both physically and chemically. The better-adapted parasites also modify host tissues to their own advantage, forming cavities, galls or giant cells. Endoparasites and sedentary ectoparasites (e.g. sheath nematodes, *Hemicycliophora*), that insert their stylets deeply and move infrequently, are able to obtain food and water, and so to develop and multiply, under external conditions that would inhibit other species.

As it is almost impossible to change the soil environment, to modify the behaviour of nematodes or to encourage enemies, the host plant remains the one component of the system that can be altered. Where land is known to be infested, susceptible crops can be withheld, for example, onions or oats following an attack of the stem eelworm (*Ditylenchus dipsaci*) on beans, to which both are sensitive, or oats, the most susceptible of the cereals to *H. avenae* where this is numerous. Manipulation of sowing dates is less effective in controlling nematode attacks than in controlling those of insects. Nevertheless, early planting may enable root systems to grow at a time when cold still inhibits nematodes. Chitwood & Feldmesser (1948) found early planting beneficial in fields infested with *H. rostochiensis* in Long Island, U.S.A., and early planting of sugar beet is recommended in fields infested with the beet cyst-nematode (*H. schachtii*) in the U.S.A. (Johnson *et al.* 1971). Early harvesting of early potatoes before *H. rostochiensis* completes its life cycle is a phenological control measure used in Belgium (van den Brande & D'Herde 1964). None of these procedures is entirely satisfactory, and the last can be disastrous to the next potato crop if the intended harvesting date is missed. Late planting into somewhat drier and warmer soils is best for crops attacked by root ectoparasitic nematodes or stem nematodes.

Crop sequences may be changed in situations where one or more of the preceding crops allows the multiplication of a species that attacks a following crop. The sequence barley, barley, sugar beet, seems to increase numbers of *Trichodorus* and *Longidorus* in coarse soils and, when there is much rain in May, these nematodes stunt sugar beet seedlings severely. Undoubtedly, more could be done by modifying crop sequences if more were known about the host preferences of the above nematodes and the many others that feed on the roots of field crops. As a rule, harmful nematodes are a residue from previous host crops or weeds and the crops that suffer most are those with small seedlings and few plants per hectare growing in weedless fields. Changing crops, however, may affect a farmer's pocket, and biologically desirable solutions may be economically disadvantageous. Populations often include several species in the same genus that seem to occupy the same ecological niche, so changing crops might change selection pressures and end by substituting one harmful species for another.

Changing the frequency of a host crop works best against species that have few hosts, and is applied especially to nematodes that persist for many years. Of these, the potato cyst-nematode is the best example. Devising rotations to avoid damage from nematodes that have several host crops is more difficult. Thus, for the beet cyst-nematode, sugar, mangolds, red beet, and almost all cruciferous crops are hosts, and none should appear on infested land more than once in a rotation. A clause to this effect is included in growers' contracts with the British Sugar Corporation and is reinforced by the Beet Eelworm Orders (Jones 1951). Devising rotations to control root-knot nematodes, *Meloidogyne* spp., is especially difficult, for most species are polyphagous. Rotations have to be tailor-made for any particular region after accurate diagnosis of the species present and careful studies of their host ranges. In East Africa, Whitehead (1965, 1969) found that *M. javanica* and *M. incognita* were the dominant species on field crops and that, for example, peppers were poor hosts for the first and cotton, var. U.K. 51, Rhodes grass, sorghum, ground nuts and maize were usually poor hosts for both. In the U.S.A., ground nuts are susceptible to *M. arenaria* and *M. hapla* and therefore cannot be used to prevent populations increasing.

Breeding for resistance to nematodes, that is, changing the genotype by incorporating genes for resistance, rather than changing to a different species, possibly in a different plant family, has met with success where the host and parasite have a specific interrelationship. Varieties resistant to races of the stem nematode (*Ditylenchus dipsaci*) have been bred in lucerne, clover and oats. Here, plants sensitive to stem nematodes are ones in which invasion leads to dissolution of the middle lamellae of parenchymatous cells, presumably due to a specific reason between the pectin of the lamella and pectinase of the nematode. Resistant plants are ones in which dissolution fails (Blake 1962). In root-knot nematodes and most cyst-nematodes, the successful development of females depends on the formation of transfer cells. Again the reaction of the plant is induced, presumably, by components in larval saliva. Genes for resistance upset the relationship, transfer cells fail to develop, and cells fed upon become necrotic. In consequence, only males develop. Effectively, genes for resistance shift the sex ratio in favour of males because the larvae of most species may become male or female according to their nutrition and only those larvae that establish proper giant cells become females (Ross & Trudgill 1969).

Resistance to *Meloidogyne* spp. has been found in many plant species (Franklin & Hooper 1959) and bred into several crops. Similarly useful resistance to *Heterodera* spp. has been found, e.g. to *H. glycines* in soya bean (Epps & Duclos 1970), to *H. avenae* in oats (Cotten & Hayes 1969) and to *H. rostochiensis* in potato (Howard 1969). Resistance is derived from one or more major genes with or without some generalized, non-specific resistance, and its inheritance does not differ from that of any other plant attribute (Hare 1965).

Some of the resistant varieties of lucerne, clover and oats have been grown commercially for years without new races of stem eelworm arising. Nevertheless, races of this nematode do interbreed on common host plants (Sturhan 1964) and new races able to circumvent resistance are a possibility. Tests with potato hybrids containing genes for resistance soon showed that populations of the potato cyst-nematode in Great Britain did not behave uniformly. Later it became clear that here two species occur, one (*H. rostochiensis* proper) with yellow females, and another, unnamed, with white females (Jones *et al.* 1970). In Britain, *H. rostochiensis* belongs to one race only, whereas in the Netherlands and Germany there are three. The unnamed species has two races in Britain and, so far as is known, only one on the Continent: tests with a larger range of differential hosts may reveal more. At first it was thought that resistant potato varieties selected races from an interbreeding continuum, but in Britain it now seems certain that the potato variety Maris Piper, and others also with a major gene derived from *S. tuberosum* ssp. *andigena*, are almost totally resistant to *H. rostochiensis* but susceptible to the unnamed species. The occasional yellow females on Maris Piper root systems are a puzzle. Are they genetically different from most other females of the same race or are they the result of some combination of circumstances, including perhaps some local change in host root systems enabling them to evoke the right response and induce giant cells? Selection of mutations within a population, or of already existent individuals with genes that would enable them to circumvent resistance, is likely to be a slow process taking many years (Jones *et al.* 1967), whereas selection of a second species which does not suffer genetic competition is rapid. In a field at Woburn where the second species was either absent, very infrequent, or at a disadvantage, a resistant variety remained resistant for more than 10 years when grown continuously. In other fields there and elsewhere, where it was present, resistance ceased to be effective within 5 or 6 years.

The evidence to date suggests that resistance against the race for which the variety was bred is long-lasting, but may fail when a second species is present for which the resistance is inoperative. As most nematode populations in fields are species mixtures, generalized resistance to a group of species or to the genus rather than to a single species may be the answer, provided suitable resistance sources can be found. The alternative is the incorporation of two or three major genes plus some background resistance from polygenes. Suitable sources of resistance to *H. rostochiensis* are known, but including them all with resistance to other diseases and with desirable commercial properties is a slow process.

Enemies

Table 2 lists the groups of organisms known to feed on or parasitize nematodes and some are illustrated in Figs. 8 and 9. Apart from the predacious fungi studied in detail by Duddington (1960) and by Cooke & Godfrey (1964), and the endogenous fungi studied by Drechsler (1937, 1941), knowledge of other groups is scanty, and mostly derived from chance observations on single species in laboratory cultures or observation boxes. Except for amoebae and nematodes, all the predatory animals listed are larger than plant nematodes, and so are unable to follow them through narrow soil pores.

Table 2. Enemies and competitors of plant nematodes (Esser & Sobers 1964, Sayre 1971).

Predators	Parasites	Competitors
Amoebae	?Viruses	Organisms
Ciliates	Rickettsias	injurious
Tardigrades	Bacteria	to
Turbellarians	Sporozoans	roots
Nematodes	Endoparasitic	
Enchytraeids	fungi	
Collembolans		
Mites	*Predacious*	
?Insects	*fungi*	

The predatory nematodes include those with large open stomas (*Anatonchus*, *Butlerius*, *Diplogaster*, *Mononchus*, *Mononchoides* and *Tripyla*) and those with stylets (*Actinolaimus*, *Discolaimus*, *Dorylaimus* and *Seinura*) but none have been studied in the detail necessary to assess their potential. Methods of rearing large numbers do not exist, and whether such work is worth attempting is debatable, considering the immense numbers needed and the fact that non-specific predators rarely exert more than a partial control. Gilmore (1970) studied soil Collembola and found that some species ate as

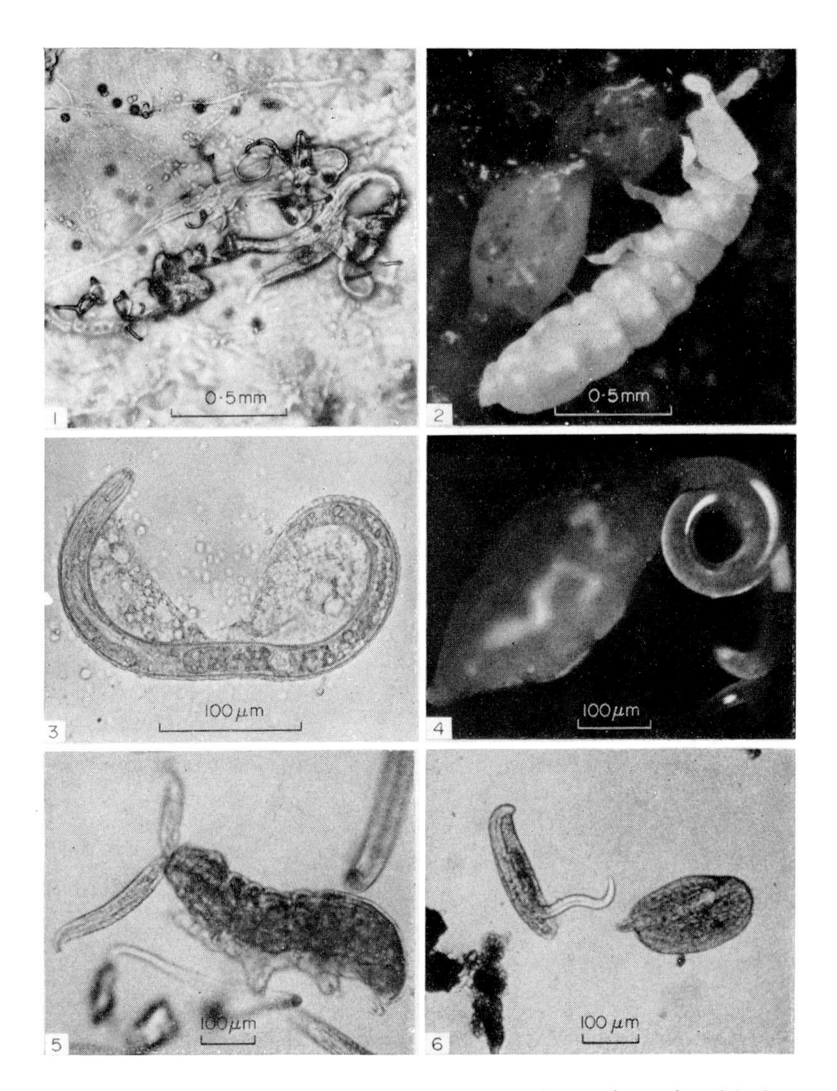

Figure 8. Some predators of soil nematodes. 1. Nematodes caught on the sticky loops of the fungus *Arthrobotrys oligospora*. 2. *Onychiurus armatus* (Collembola) feeding on female *H. trifolii*. 3. *Theratromyxa weberi* (Proteomyxa) having caught and beginning to coil larva of *H. rostochiensis*. 4. *Aporcelaimus* sp. feeding on the contents of a female *H. trifolii*. 5. *Macrobiotus vichtersii* (Tardigrada) feeding on free-living soil nematode. 6. *Urostyla* sp. (Ciliata) with partly ingested nematode (left) and showing damage caused by ingesting a nematode (right). *Urostyla* could not digest nematode 'prey'. (Photographs by C. C. Doncaster, Rothamsted Experimental Station.)

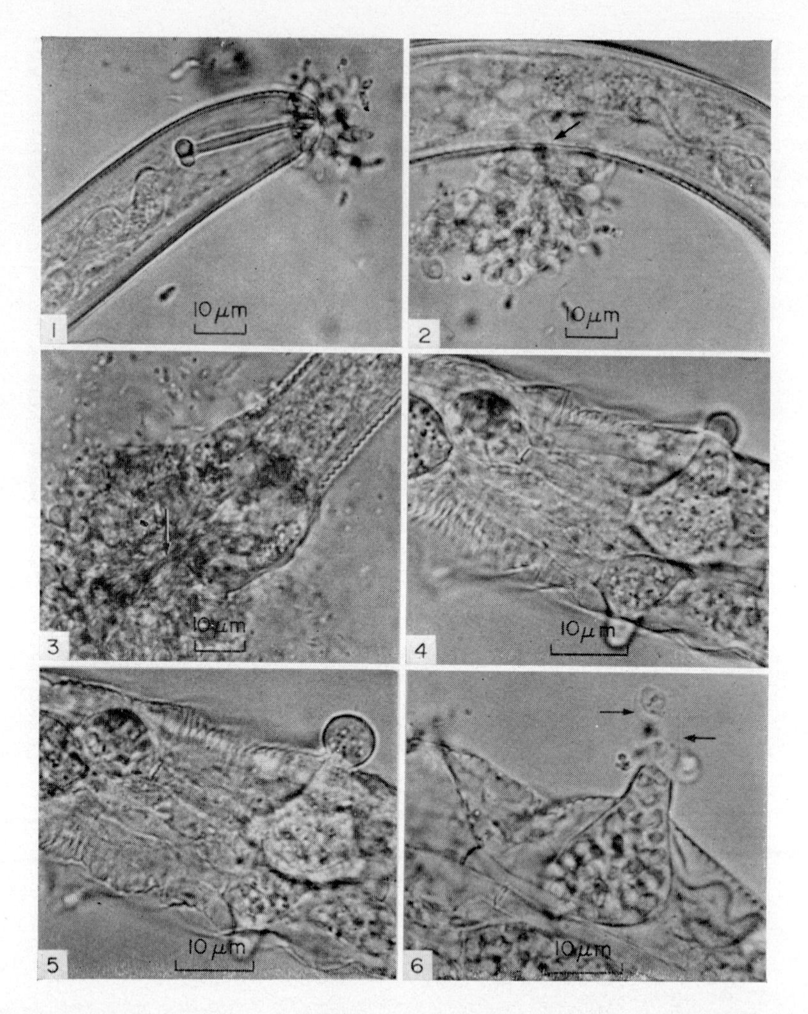

Figure 9. Stages in the life cycle of a fungus predator (probably *Catenaria anguillulae*) of *H. rostochiensis* males. 1, 2. zoospores clustering around the mouth and excretory pore (arrowed). 3. Swelling at tail end of male with filamentous hyphae (arrowed). 4, 5, 6. Stages in development of sporangium. 6. Released zoospores with flagella (arrowed). (Evans 1968, photography by C. C. Doncaster, Rothamsted Experimental Station.)

many as 1000 nematodes a day and completed their life cycles on them. Comparing the great numbers of Collembola that occur in soil with the greater numbers of nematodes, he concluded that, at the feeding rates he observed, Collembola were potentially able to effect great reductions in nematode populations. But in his laboratory cultures the Collembola had no choice of food and the nematodes no means of escape: in field soils alternative food is abundant and the nematodes can move through spaces the Collembola cannot penetrate.

Much work has been done on predacious fungi that capture nematodes by sticky branches, networks, or knobs or by constricting and non-constricting rings. Apparently, capturing nematodes is a side-line, used only when the substrates they normally use are depleted. Cooke (1968) reviewed the nutritional requirements of nematode-trapping fungi and several people have developed methods for studying them, notably Cooke (1961), Eren & Pramer (1966) and Hayes & Blackburn (1966). Organic matter only temporarily increased the fungistatic properties of soil (Mankau 1962). Cooke (1963) found there was a succession of species acting predaciously from 7 to 11 weeks after the addition of organic matter. Their activity seems to bear no relation to the numbers of nematodes in soil, and it is therefore not surprising that in most outdoor experiments no worthwhile reductions in nematode numbers have followed attempts to increase the activity of these fungi by applying large amounts of organic matter (e.g. Duddington, Jones & Williams 1956, Duddington *et al.* 1956).

Although Drechsler listed many fungi that were endogenous parasites of nematodes, few have studied the details of their life cycles. Van der Laan (1956) studied the fungi that attack within *Heterodera rostochiensis* cysts. He showed that some of these fungi reduced the numbers of females (cysts) produced on potato roots in plots. Work at Cambridge (Jones 1945, Tribe unpublished) indicates that the beet cyst-nematode is often parasitized by a range of fungi.

Much more work is needed on life cycles, frequency of occurrence, effectiveness and means of culturing spore forms that can be distributed, before attempts can be made to use endoparasitic fungi as control agents. The same applies to protozoan parasites, of which only *Duboscquia penetrans* seems to have received more than comment (Williams 1967). Although it is almost certain that there are virus and other diseases of nematodes, e.g. rickettsias (Shepherd 1971), no means have yet been devised of screening for them. In short there are no immediate or long-term prospects of using biological agents against nematodes. The many that already exist in soil are ineffective against pest species and whether they suppress less common species that might otherwise cause trouble is unknown. No examples have yet been brought to light by 'flare backs' following the commercial use of nematicides, although formalin used in experiments revealed an unknown agent controlling *H. avenae* in Britain (see page 258). The increases in potato cyst-nematode after applying nematicides seem to be from the better growth of roots: similar increases follow short rotations or any control method that is only partly effective, all of which tend to bring numbers to the peak of the population curve (Fig. 6).

Integrating control methods

The methods used to control most harmful nematodes before planting rarely kill so many that the species cannot recover its numbers by harvest. Some slow-breeding ectoparasitic nematodes are an exception. Figure 10 shows

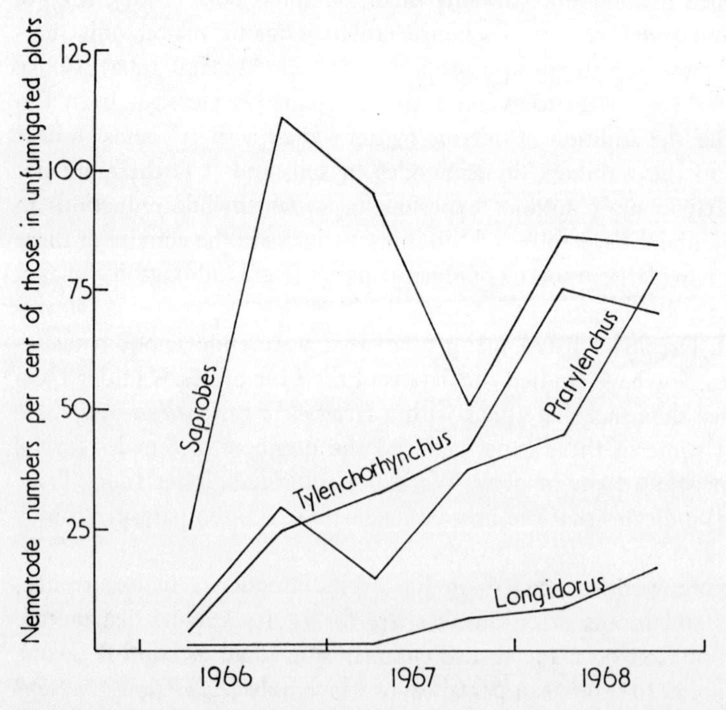

Figure 10. Changes in nematode numbers following fumigation with DD before planting sugar beet in 1966. (Redrawn from Cooke & Hull 1972.)

post-fumigation population increases of two ectoparasitic nematodes (*Longidorus*, *Tylenchorhynchus*), one endoparasitic (*Pratylenchus*) and the total of saprobic species. After fumigation, the last group multiply rapidly, presumably on the nutrients provided by dead bodies of soil organisms of many kinds killed by fumigation and, possibly, freed from some of their enemies. *Pratylenchus* and *Tylenchorhynchus* take 3 years to recover and *Longidorus* longer. Whether a nematode can recover its numbers within a crop season depends on the kill achieved and on the nematode's maximum rate of multiplication between planting and harvest. Thus a kill of 90% is balanced by a reproductive rate of ten times and one of 99% by a rate of a hundred times (Jones 1969a, 1969b). For effective control of nematodes able to multiply between thirty and a hundred and fifty times, kills better than 99% are required. Most available methods, although giving large yield increases, are ineffective on this criterion, and it is only by combining methods that effective kills can be obtained (Table 3). In the Netherlands, to protect

Table 3. Integration of control methods for *H. rostochiensis*

Control method(s)	Resulting population (% initial population)	Kill (%)	Population after growing and harvesting a susceptible variety, calculated at two assumed nematode multiplication rates.* (% initial population)	
			30×	70×
1. 4 years without potatoes	3	97	90	>100
2. 1 year with resistant potatoes	20	80	>100	>100
3. Nematicide treatment	25	75	>100	>100
4. 1 and 2	0·6	99·4	18	42
5. 1 and 3	0·75	99·25	22·5	52·5
6. 2 and 3	5	95·0	>100	>100
7. All three methods	0·15	99·85	4·5	10·5

* The observed maximum reproductive rate lies between 30 and 70 times.

the potato starch industry in the Groningen area from *H. rostochiensis*, these principles are being applied on a large scale (Nollen & Mulder 1969). DD (dichloropropane–dichloropropene mixture) is the nematicide being employed. Beneficial side effects include control of *H. avenae*, a decrease in the incidence of tobacco rattle virus through the control of its nematode vector (*Trichodorus*), and the killing of some weed seeds. Harmful side effects include tuber taint (unimportant in starch potatoes) and damage to wheat ears. Because of the large amounts of nematicides to be applied, there are objections by conservationists. Whether their objections are valid is debatable, for DD has been used extensively in the U.S.A. for many crops, and is recommended combined with crop rotation for the control of *H. schachtii* in sugar beet. DD does not persist long, does not kill the soil fauna and flora totally, and its breakdown products are innocuous.

Integrating control methods with the activities of enemies and competitors of nematodes is impossible until more is known about them, but those experimenting with nematicides seek to apply the minimum effective dose. Commercial pressures to cheapen costs operate strongly in the same direction.

References

BLAKE C.D. (1961) Importance of osmotic potential as a component of the total potential of the soil water on the movement of nematodes. *Nature, Lond.* **192**, 144–5.

BLAKE C.D. (1962) The etiology of tulip-root disease in susceptible and in resistant varieties of oats infested with the stem nematode *Ditylenchus dipsaci* (Kühn) Filipjev II. *Ann. appl. Biol.* **50**, 713–22.

BRANDE J. VAN DEN & J. D'HERDE (1964) Phenological control of the potato root eelworm (*Heterodera rostochiensis* Woll.) *Nematologica* **10**, 25–8.

CHITWOOD B.G. & J. FELDMESSER (1948) Golden nematode population studies. *Proc. helm. Soc. Wash.* **15**, 43–55.

CHRISTIE J.R. (1959) *Plant Nematodes, their Bionomics and Control.* Gainesville Agric. Exp. Stn. Univ. Fla. 256 pp.

CLARKE A.J. (1972) Sex attractants and hatching factors. *Rep. Rothamsted epx. Stn.* for 1971 Pt. 1, 165.

COOKE D.A. & R. HULL (1972) The effects of soil fumigation with DD on the yields of sugar beet and other crops. *Ann. appl. Biol.* **71**, 59–67.

COOKE R. (1963) Succession of nematophagous fungi during decomposition of organic matter in the soil. *Nature, Lond.* **197**, 205.

COOKE R.C. (1961) Agar disk method for the direct observations of nematode-trapping fungi in the soil. *Nature, Lond.* **191**, 1411–12.

COOKE R.C. (1968) Relationships between nematode-destroying fungi and soil borne phytonematodes. *Phytopathology* 58, 909–13.

COOKE R.C. & B.E.S. GODFREY (1964) A key to the nematode-destroying fungi. *Trans. Br. mycol. Soc.* 47, 61–74.

COTTEN J. & J.D. HAYES (1969) Genetic resistance to cereal cyst-nematode (*Heterodera avenae*). *Heredity, Lond.* **24**, 593–600.

DAULTON R.A.C. (1965) Soil fumigants in Rhodesian tobacco. *Rhodesian Farmer* 36, May, 2 pp.

DRECHSLER C. (1937) Some Hyphomycetes that prey on free-living terricolous nematodes *Mycologia* **29**, 447–552.

DRESCHLER C. (1941) Predacious fungi. *Biol. Rev. Cambridge Phil. Soc.* **16**, 265–90.

DUDDINGTON C.L. (1960) Biological control—predaceous fungi. In *Nematology*, eds. J.N. Sasser and W.R. Jenkins. University of North Carolina Press, Chapel Hill.

DUDDINGTON C.L., F.G.W. JONES & F. MORIARTY (1956) The effect of predacious fungus and organic matter upon the soil population of beet eelworm *Heterodera schachtii* Schm. *Nematologica* **1**, 344–8.

DUDDINGTON C.L., F.G.W. JONES & T.D. WILLIAMS (1956) An experiment on the effect of a predacious fungus upon the soil population of the potato root eelworm *Heterodera rostochiensis* Woll. *Nematologica* **1**, 341–3.

EMPSON D.W. & P.J. JAMES (1966) An economic approach to the potato root eelworm problem. *N.A.A.S. Quarterly Review* No. 73, 22–9.

EMPSON D.W. & P.J. JAMES (1967) A further note on the economics of the potato root eelworm. *N.A.A.S. Quarterly Review* No. 76, 160–5.

EPPS J.M. & L.A. DUCLOS (1970) Races of the soybean cyst nematode in Missouri and Tennessee. *Pl. Dis. Reptr* 54, 319–20.

EREN J. & D. PRAMER (1966) Application of immuno-fluorescent staining to studies of the ecology of soil microorganisms. *Soil. Sci.* **107**, 39–45.

ESSER R.P. & E.K. SOBERS (1964) Natural enemies of nematodes. *Proc. Soil Crop Science Soc. Fla.* **24**, 326–53.

EVANS K. (1968) Influence of some factors on the reproduction of *Heterodera rostochiensis*. Ph.D. Thesis, London University, 218 pp.

FRANKLIN M.T. & D.J. HOOPER (1959) *Plants recorded as resistant to root-knot nematodes* (*Meloidogyne* spp.). *Tech. Commun.* No. 31 *Commonw. Bur. Helminth.* 33 pp.

GILMORE S.K. (1970) Collembola predation on nematodes. *Search Agriculture*, *Entomology*, *Limnology* **1**, 1–12.

GREEN C.D. (1971) Mating and host finding behaviour of plant nematodes. In *Plant Parasitic Nematodes*, eds. B.M. Zuckerman, W.F. Mai and R.A. Rohde. Academic Press, New York & London. Vol. 2, 247–66.

GREET D.N. (1972) Sex attractants and hatching factors. *Rep. Rothamsted exp. Stn* for 1971 Pt. 1, 164.

HARE W.W. (1965) The inheritance of resistance to plant nematodes. *Phytopathology* **55**, 1162–7.

HAYES W.A. & F. BLACKBURN (1966) Studies on the nutrition of *Arthrobotrys oligospora* Fres. and *A. robusta* Dudd. II. The predaceous phase. *Ann. appl. Biol.* **58**, 51–60.

HOWARD H.W. (1969) Breeding potatoes resistant to cyst-nematode. *Proc. 5th Br. Insectic. Fungic. Conf. Brighton 1969* **1**, 159–63.

JOHNSON R.T., T.T. ALEXANDER, G.E. RUSH & G.R. HAWKES (1971) *Advances in Sugar Beet Production*, *Principles and Practices*. University Press, Ames, Iowa. 470 pp.

JONES F.G.W. (1945) Soil populations of beet eelworm (*Heterodera schachtii* Schm) in relation to cropping. *Ann. appl. Biol.* **32**, 351–80.

JONES F.G.W. (1951) The sugar beet eelworm order 1943. *Ann. appl. Biol.* **38**, 535–7.

JONES F.G.W. (1969a) Some reflections on quarantine, distribution and control of plant nematodes. In *Nematodes of Tropical Crops*, ed. J.F. Peachey, Farnham Royal, Commonw. Agric. Bur. 67–80.

JONES F.G.W. (1969b) Integrated control of the potato cyst-nematode. *Proc. 5th Br. Insectic. Fungic. Conf. Brighton 1969* **3**, 646–56.

JONES F.G.W. (1970) The control of the potato cyst-nematode. *Jl R. Soc. Arts* **118**, 179–196.

JONES F.G.W., J.M. CARPENTER, D.M. PARROTT, A.R. STONE & D.L. TRUDGILL (1970). Potato cyst-nematode: one species or two? *Nature*, *Lond.* **227**, 83–4.

JONES F.G.W., D.W. LARBEY & D.M. PARROTT (1969) The influence of soil structure and moisture on nematodes, especially *Xiphinema*, *Longidorus*, *Trichodorus* and *Heterodera* spp. *Soil & Biochemistry* **1**, 153–65.

JONES F.G.W. & D.M. PARROTT (1969) Population fluctuations of *Heterodera rostochiensis* Woll. when susceptible potato varieties are grown continuously. *Ann. appl. Biol.* **63**, 175–81.

JONES F.G.W., D.M. PARROTT & G.J.S. ROSS (1967) The population genetics of the potato cyst-nematode, *Heterodera rostochiensis*. Mathematical models to simulate the effects of growing eelworm-resistant potatoes bred from *Solanum tuberosum* ssp. *andigena*. *Ann. appl. Biol.* **60**, 151–71.

LAAN P.A. VAN DER (1956) Onderzoeken over schimmels, die parasiteren op de cysteinhoud van het aardappelcystenaaltje (*Heterodera rostochiensis* Wollenw.). *Tyschr. Pl Ziekten* 62, 305–21.

MANKAU R. (1962) Soil fungistasis and nematophagous fungi. *Phytopathology* 52, 611–15.

MOUNTAIN W.B. & H.R. BOYCE (1958) The peach replant problem in Ontario. *Can. J. Bot.* 36, 125–34.

NOLLEN H.M. & A. MULDER (1969) A practical method for economic control of potato cyst-nematodes. *Proc. 5th Br. Insectic. Fungic. Conf. Brighton 1969* 3, 671–4.

OOSTENBRINK M. (1971) Quantitative aspects of plant-nematode relationships. *Indian J. Nematol.* 1, 68–74.

RAPOPORT E.H. & M. TESCHAPEK (1967) Soil water and soil fauna. *Rev. Ecol. Biol. Sol.* 4, 1–58.

ROSS G.J.S. & D.L. TRUDGILL (1969) The effect of population density on the sex ratio of *Heterodera rostochiensis*, a two dimensional model. *Nematologica* 15, 601–7.

RUSSELL E.J. (1960) *Soil Conditions and Plant Growth* (8*th* Ed. by E.W. Russell). Longmans Green, London. 635 pp.

SAYRE R.M. (1971) Biotic influence in soil environment. In *Plant Parasitic Nematodes*, eds. B.M. Zuckerman, W.F. Mai and R.A. Rohde, 235–56. Academic Press, New York & London.

SCHMIDT C.T. (1964) Sustained control of eelworm disease in pineapple, a permanent crop of tropical Hawaii. *Nematologica* 10, 62 [Abstract].

SEINHORST J.W. (1970) Dynamics of populations of plant parasitic nematodes. *Ann. Rev. Pl. Path.* 11, 131–56.

SHEPHERD A.M. (1971) Rickettsia-like organisms in cyst-nematodes. *Rep. Rothamsted exp. Stn* for 1970, 147.

SHEPHERD A.M. & A.J. CLARKE (1971) Molting and hatching stimuli. In *Plant Parasitic Nematodes*, eds. B.M. Zuckerman, W.F. Mai and R.A. Rohde. Academic Press, New York & London. Vol. 2, 267–87.

STURHAN D. (1964) Kreuzungsversuche mit biologischen Rassen des Stengelälchens (*Ditylenchus dipsaci*). *Nematologica* 10, 328–34.

SUIT R.F. & E.P. DU CHARME (1957) Spreading decline of citrus. *State Plant Board of Florida*, Vol. 2, Bull. 11, 1–24.

TOWNSHEND J.L. & L.R. WEBBER (1971) Movement of *Pratylenchus penetrans* and the moisture characteristic of three Ontario soils. *Nematologica* 17, 47–57.

WALLACE H.R. (1963) *The Biology of Plant Parasitic Nematodes*. Edward Arnold, London. 280 pp.

WHITEHEAD A.G. (1965) Taxonomy, distribution and host-parasite relationships of the genus *Meloidogyne* Goeldi. Ph.D. Thesis, University of London, 321 pp.

WHITEHEAD A.G. (1969) The distribution of root-knot nematodes (*Meloidogyne* spp.) in tropical Africa. *Nematologica* 15, 315–33.

WHITEHEAD A.G., R.A. DUNNING & D.A. COOKE (1971) Docking disorder and root ectoparasitic nematodes of sugar beet. *Rep. Rothamsted exp. Stn* for 1970, Pt 2, 219–36.

WILLIAMS J.R. (1967) Observations on parasitic protozoa in plant parasitic nematodes. *Nematologica* 13, 336–42.

WILLIAMS T.D. (1969) The effects of formalin, nabam, irrigation and nitrogen on *Heterodera avenae* Woll., *Ophiobolus graminis* Sacc. and the growth of spring wheat. *Ann. appl. Biol.* 64, 325–34.

Integrated control of pests and diseases of sugar beet

R. HULL *Broom's Barn Experimental Station, Higham, Bury St. Edmunds, Suffolk*

The circumstances in which arable crops are grown differ greatly from those for forest, orchard, glasshouse or horticultural crops. A few arable crops are biennials but most are annuals, grown in rotation and consequently the land is frequently laid bare and ploughed. Cropped fields are large and tending to become larger. Since only few species of plants are suitable for arable cropping, extensive areas are cropped with each. The environment of arable fields ranges from extensive tracts of land with few or no undisturbed habitats, to those latticed with hedgerows and chequered with copses, surrounded by forest or interspersed with urban and industrial developments. Intensive agriculture tends to unify the environment and make it suitable for the range of crops cultivated. This, of course, is the farmer's objective and his success is measured by the size and regularity of his crop yields and the profit he makes from them. The more intensive his culture the greater the risk of pests and diseases becoming epidemic on his crops, so his continued success depends on his skill in avoiding them and controlling them. His objective is to make a profit, so the methods employed will be determined to a great extent by their cost in relation to the amount of yield saved or augmented.

The circumstances in which crops are grown determines the incidence of pests and diseases. The grower is aware of many of the factors leading to epidemics, and avoids them. Often he is depending unconsciously on biotic factors to help to restrict pest and disease incidence. In this paper I propose to discuss some of the factors influencing the incidence of pests and diseases in the sugar-beet crop and how farming practices and the use of pesticides can be integrated to get most benefit from the limitations exerted by nature on the development of populations of organisms that become pests. This seems to me to be biological control in its widest sense and to have meaning and application in arable agriculture. Integrated control of pests and diseases of sugar beet, including the contribution of biotic factors, was the subject of the

34th Winter Congress of the International Institute of Sugar Beet Research. A summary of the conclusions and many of the papers given there have now been published (Hull 1972).

Crop rotation

All sugar beet are grown on contract with the British Sugar Corporation and a clause in the contract specifies that growers will not sow beet on land which in the two previous years was under sugar beet and related crops or Brassicae (which are susceptible to beet cyst nematode). This enforces a rotation of one crop in three years, and because growers cannot get contracts for as much acreage as they want, most grow beet once in five years. This enforced rotation has had a profound effect on the spread of cyst nematode (*Heterodera schachtii*) and pygmy beetle (*Atomaria linearis*), two pests that are serious where land is cropped continuously with beet (Jones & Dunning 1969). Neither is a serious problem in England. The control of cyst nematode is reinforced by the Beet Eelworm Orders, 1960 and 1962, which enforce stricter control of cropping of infested land. Pygmy beetle is a sporadic pest of slow-growing seedlings that are attacked by immigrant beetles, or occasionally of edges of crops abutting on land cropped with beet the previous year where beetles have hibernated in buried beet debris.

A consequence of cultivation of the soil and of crop rotation is that the composition of the weed flora changes from year to year, as does the environment produced by the different crops and so presumably does the composition of the edaphic fauna. This unstable community militates against any effort to establish a soil fauna of resident predators that would control pests.

Crop hygiene

Our first efforts in the 1940s at controlling virus yellows, which devastated the sugar-beet crop, were largely by hygiene measures. A study of the epidemiology of the disease showed that *Myzus persicae* carried the virus to the young root crop from infected plants that had survived the winter (Watson *et al.* 1951). The most important sources were the biennial seed crops of sugar beet and mangold that had been infected with the virus as young plants (stecklings). This cycle was broken by growing only virus-free seed crops and separating seed and root crops as far as practicable. Virus-free stecklings were raised in isolated areas with climates unfavourable for aphids, by careful choice of sowing date to avoid the main aphid migrations, by spraying with insecticide, and by raising stecklings under cover crops

Figure 1. Spread of yellows through a sugar-beet crop from aphid-infested mangolds which were clamped on the round concrete platform. The lighter area and spots contain the yellows infected plants.

(Hull 1967). The last proved for many years the most successful and practical. The success of these measures was monitored by inspecting the stecklings, and only plants from beds certified as satisfactorily healthy were grown-on as seed crops. These measures stopped the spread of the disease between seed and root crops. More recently many seed crops have been grown right away from root crops, in the Cotswold area and the upper Thames Valley. This has been made possible by developing methods of growing and harvesting the crops without hand labour so that it fits into the farming system in these areas. Separation of seed and root crops stopped the spread between them of several other air-borne pathogens such as *Peronospora schachtii*, *Erysiphe*, *Uromyces* and mosaic virus.

Mangolds stored in clamps through the winter for feeding cattle are often infested with aphids. These aphids feed and multiply on the sprouts that develop during the winter, and the alatae carry virus from the mangold sprouts to young sugar beet in spring (Heathcote & Cockbain 1966). Growers have been encouraged to use up their mangolds before the beet crop has germinated and to destroy unwanted clamps. This problem has largely been solved for us by a change in farming practice. The mangold crop has declined from 120,000 ha in 1944 to less than 40,000 ha, and instances, once common, where yellows has obviously spread from a clamp to beet crops are now rare (Fig. 1).

Other sources of virus and aphids are dealt with as they are found, and sugar factory fieldmen are encouraged to seek them. Winter spinach, derelict, unharvested beet or mangold crops on abandoned land, groundkeeping beets and volunteer beets or beet clamp sites have from time to time needed removing. Other virus sources such as susceptible weeds, garden crops of seakale beet or stored red beets cannot be eliminated, so some root crops always remain at risk, and measures other than hygiene must be taken to restrict the incidence of yellows. Figure 2 shows that when conditions favoured the spread of yellows in 1957, hygiene measures alone failed to control it.

Sowing date

The sowing date of annual crops may be adjusted to prevent the most susceptible stages of growth coinciding with peak pest populations. Sugar beet sown in England in March get fewer aphids and less yellows than when sown in May; the reverse occurs in central California. Sugar-beet stecklings sown before mid-July get infested with aphids and viruses, whilst those sown at the end of July often escape infestation (Heathcote 1970). However, stecklings sown in June often escape downy mildew because they are at their most susceptible stage of growth when the weather is too hot and dry for mildew to spread; those sown in August get infected as seedlings in the cooler, dewy nights of September and October. In England, sowing sugar

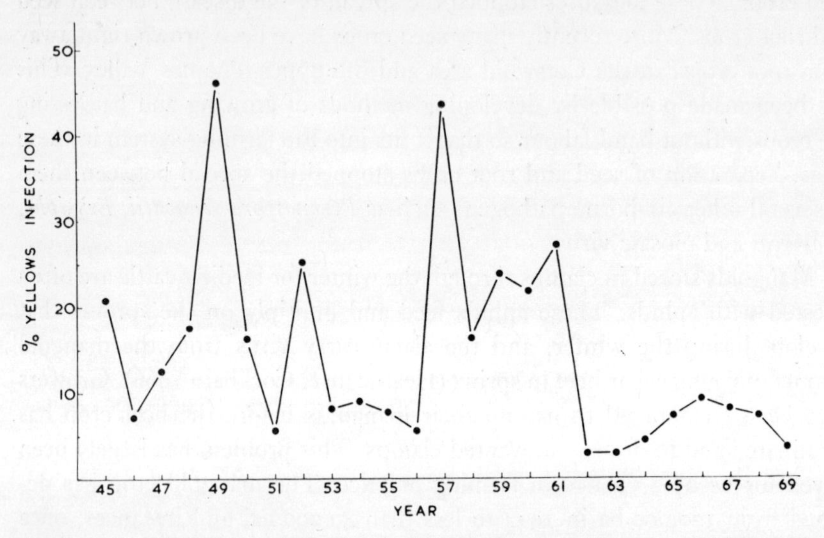

Figure 2. Graph showing the average incidence of virus yellows in sugar-beet crops in Great Britain, 1945–69. Intensive hygiene measures have been taken since 1953 and the spray warning scheme has been operated since 1959.

beet in March gives the greatest yield as well as avoiding yellows, but in California crops can be sown late to avoid yellows only at the loss of much yield. Although manipulation of the sowing date can be a help, it is no panacea.

Plant spacing and cover crops

Aphids prefer widely spaced sugar beet to dense stands, and consequently thin, gappy sugar-beet crops suffer more from yellows than do uniform stands of about 75,000 plants/ha. In contrast, leaf spots spread more in dense than in sparse stands. In England, dense stands of sugar-beet seed plants are often defoliated by *Ramularia* whilst sparser stands retain their leaves. Sowing stecklings in April under cover crops gives excellent protection against aphids and yellows and against downy mildew, whereas sowing under mustard cover in July results in much mildew infection but little yellows. These diverse effects of cover have several causes. If there are fewer plants per unit area, mobile pests like aphids will congregate on them. There is evidence that fewer alate aphids land on a uniform cover of foliage than on a mosaic of foliage and bare soil. Cover crops of tall cereals over stecklings are a physical barrier to aphids and prevent them reaching the stecklings at their base. The humid microclimate in dense vegetation that encourages fungi (*Entomophthora* spp.) to parasitize aphids also encourages sporulation and spread of the leaf spot fungi. These interactions of environment with crop and pest are complicated. Some can be exploited by the grower, but obviously he has to be knowledgeable of the biology of both crop and pest if he is to be successful and avoid the pitfalls.

Methods of establishing the sugar-beet crop are changing rapidly (Hull & Jaggard 1971). Instead of sowing seed thickly and singling the seedlings, growers now sow thinly, and in 1971 more than a third of the sugar-beet crop in England was sown to stand, without subsequent hand-work, at a seed spacing of 12·5 cm or more. This thin stand of seedlings is prone to damage from seedling pests, and much concern has been caused by damage from birds, believed to be mainly larks, grazing on the emerging seedlings. This problem gives great scope for biological control!

Variety

The value of resistant varieties has already been fully discussed, and my only comment will concern the policy for breeding. When a dominant pest or disease regularly devastates a crop, the case for breeding and sowing resistant varieties is obvious; such an instance is the American sugar-beet crop which was established and is maintained in the west only by breeding varieties resistant to curly top. Most of our pests and diseases in England occur sporadically, locally and often unpredictably. Resistant varieties are not a great help because resistance is usually achieved only at the loss, in the absence of the pest, of some desirable character of yield or quality. The objective of breeding should be yield in normal circumstances, avoiding excessive susceptibility to sporadically occurring pathogens. With out-pollinated sugar beet, using many different genotypes as the constituents of varieties, susceptibility may be augmented unconsciously in the breeding programme in the absence of the pest or disease in the selection fields. All breeding material should be screened from time to time for susceptibility to the most important pathogens, and this means either developing screening methods under glass, or providing local epidemics artificially in the field. We have done this now for several years, for downy mildew of sugar beet, in co-operation with the National Institute of Agricultural Botany at their centre at Trawscoed, Cards.

Pesticides

Pesticides have benefited agriculture enormously and their use is essential to control pests and maintain yields. If they are to be and to remain effective their use has to be integrated into the agricultural ecological system, and an example of how this has been done is the control of aphids and yellows in sugar beet.

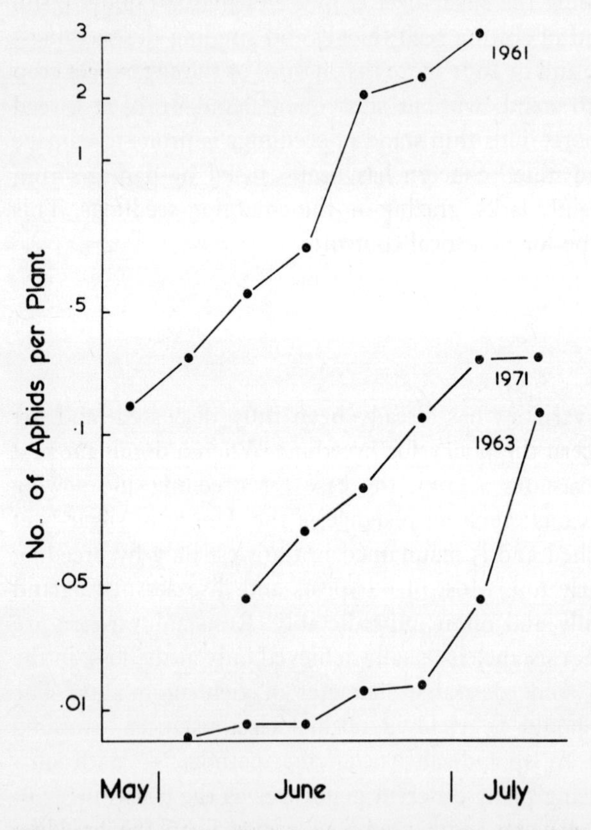

Figure 3. Graphs showing the development of green aphid infestation on sugar beet, averaged throughout England, in three contrasting years. The early and heavy infestation in 1961 resulted in an average incidence of 27% of plants with yellows at the end of August and justified widespread spraying with insecticide. The later and smaller infestations in 1963 and 1971 gave only 2–3% yellows.

The first essential is an effective pesticide, and, for the control of aphids and yellows, demeton-methyl was a fortunate discovery. Its fumigant, contact and persistent systemic insecticidal properties made it effective for long enough against the small populations of aphids that must be controlled to restrict virus spread. The second task was to establish the relationship between pathogen intensity and loss of yield (Watson *et al.* 1946, Hull 1953). Results from field experiments and surveys showed that populations of one *Myzus persicae* per 4 sugar beet in May or June could lead to more than 20% of plants having yellows by the end of August, and at this incidence spraying would augment yield enough to pay for the treatment. Routinely now, sugar factory fieldmen examine crops for aphids from the beginning of May, and when infestations start to develop in a district, spray warnings are sent to growers there. Thus, the spray recommendations are based on factual justification that treatment is likely to be economical and necessary (Hull 1968, von Steudel 1971). Figure 2 shows that since the introduction of the spray warning scheme in 1959 (Hull 1968), yellows incidence has remained at a very low level.

Insecticidal sprays do not necessarily prevent yellows, but delay and restrict its spread and thereby increase yield. Spraying halves incidence, say from 60% to 30%. If infection pressure is so great that every plant has chances of being infected on several occasions, sprayed crops may end up with most plants infected and, although spraying may increase yield by delaying infection, the loss can still be large. There is limited satisfaction in increasing yield from 8 to 12 tons when a healthy crop would yield 16 tons. Insecticidal sprays successfully control yellows only when the infective pressure of the pathogen is not too great, so restriction of the disease pressure by hygiene, sowing date, plant population and variety are integral parts of the control system. Predators help to restrict aphid infestations decreased by spraying with organophosphorus insecticide. The sprays temporarily affect the population of coccinellids, more by removing their food supply than by killing the beetles.

Predators and parasites

Occasionally, as in 1971, coccinellids are numerous among young sugar beet and appear to prevent winged aphids establishing appreciable populations. Usually, however, they and other migratory predators become numerous in the crop only when there is an adequate population of aphids on which to feed. This is useless for controlling the spread of virus by aphid vectors. Early colonization of beet fields by predators usually occurs when a large population has built up in the previous year and over-wintered in shelter near the field. To attempt to repeat this as a specific control measure seems neither practical, economic nor politic. Growers tend to clean up shelter around their arable fields when they can, because experience shows that numerous pests, such as capsids, flea beetles, mice and rabbits, are likely to invade their crops from it.

Most investigations of biological control of aphid populations seem to have concentrated on migratory predators (van Emden 1966) that are dependent on the pest for food, and only establish on populations large enough to damage the crop. When sugar beet are sprayed with persistent insecticides (e.g. DDT) that are relatively ineffective against aphids, the numbers of aphids increase more rapidly on the sprayed than on unsprayed plots even though migratory predators may not be numerous (Hull & Gates 1953). This suggests that other resident non-specialized predators may be restricting the aphid population. A parochial population of predators resident in the soil would be much more amenable to manipulation and would seem to be a more profitable line to investigate, and a more likely way of increasing the biological contribution to aphid control.

References

VAN EMDEN H.F. (1966) The effectiveness of aphidophagous insects in reducing aphid populations. In *Ecology of Aphidophagous Insects*, ed. Ivo Hodek, 227–35. Academia, Prague.

HEATHCOTE G.D. (1970) Effect of plant spacing and time of sowing of sugar beet on aphid infestations and spread of virus yellows. *Pl. Path.* 19, 32–9.

HEATHCOTE G.D. & A.J. COCKBAIN (1966) Aphids from mangold clamps and their importance as vectors of beet viruses. *Ann. appl. Biol.* 57, 321–36.

HULL R. (1953) Assessments of losses in sugar beet due to virus yellows in Great Britain, 1942–52. *Pl. Path.* 2, 39–43.

HULL R. (1967) Aphid borne viruses of sugar beet; a retrospective exercise in integrated control. *Proc. 4th Brit. Insect. Fung. Conf. 1967*, pp. 472–7.

HULL R. (1968) The spray warning scheme for control of sugar beet yellows in England. Summary of results between 1959–66. *Pl. Path.* 17, 1–10.

HULL R. (1972) Conclusions of the symposium on integrated control of pests and diseases of sugar beet. *J. int. Inst. Sug. Beet Res.* 5, 191–8.

HULL R. & L.F. GATES (1953) Experiments on the control of beet yellows virus in sugar-beet seed crops by insecticidal sprays. *Ann. appl. Biol.* 40, 60–78.

HULL R. & K.W. JAGGARD (1971) Recent developments in the establishment of sugar-beet stands. *Field Crop Abstracts* 24, 381–90.

JONES F.G.W. & R.A. DUNNING (1969) Sugar Beet Pests. Bull. No. 162 of Min. of Agric. Fish. & Fd., H.M.S.O.

VON STEUDEL W. (1971) Entwicklung und Notwendigkeit von Schadvoraussagen und Spritzwarnsystemen bei der Zuckerrübe, speziell bei Viruskrankheiten. *Zucker* 24, 465–70.

WATSON M., R. HULL, B. HAMLYN & J.W. BLENCOWE (1951) The spread of beet yellows and beet mosaic viruses in the sugar beet crop. I. Field observations on the virus diseases of sugar beet and their vectors, *Myzus persicae* Sulz., and *Aphis fabae*, Koch. *Ann. appl. Biol.* 38, 743–58.

WATSON MARION A., D.J. WATSON & R. HULL (1946) Factors affecting the loss of yield of sugar beet caused by beet yellows virus. *J. agric. Sci., Camb.* 36, 151–66.

Role of biology in the control of pests and diseases of vegetable crops

G. A. WHEATLEY *National Vegetable Research Station, Wellesbourne, Warwick*

Before discussing contributions that biology can make and is making to the solution of pest and disease control problems encountered in the production of outdoor vegetable crops in the U.K., I should like to dispel any illusions that the use of pesticides for this purpose has been a failure. Viewed as a whole, their use has been very successful and they now play an essential role in present-day vegetable culture, even though they have short-comings and have suffered temporary set-backs.

There are difficulties with insecticides at the present time (Gair 1971) but these are being aggravated by other contributory factors (Wheatley 1971a). Only artificial curtailment of their use will prevent insecticides continuing as the main method of pest control for at least the next 15 to 20 years while new methods are being devised and developed. There is as yet no evidence that the new methods at present envisaged will be more durable or more appropriate than insecticides for protecting outdoor vegetables from pests in intensive systems of culture.

The control of fungal diseases of crop plants has entered a phase where increasing reliance is being placed on chemicals, and the specialists concerned must quickly become familiar with the range of likely problems and side-effects now well-exemplified by the histories of insecticide and herbicide usage. History should not be allowed to repeat itself with a succession of problems analogous to those that have beset the applied entomologist during the past 20 years. The consequences of over-dependence on chemicals at the expense of soundly-based biological principles are now fully evident.

Present difficulties with pest and disease control in outdoor vegetable crops are not all attributable to pesticides. Rapid changes in agronomic practices have been occurring in the industry, some simply making it more difficult to use the chemicals effectively, while others may be actively encouraging pests and diseases (Clements 1971, Wheatley 1971b). In some instances the standards of pest control demanded are proving impossible to meet by chemical methods, and the biologically-based methods at present under consideration show little promise of being any more effective. There is, however, a reasonable chance that pesticides combined with the appropriate biologically-based techniques may succeed at least temporarily until more permanent methods can be developed. While there is dependence on pesticides, however, they will have to be used rationally, and situations which will foreshorten their useful life, or make them less effective, must not be allowed to develop before satisfactory alternatives are found. Blending together relevant methods to meet the high standards now being set will be one of the principal roles for applications of biology to pest and disease control in the immediate future.

The time-scale for the development of biologically-based methods of control is not likely to be greatly contracted simply by increasing effort. It is partly dictated by the annual cycles of events and there are deep political implications in the widespread application of pest/disease management systems. The biologist working on problems of outdoor field crops would make better progress if mathematical models could be developed to simulate practical events. Meanwhile there is a real risk workers will be encouraged to 'cut corners' in an attempt to make faster progress and, in doing so, they may make expensive and discouraging mistakes.

Assuming that pesticides continue to be developed along present lines, namely the production of relatively non-selective compounds with fairly wide ranges of potential usage, in what ways can biology contribute? The following seem to be the main roles:

(a) To provide ways and means of monitoring the effects of rapid changes in production methods so that agronomists can be warned promptly when adverse situations *begin* to develop.

(b) To extend and quantify knowledge of all aspects of the biology and behaviour of pest and disease organisms and of their relationships with other organisms and factors in their environment.

(c) To research and develop new methods for controlling pests and diseases.

(d) To research and develop efficient methods enabling Technical Services to provide precise advice on a very local basis for the individual farm or even a field (Gair 1974, this volume).

These roles are illustrated by describing selected examples, but they will first be set in context by a consideration of the changes influencing vegetable production.

Features of modern vegetable production

Production systems are largely determined by economic factors deriving from social and political influences as well as by agronomic considerations. Individual methods for pest and disease control have to fit reasonably conveniently within this framework of restraints to be acceptable, as is evident from Gair (1974, this volume). The situation is not static, as Strickland (1974, this volume) points out in discussing the principal trends in British farming. These trends have important influences on the vegetable industry and affect its outlook towards pest and disease control.

The output value of vegetables in England and Wales is at present about £80M, if potatoes, tomatoes, mushrooms, cucumbers and rhubarb are excluded; this mainly comprises brassicas (£36M), peas and beans (£20M), carrots, parsnips and celery (£11M) and lettuce (£7M). A major shift from horticultural to farm-scale production of vegetables has been occurring (Clements 1971). Fewer growers are now involved in vegetable production than hitherto, and they are increasingly more technologically-minded and receptive to innovation provided that it is effective, economical to apply and works reliably. Vegetable production is now a highly competitive and semi-industrialized industry. The grower has to produce the right type of crop at the right time, or in regular supply, and get economic yields to secure his income. To do this, he has developed efficient production systems and he is forward-contracting an increasing proportion of his crop (Kovachich 1969, Strickland 1974, this volume).

A careful watch needs to be kept to detect adverse influences of production trends on local ecology, not only from the agronomic viewpoint, but also for their possible impact on conservation or other environmental problems.

Some trends influencing pests and diseases

Certain trends in production systems are contrary to the best practices for combating pests and diseases. Growers are specializing in fewer crops. Sly (1969) found on average only 2·4 different vegetable crops per holding in Norfolk compared with 5·4 to 8·6 on holdings in four other counties surveyed. This means shorter rotations, placing greater emphasis on the need to avoid crops with similar susceptibilities to pests and diseases. The length and composition of rotations mainly influence the incidence of relatively static organisms such as soil-inhabiting nematodes, or diseases of which club root (*Plasmodiophora brassicae*) and violet root rot (*Helicobasidium purpureum*) are examples. For mobile or airborne organisms, the key factor is often the distances between sites of successive annual host-crops within a locality.

K

One organization with 600 ha of vegetables has rotations which include carrots, celery, redbeet, onions and potatoes, and an acute cutworm (Noctuidae) problem has developed during the past few years. The pest has apparently taken advantage of the succession of susceptible crops in seasons which were in any case favourable to it.

Allied to problems of inadequate rotations, localization of crop production also has important ecological implications. This is not a new trend. Even by 1961 about 75% of our celery was grown within 25 km of Littleport, Isle of Ely, Cambridgeshire, about half our Brussels sprouts came from within a 20 km radius of Biggleswade in Bedfordshire, more than one-quarter of the parsnip production was concentrated in Norfolk, two-thirds of the cauliflower production was confined to small coastal districts in Lincolnshire, Kent, Cornwall and Lancashire, and a similar proportion of our carrots were from Cambridgeshire, Huntingdonshire, East Norfolk and Suffolk (Anon. 1967). This concentration of cropping often (but not always) exploits the natural advantages of certain areas, for example the overwintering of carrots in deep fen soil which protects them from frost. The trend has intensified in recent years because of the need to grow certain crops near to a processing factory or packaging organization. Kovachich (1969) stated that it is not uncommon for 6,000 to 8,000 ha of a crop to be grown within 20 to 30 km of a factory, and he recognized that this may induce local pest and disease problems requiring specific solutions. Technical services will thus have to operate more effectively on a local level than hitherto.

Increased mechanization, especially the development of equipment for once-over harvests, makes uniform growth and maturation of plants essential to maximize the amount of marketable produce. Consequently freedom from random or irregular attacks by pests and diseases is important at all stages of crop development, quite apart from the need for the produce to be of a high quality.

Cultural methods

The ready acceptance of innovation effectively prevents new practices from being comprehensively investigated *before* they are used by growers, who often undertake the final development themselves. Herbicides have become the key to modern production methods, enabling the grower to discard many traditional restraints on growing systems. With effective herbicide protection, crops can be precision-sown to a stand, making it important to achieve a better-than-95% survival of seedlings. Pests such as wireworms (Elateridae), leatherjackets (Tipulidae) or cutworms (Noctuidae), then become important at such low population densities that they are difficult to detect

(Strickland 1970). Irrigation affects microclimates and moisture stress in crops and hence the susceptibilities to pests such as aphids, and to several diseases (Clements 1971). Natural or artificial barriers may protect crops from wind and so increase yields, but they also affect the local distribution of airborne pest and disease organisms (Lewis & Dibley 1970). New cultivars are being introduced for specific purposes but their susceptibilities to pests and diseases are not known relative to the older cultivars which they displace.

Plant density

By altering the plant density, a grower can produce crops to particular size specifications, a freedom gained by the availability of suitable herbicides. Carrots, for instance, may be sown in many row-spacing arrangements at plant densities ranging from about 50 to 500 carrots/m^2, each giving different patterns of foliage cover and microclimates affecting pest distribution. Way & Heathcote (1966) and Heathcote (1970), found that individual sugar-beet plants were less heavily attacked by *Aphis fabae* and showed less yellows virus at high than at low plant densities, the high densities apparently ameliorating the effects of the pest. This seems also to occur with other pests such as the cabbage root fly (*Erioischia brassicae*), or willow-carrot aphid (*Cavariella aegopodii*), but not invariably so with carrot fly (*Psila rosae*), on carrots (Hardman & Wheatley 1971). However, experiments have also revealed that the absolute population numbers of cabbage root fly, carrot fly or willow-carrot aphid can be greater at high than at low plant densities. There is thus a risk that high-density systems may induce heavier attacks on subsequent crops on nearby sites. This would automatically reduce the practical effectiveness of chemical methods of control, the efficiencies of which are essentially independent of pest-density.

There is evidence that high-density cropping also encourages certain diseases. Clements (1971) cited increasing mildew (*Peronospora parasitica*) on sprouts associated with higher water stress, increased halo blight (*Pseudomonas phaseolicola*) of beans following irrigation, increased white rot (*Sclerotium cepivorum*) of onions and increased weeds, cucumber mosaic virus and downy mildew (*Bremia lactucae*) in high-density lettuce crops. The status of both pest and disease problems can therefore be affected by changing plant densities, particularly if this also interferes with the application and hence the efficiency of pesticides, for instance applying sprays to control cabbage aphid (*Brevicoryne brassicae*) on closely-spaced Brussels sprouts.

New crops

Examples of new types of crop developed in recent years are calabrese for quick-freezing and the mini-cauliflower primarily for packaging fresh or freezing. The mini-cauliflower cropping system embodies several features typifying present-day methods. It is an ultra-high density brassica crop, commonly with 4 to 5 $\times$ 10^5 plants/ha, 10 to 40 times the conventional plant-density for brassicas. About every ten days from March until July a batch of several varieties is sown in close-row beds. If varieties with differing maturation rates are selected, one variety of one or other of the batches can be expected to be ready to harvest on any one day from June to November. Uniformity of growth is vital because there is a latitude of only $\pm$ 1 day for harvesting each crop to ensure the maximum yield of curds in the desired size-range of about 3 to 8 cm diameter. The status of pest and disease problems is likely to be different for each sowing and each cultivar may differ in its susceptibility or tolerance. To a lesser degree, other crops such as Brussels sprouts, peas, dwarf beans and lettuce may encounter similar difficulties. Clearly it is impossible to investigate all combinations or permutations of circumstances that could occur, and so it has become essential to understand the basic factors encouraging or discouraging pests and diseases in order to predict both the problems and their likely solutions.

By monitoring cropping systems, adverse influences could be detected at an early stage to warn the agronomist that advice on suitable corrective action will be needed. One objective of future research must be to provide the knowledge and the methods for this to be done by the competent, but not necessarily highly-informed, Technical Officer in the course of his many other duties.

Applications of biological studies of pests and diseases

A good understanding of the biology and behaviour of pest and disease organisms usually reveals logical ways to attack them. The realization of this in the mid-nineteenth century led to improved methods of pest and disease control. Since then biological foundations have been built up for virtually all methods of control. This includes the use of pesticides, although the layman can be excused for imagining otherwise; the underlying biological principles are not usually stated explicitly on the label prescribing how a product should be used. They are a hidden contribution.

Classically, the roles of biology have included identification of the causal organisms, elucidation of their life cycles, habits, host-ranges, modes of dispersal, migration and epidemiology, their physiology and innate behaviour, host-plant relationships and, ultimately, an understanding of the factors controlling their incidence. The applications of pest and disease biology are now complex and varied, as illustrated by the following examples.

Viruses

New virus problems are continually being recognized on vegetables. Two examples illustrate the application of different methods for reducing virus incidence.

Cucumber mosaic virus on lettuce

The importance of interdisciplinary investigations can be illustrated by reference to cucumber mosaic virus (CMV) affecting lettuce. Infected plants tend to be stunted and to have yellowish leaves, symptoms resembling those of nitrogen deficiency. Tomlinson & Carter (1969) demonstrated that CMV was present in the sap of diseased plants and subsequently showed that infecting healthy plants with the virus could produce the range of symptoms observed. The vector of this non-persistent virus, the peach-potato aphid (*Myzus persicae*), becomes infected within seconds of probing into a diseased plant. The virus therefore spreads readily, and killing the aphid with relatively slow-acting insecticides does not effectively limit the introduction of the virus into the crop, nor its subsequent spread. Several common weeds of arable land, for instance chickweed (*Stellaria media*), annual nettle (*Urtica urens*), groundsel (*Senecio vulgaris*) and scentless mayweed (*Tripleurospermum maritimum* ssp. *inodorum*) are virtually symptomless carriers of CMV (Tomlinson *et al.* 1970). Tomlinson & Carter (1970) then found that the virus was seed-transmitted in chickweed, and the seeds are now known to remain infective for at least two years in the soil. A population of 12,000,000 chickweed seeds/ha is not uncommon in arable land (Roberts & Stokes 1966); if only 1% are infected and 10% germinate in any one year, then there will be at least one infective seedling/m^2. The biological investigation showed that the solution to the problem lay not only in entomology or plant pathology, but also in weed control using pre-emergence herbicides such as chlorpropham (CIPC) or pronamide.

Rhubarb viruses

One purely biological solution to a pest or disease problem is the rejuvenation of rhubarb stocks chronically infected with several viruses that severely restrict plant vigour, and reduce yield and also the useful life of the stocks. By use of an apical meristem culture technique (Walkey 1968), healthy stocks have now been produced for distribution to growers through their co-operative organizations.

Host-plant resistance

Compared with advances in developing cultivars of cereals (Lupton 1974, this volume) or fruit trees (Knight & Alston 1974, this volume) resistant to pests and/or diseases, progress with this approach for protecting vegetable crops has been slow. Host-plant resistance has not yet received the detailed attention that it merits, and experiences to date have shown clearly how complicated interactions of extraneous factors can affect the usefulness of the end-product, the resistant cultivar, and effectively negate research.

Lettuce resistant to downy mildew and lettuce root aphid

Outdoor summer lettuce suffers from downy mildew (*Bremia lactucae*) and it is also sporadically attacked by the lettuce root aphid (*Pemphigus bursarius*). By 1932, at least five races of downy mildew were recognized in California and a plant breeding programme was begun to develop resistant cultivars (Whitaker *et al.* 1958). Several lines from this programme survived a heavy attack by lettuce root aphid in a variety trial at Wellesbourne in 1955 (Dunn 1956) and were subsequently proved to be highly resistant, especially line 45634M (Dunn 1960). Meanwhile Watts & George (1964) used line 45634M to develop the downy mildew resistant cultivars Avoncrisp and Avondefiance which were released to growers in 1966. Independently, another mildew-resistant cultivar, Mildura (Tozer Ltd.), had been produced, and all three cultivars were included in a trial heavily attacked by the root aphid in 1967 at the National Institute of Agricultural Botany, Cambridge. Avoncrisp and Avondefiance, but not Mildura, survived, showing that the root aphid resistance of 45634M had been retained in the Avon-cultivars without deliberate selection for this character, a useful bonus for the plant breeder. Several interrelated strains of downy mildew then began to be encountered in the U.K. and in Holland (Channon & Higginson 1971), and there is now no suitable butterhead-type lettuce to resist both downy mildew strain W.2 and the lettuce root aphid (Table 1).

The full advantages of having root aphid and downy mildew resistance in the same cultivars have thus been quickly lost by the development of local races of mildew, illustrating the need to consider pest and disease resistance problems simultaneously and not independently. For this approach to pest and disease control to be a lasting success, a continuing technological programme to develop resistant cultivars is necessary, a practice adopted for

Table 1. Resistance of lettuce cultivars to races of downy mildew and to lettuce root aphid. R = resistant. S = susceptible

Cultivar	Type	Race of mildew*				Lettuce root aphid
		W.4→	*W.1*→	*W.2*	*W.3*	
Avondefiance	Butterhead	R	R	S	S	R
Avoncrisp	Crisp	S	R	R	S	R
Mildura	Butterhead	S	S	S	R	S

* Mildew race W.4 is probably the precursor of W.1, which is similarly able to give rise to W.2. Race W.3 seems to be a separate strain.

controlling Hessian fly on cereals in California (Holcomb 1970). A sufficient range of cultivars of the required types has to be developed and maintained, so that they can be changed, preferably as a planned strategy, before they fail completely.

Resistance of parsnips to canker

A different saga in the development of disease-resistant cultivars is illustrated by the parsnip cv. Avonresister which is resistant to canker, caused by at least two fungi, *Itersonilia pastinacae* and *Phoma* spp. (Channon 1963). Avonresister is a selection from an Offenham-type parsnip and its resistance to canker seems to depend on its relatively small, compact roots being less prone to splitting than those of most other cultivars. Unfortunately, its usefulness is limited largely by lack of aesthetic appeal. It is too distinctly 'shouldered', has a yellowish skin and consequently tends not to be favoured by customers in retail chain-stores, an outlet for which it would otherwise seem to be suitable. Resistance must therefore be combined with acceptable market qualities, whether technical or aesthetic.

Present research

Present research on biological aspects of pest and disease control appears to have become centred on the interrelations between the host-plant and the parasitic organisms and on detailed investigations of their biology and behaviour, in the expectation that further elucidation will suggest novel methods for dealing with the problems.

For many years both plant pathologists and plant breeders have been aware of the value of considering disease resistance either as a primary or as a subsidiary objective of vegetable breeding programmes. Selection for disease resistance tends to occur automatically since diseased plants compare unfavourably with healthy plants and are therefore discarded. This process has undoubtedly contributed to a measure of disease-resistance in some present-day cultivars.

There is also a long, well-documented history of development of disease resistance in field crops, of which resistance of potato cultivars to wart disease (*Synchytrium endobioticum*) is one of the best known and most successful examples. Studies have been undertaken to develop peas resistant to *Fusarium* wilt and brassicas resistant to club root disease. At the present time selected lines of outdoor tomatoes resistant to *Didymella* disease are nearing final trials, and the development of spinach resistant to cucumber mosaic virus is well advanced. Experience with disease-resistant cultivars is therefore considerable and reveals many possible pitfalls, as already indicated, a warning against over-confidence in the outlook for resistance against insect pests.

In the case of insect pests, however, insecticides have generally been used to preserve plant breeding material, so preventing inadvertent selection for resistance. The overall success of insecticides has also effectively discouraged interest in host-plant resistance, and only in recent years have its possibilities been seriously considered for controlling insect pests of vegetables in the U.K.

Radcliffe & Chapman (1966a) observed that the relative resistance of cabbage cultivars to the cabbage aphid (*Brevicoryne brassicae*) in Wisconsin was determined more by the suitability of the host-plant than by any preferential selection of plants by the aphids. In the U.K., Dunn & Kempton (1971) showed that some Brussels sprout cultivars both tolerated aphid infestations better than others, and also recovered more rapidly when the aphid populations declined during the autumn. By vegetatively propagating clones from selected aphid-resistant sprouts, Dunn & Kempton (personal communication) have shown that both antibiosis and non-preference contribute to the resistance and that biotypes of the aphid occur in different localities. Seven have been discovered in England to date and none of the selected sprout clones are resistant to all of the biotypes. The benefits accruing from the development of a cabbage aphid-resistant cultivar may thus be limited unless one can be found with universal resistance.

Doane & Chapman (1962) reported that the cabbage root fly laid eggs on rutabagas (swedes) and turnips in preference to radish or mustard, cauliflower being the least preferred of the crops tested. Differences in the susceptibilities of cultivars of cruciferous crops to cabbage root fly and several other pests were subsequently recorded (Radcliffe & Chapman 1966b), the relative resistances to different pest species apparently being independent. Antibiosis in cabbage cultivars was considered by Radcliffe & Chapman (1966a) to be more important than host-selection, but Coaker (1970) found egg-laying preferences for strains of rape originating from parents with high as compared with low thioglucoside contents. Studies at Wellesbourne (Ellis & Hardman 1972) have confirmed that Brussels sprout, cabbage and radish cultivars differ in their average attractiveness to cabbage root fly for egg-laying (Table 2) and have shown that there are large differences in the

Table 2. Differences in the distribution of egg-laying preference of cabbage root fly for radish cultivars in the laboratory (Ellis & Hardman, unpublished)

	Frequency (%) of plants with			
Eggs/plant	< 10	11–50	51–250	> 250
Cultivar				
China Rose	0	0	95	5
French Breakfast	0	11	63	26
Long White Icicle	15	25	35	25
Improved French Breakfast	0	50	50	0
Crimson French Breakfast	15	40	45	0

attractiveness of individual plants within cultivars of these crops, and also of cauliflower (Table 3). The existence of differences in the resistances of

Table 3. Consistency of differences in the egg-laying preferences of cabbage root fly for individual cauliflower plants (cv. Early Mechels) grown outdoors, Wellesbourne, 1971 (Ellis & Hardman, unpublished)

Date	Plant A	B	C	D	Total
	Numbers of eggs/plant				
17 May	0	8	11	22	41
19 May	0	8	24	8	40
21 May	5	0	16	23	44
24 May	0	5	7	53	65
26 May	1	3	11	29	44
28 May	6	6	28	31	71
Total	12	30	97	166	305
Mean	2	5	16	28	51

cruciferous crop-plants to cabbage root fly, due to both host non-preference and antibiosis, has therefore been established.

At least twofold differences have now been found in the degree to which carrot cultivars are attacked by the carrot fly (Hardman *et al.* 1971) but intrinsic differences in susceptibilities are confounded with the shelter-effects of different amounts of foliage in experiments where a choice is offered. Although it is arguable whether such results can be extrapolated to a field-scale situation where no choice exists, Gair (private communication) has observed different levels of carrot fly attack among three cultivars growing in the same field. Cv. Danvers, which characteristically produces a large amount of foliage, was the most heavily attacked. Effects observed in small-plot experiments may therefore still occur when the pest has little or no choice of host-crop. Even the modest twofold differences in susceptibility observed so far would be important, because of the present difficulty in controlling carrot fly with insecticides to the required standards.

The results of more than a century of observation and study of the cabbage root fly were reviewed recently by Coaker & Finch (1971), and a major project has been in progress at Wellesbourne since 1962 to investigate the biology and behaviour of the adult fly with a view to devising new methods for suppressing its populations and thereby reducing the damage which it causes to cruciferous crops. The study has shown that the male flies mate at any time after they are two days old. They tend to aggregate near feeding sites, to disperse relatively slowly and to be attracted to traps only over distances of about 1 m. Females mate only once, when they are about 4 days old, usually at or near to the emergence site. They have to feed on carbohydrate to mature their first batch of 40–60 eggs and, although they can lay several more batches, few get protein to enable them to do so (Finch 1971). Immature females tend to remain with the males, but once they are about six days old they begin to move into the crop to oviposit, returning periodically to the food sources. Gravid females will orientate upwind towards brassica crops at least 30 m away (Hawkes 1971) and some females may move 1 km or further from their site of emergence (Finch & Skinner 1972).

Three possible new ways of limiting cabbage root fly attacks on brassicas have now been explored (Finch & Skinner 1971, 1972). The first, removing hedgerow plants near a brassica crop, to eliminate food sources for the adult fly, failed to reduce egg-laying. The second, using baited lures containing a chemosterilant, was effective in large field cages in which 94 to 96% of the eggs laid were sterile, but not in the open field where only 16 to 31% egg-sterility could be obtained. In both instances the lack of success was attributable to the immigration of fertile females into the experimental areas. During 1971, Wallbank (1972) discovered that allylisothiocyanate was an effective attractant outdoors, despite a negative report by Muller (1971), and Finch & Skinner (1972) have subsequently been able to trap flies in sufficiently large numbers to depress local populations. This third approach seems promising and is now being further investigated.

Technical services

All aspects of production costs for vegetables are now coming under scrutiny, and this is creating an incentive to optimize expenditure on pest and disease control measures and to eliminate unnecessary operations. To achieve this, accurate quantitative advice will be needed.

The need to set technical objectives for individual pest or disease/crop/market problems has been discussed by Wheatley & Coaker (1969) who proposed a simple model whereby pest damage could be specified for this purpose. Well-defined objectives would enable pest and disease control to be optimized, as well as providing targets to be achieved when developing new methods. For example, Zink *et al.* (1956) suggested that lettuce seed should contain $< 0{\cdot}1\%$ of seeds infected with lettuce mosaic virus to enable the disease to be satisfactorily controlled in the field. This is not easily achieved, and up to 5% may be infected in commercial seed stocks (Tomlinson 1962). A Brussels sprout crop would normally be rejected for quick freezing if more than 0·4% of the sprouts contained cabbage root fly maggots. This is one of the most critical requirements for pest control at the present time, and it implies that the 70–80% efficiency of recommended trichlorphon sprays (Coaker 1967, Coaker & Ensor 1967) would not be sufficiently effective if more than 2% of the buttons were attacked. In contrast, early summer cauliflowers can tolerate up to about 30% of the root system being damaged by the time of harvest, provided they are well protected during their first few weeks of growth (Coaker 1969, Rolfe 1969). Wheatley (1969) suggested that 10% of carrots damaged by carrot fly larvae should normally be acceptable for the fresh market, whereas a tolerance of 2% would probably be necessary in a crop destined for canning, and this would normally pose grading difficulties. Studies of carrot fly infestations have enabled the efficiency of control measures to be estimated (Wheatley 1969), and also the relationship between this efficiency, the level of attack and residual damage after treatment to be determined (Wheatley & Coaker 1969, Wheatley 1971b), so that requirements can be specified unambiguously.

Damage assessment studies to determine economic injury levels or levels to which pests can be permitted to develop before control measures are applied are typified by the work of the Conference of Advisory Entomologists (1957), which objectively assessed the importance of cabbage aphid infestations. There is scope for developing simple sampling procedures to reduce the work burden of such undertakings, and also to facilitate surveys aimed at revealing what is happening in practice. Feed-back of this information is important to aid management of applied research, as shown by recent limited surveys of carrot fly infestations and control measures on carrots, and of cabbage root fly damage on brassicas. In the Vale of Evesham, for instance (Dennis 1971, personal communication), of 62 brassica fields sampled in 1971, 22 had not been treated with insecticide for control of cabbage root fly, and of these 15 had yielded apparently satisfactory crops. More than half of the summer brassica crops visited were growing either on land which had carried a cruciferous crop during the previous year or in an adjacent field.

The ability to forecast the time and intensity of pest and disease incidence will be of increasing importance in the future. The intensity of pest attacks cannot yet be predicted reliably, but attempts to identify the chief factors dynamically regulating populations may assist in this way. Factors operating on *Aphis fabae* populations on their summer host crops seem to be the dominant regulatoıy agencies affecting this aphid, rather than loss of over-wintering eggs on spindle, *Euonymus europaeus* (Way & Banks 1964). The two-year cycle of abundance of this aphid seems to be linked to activities of natural enemies on spindle and on the summer hosts interacting with weather conditions. Potential outbreak years should be predicatable at least one year in advance, but the widespread use of aphicides may be masking this by interfering with the normal impact of natural enemies and consequently the basic regulatory mechanisms (Way & Banks 1968).

Attempts to forecast the beginning of particular pest activities have also met with only mixed success in the variable climate of the British Isles. Wheatley & Dunn (1962) reported that emergence of the pea moth (*Laspeyresia nigricana*) could be predicted to within about ± 3 days by measurements of soil temperature. However, this was not sufficiently precise for practical purposes, and the method adopted in practice is to sweep moths or to observe when the first eggs are laid on pea crops. Coaker & Wright (1963) were able to predict emergence of the cabbage root fly adults during April and early May in the centre of England by computing day-degrees above 5·6° C from meteorological-screen temperatures. Depending on subsequent weather conditions, the first eggs would be laid 5 or 6 days later. They had also observed that this event was closely correlated with the appearance of the first flowers on hedge parsley (*Anthriscus sylvestris*), a very useful phenological guide. ADAS entomologists of Eastern Region issue warnings of the appearance of pests such as the cabbage root fly, carrot fly, pea moth, pea midge and noctuid (cutworm) moths on a sub-regional basis, and similar services are provided by other ADAS Regions. Better methods of forecasting are needed, but they must be simple and not be wasteful of labour.

References

ANON. (1967) *Horticulture in Britain*. Part 1: Vegetables. Min. of Agric. Fish. & Fd., H.M.S.O., p. 45.

CHANNON A.G. (1963) Studies on parsnip canker. I: The causes of the diseases. *Ann. appl. Biol.* 51, 1–15.

CHANNON A.G. & Y. HIGGINSON (1971) Studies on a further race (W.4) of *Bremia lactucae* Regel. *Ann. appl. Biol.* 68, 185–92.

CLEMENTS R.F. (1971) Developments in husbandry techniques in crop production (horticultural crops). *Proc. 6th Br. Insectic. Fungic. Conf.* 3, 637–42.

COAKER T.H. (1967) Insecticidal control of cabbage root fly (*Erioischia brassicae* (Bouché)) in the axillary buds of Brussels sprouts. *Ann. appl. Biol.* 59, 339–47.

COAKER T.H. (1969) Plant tolerance to cabbage root fly damage. *Rep. natn. Veg. Res. Stn* for 1968, p. 73.

COAKER T.H. (1970) Host-plant preferences by the cabbage root fly. *Rep. natn. Veg. Res. Stn* for 1969, pp. 97–8.

COAKER T.H. & H.L. ENSOR (1967) Control of cabbage root fly attack on Brussels sprout buttons. *Proc. 4th Br. Insectic. Fungic. Conf.* 1, 226–8.

COAKER T.H. & S. FINCH (1971) The cabbage root fly. *Rep. natn. Veg. Res. Stn* for 1970, pp. 23–42.

COAKER T.H. & D.W. WRIGHT (1963) The influence of temperature on the emergence of the cabbage root fly (*Erioischia brassicae* (Bouché)) from overwintering pupae. *Ann. appl. Biol.* 52, 337–43.

CONFERENCE OF ADVISORY ENTOMOLOGISTS (1957) Cabbage aphid assessment and damage in England and Wales, 1946–55. (Summarized by A. H. Strickland.) *Pl. Path.* 6, 1–9.

DOANE J.F. & R.K. CHAPMAN (1962) Oviposition preference of the cabbage maggot *Hylemya brassicae* (Bouché). *J. econ. Ent.* 55, 137–8.

DUNN J.A. (1956) Lettuce root aphid. *Rep. natn. Veg. Res. Stn* for 1955, pp. 48–9.

DUNN J.A. (1960) Varietal resistance of lettuce to attack by the lettuce root aphid, *Pemphigus bursarius* (L.). *Ann. appl. Biol.* 48, 764–70.

DUNN J.A. & D.P.H. KEMPTON (1971) Difference in susceptibility to attack by *Brevicoryne brassicae* on Brussels sprouts. *Ann. appl. Biol.* 68, 121–34.

ELLIS P.R. & J.A. HARDMAN (1972) Resistance of brassicas and radish to cabbage root fly. *Rep. natn. Veg. Res. Stn* for 1971, pp. 72–3.

FINCH S. (1971) The fecundity of the cabbage root fly *Erioischia brassicae* (Bouché) under field conditions. *Entomologia exp. appl.* 14, 147–60.

FINCH S. & G. SKINNER (1971) Studies on the adult cabbage root fly: removal of adult feeding sites: chemosterilisation. *Rep. natn. Veg. Res. Stn* for 1970, pp. 95–6.

FINCH S. & G. SKINNER (1972) Studies on the adult cabbage root fly: alternative methods of population control. *Rep. natn. Veg. Res. Stn* for 1971, pp. 70–1.

GAIR R. (1971) Organochlorine alternatives—a review of the present position in the U.K. *Proc. 6th Br. Insectic. Fungic. Conf.* 3, 765–70.

GAIR R. (1974) The future of biological control in Britain: a grower's view of the short term. In *Biology in Pest and Disease Control*, eds. D. Price Jones and M.E. Solomon, 315–20. Blackwell, Oxford.

HARDMAN J.A. & G.A. WHEATLEY (1971) Observations on carrot fly infestations: carrot densities and levels of infestation. *Rep. natn. Veg. Res. Stn* for 1970, p. 99.

HARDMAN J.A., G.A. WHEATLEY & P.R. ELLIS (1971) Observations on carrot fly infestations: susceptibility of carrot varieties to attack. *Rep. natn. Veg. Res. Stn* for 1970, pp. 99–100.

HAWKES C. (1971) Behaviour of the adult cabbage root fly: field dispersal and behaviour. *Rep. natn. Veg. Res. Stn* for 1970, p. 89.

HEATHCOTE G.D. (1970) Effect of plant spacing and time of sowing of sugar beet on aphid infestation and spread of virus yellows. *Pl. Path.* 19, 32–9.

HOLCOMB R.W. (1970) Insect control: alternatives to the use of conventional pesticides. *Science* 168, 456–8.

KOVACHICH W.G. (1969) Future developments in growing crops for processors. In *Technological Economics of Crop Protection and Pest Control. S.C.I. Monogr. No. 36*, pp. 665–70.

KNIGHT R.L. & F.H. ALSTON (1974) Pest resistance in fruit plant breeding. In *Biology in Pest and Disease Control*, eds. D. Price Jones and M.E. Solomon, 73–86. Blackwell, Oxford.

LEWIS T. & G.C. DIBLEY (1970) Air movement near windbreaks and a hypothesis of the mechanism of the accumulation of airborne insects. *Ann. appl. Biol.* 66, 477–84.

LUPTON F.G.H. (1974) Plant breeding for disease resistance. In *Biology in Pest and Disease Control*, eds. D. Price Jones and M.E. Solomon, 87–96. Blackwell, Oxford.

MULLER H.P. (1971) Das Autozidverfahren und sein Erfolgsaussichten bei der Bekämpfung der Kleinen Kohlfliege (*Phorbia brassicae*) Bouché. *Z. angew. Ent.* 67, 119–33.

RADCLIFFE E.B. & R.K. CHAPMAN (1966a) Plant resistance to insect attack in commercial varieties. *J. econ. Ent.* 59, 116–20.

RADCLIFFE E.B. & R.K. CHAPMAN (1966b) Varietal resistance to insect attack in various cruciferous crops. *J. econ. Ent.* 59, 120–5.

ROBERTS H.A. & F.G. STOKES (1966) Studies on the weeds of vegetable crops. VI. Seed populations of soil under commercial cropping. *J. appl. Ecol.* 3, 181–90.

ROLFE S.W.H. (1969) Co-ordinated insecticide evaluation for cabbage root fly control. *Proc. 5th Br. Insectic. Fungic. Conf.* 1, 238–43.

SLY J.M.A. (1969) Insecticide usage on vegetable crops. *Proc. 5th Br. Insectic. Fungic. Conf.* 3, 657–64.

STRICKLAND A.H. (1970) Some economic principles of pest management. In *Concepts of Pest Management*, eds. R.L. Rabb and F.E. Guthrie, 30–43. University of North Carolina Press, Chapel Hill.

STRICKLAND A.H. (1974) The future of biological control in Britain: a grower's view of the long term. In *Biology in Pest and Disease Control*, eds. D. Price Jones and M.E. Solomon, 321–31. Blackwell, Oxford.

TOMLINSON J.A. (1962) Control of lettuce mosaic by the use of healthy seed. *Pl. Path.* 11, 61–4.

TOMLINSON J.A. & A.L. CARTER (1969) Lettuce viruses. *Rep. natn. Veg. Res. Stn* for 1968, p. 87.

TOMLINSON J.A. & A.L. CARTER (1970) Studies on the seed transmission of cucumber mosaic virus in chickweed (*Stellaria media*) in relation to the ecology of the virus. *Ann. appl. Biol.* 66, 381–6.

TOMLINSON J.A., A.L. CARTER, W.T. DALE & C.J. SIMPSON (1970) Weed plants as sources of cucumber mosaic virus. *Ann. appl. Biol.* 66, 11–16.

WALKEY D.G.A. (1968) The production of virus-free rhubarb by apical tip-culture. *J. hort. Sci.* 43, 283–7.

WALLBANK B.E. (1972) Studies on the adult cabbage root fly: host-plant volatiles. *Rep. natn. Veg. Res. Stn* for 1971, pp. 69–70.

WATTS L.E. & R.A.T. GEORGE (1964) Lettuce: downy mildew disease. *Rep. natn. Veg. Res. Stn* for 1963, p. 21.

WAY M.J. & C.J. BANKS (1964) Natural mortality of eggs on the black bean aphid, *Aphis fabae* Scop., on the spindle tree, *Euonymus europaeus* L. *Ann. appl. Biol.* **54**, 255–67.

WAY M.J. & BANKS C.J. (1968) Population studies on the active stages of the black bean aphid, *Aphis fabae* Scop., on its winter host, *Euonymus europaeus* L. *Ann. appl. Biol.* **62**, 177–97.

WAY M.J. & G.D. HEATHCOTE (1966) Interactions of crop density in field beans, abundance of *Aphis fabae* Scop., virus incidence and aphid control by chemicals. *Ann. appl. Biol.* **57**, 409–23.

WHEATLEY G.A. (1969) The problem of carrot fly control on carrots. *Proc. 5th Br. Insectic. Fungic. Conf.* **1**, 248–54.

WHEATLEY G.A. (1971a) Pest control in vegetables: some further limitations in insecticides for cabbage root fly and carrot fly control. *Proc. 6th Br. Insect. Fungic. Conf.* **2**, 386–95.

WHEATLEY G.A. (1971b) The role of pest control in modern vegetable production. *World Rev. Pest Control* **10**, 81–93.

WHEATLEY G.A. & T.H. COAKER (1969) Pest control objectives in relation to changing practices in agricultural crop production. In *Technological Economics of Crop Protection and Pest Control, S.C.I. Monogr. No. 36*, pp. 42–55.

WHEATLEY G.A. & J.A. DUNN (1962) The influence of diapause on the time of emergence of the pea moth *Laspeyresia nigricana* (Steph.). *Ann. appl. Biol.* **50**, 609–11.

WHITAKER T.W., G.W. BOHN, J.E. WELCH & R.G. GROGAN (1958) History and development of head lettuce resistant to downy mildew. *Proc. Amer. Soc. hort. Sci.* **72**, 410–16.

ZINK F.W., R.G. GROGAN & J.E. WELCH (1956) The effect of the percentage of seed transmission upon subsequent spread of lettuce mosaic virus. *Phytopath.* **46**, 662–4.

Progress towards biological control under glass

I. J. WYATT *Glasshouse Crops Research Institute, Littlehampton, Sussex*

The main purpose of this paper is to provide an account of what has been achieved at the Glasshouse Crops Research Institute in the development of biological pest control, and to describe the present programme of research. However, by way of introduction, and particularly in the light of other contributions to this Symposium, it is relevant to discuss the reasons for the success of recent biological control ventures under glass compared with the limited achievements on field crops in Britain.

Glasshouse conditions favourable to pests and natural enemies

Pests

The conditions of glasshouse culture are regulated to provide an optimum environment for crop growth. They, in turn, provide an ideal situation for the multiplication of many pests. This suitability can be attributed to several factors, seven of which will be discussed here.

1 Space in a glasshouse is at a premium and, therefore, dense stands of crops are grown in pure monoculture. Thus, once a pest is established, there is little (apart from the grower's attempted control measures) to prevent it from achieving its maximum reproductive potential and spreading freely through the dense crop. The ensuing rapid increase is ultimately terminated either by intraspecific competition or the collapse of the crop. With reference to Dempster & Coaker (1974, this volume) and Way (1974, this volume), this is an example of low diversity accompanied by extreme instability.

2 The equable controlled environment of a glasshouse enables any pest

adapted to such conditions to multiply without the normal outdoor hazards of variable weather.

3 The glasshouse structure prevents the ingress of predators and parasites which would normally take a heavy toll of a pest population.

4 In the same way, the pest is enclosed on the crop, and even when dispersal mechanisms come into play as a result of overpopulation, the migrants are usually forced to recolonize the crop.

5 Overwintering of pests is encouraged by the existence of hiding places in the structure, by continued heating for winter catch crops, and even by continuous year-round culture of crops such as chrysanthemums.

6 Distribution of pests is often fostered by the raising of seedlings in propagation houses before subsequent planting in growing houses. Specialist propagators of several crops, particularly chrysanthemums, distribute cuttings and seedlings throughout the country, frequently accompanied by highly adapted pests.

7 Finally, because of the high value of the crops, growers are forced to adopt intensive spray programmes. Owing to the dense stand, chemical treatments are seldom thorough, with the result that insecticide-resistant pest strains are liable to be selected.

Natural enemies

It was the appearance of such resistant strains and the consequent difficulty of controlling them that led the entomology staff of the Glasshouse Crops Research Institute to divert most of their efforts to devising practical means of biological control for the use of glasshouse growers. Once attempts were made, it was soon discovered that the same factors which favour pests in glasshouses also favour their natural enemies and so provide suitable conditions for biological control. The same seven aspects can be enumerated.

1 The dense monoculture enables natural enemies to move rapidly and search effectively for their hosts. Here again the low degree of diversity creates conditions of extreme instability and a catastrophic reduction of the pest results. Total elimination of both pest and natural enemy may occur, but this is seldom of any serious consequence, owing to the short duration of many glasshouse crops. In fact it must be concluded that biological control under glass often aims at exploiting the instability of the situation, whereas in outdoor crops the aim is usually to increase stability while maintaining the pest below economic densities.

2 The controlled conditions in a glasshouse make it possible to define and select natural enemies most suited to that environment and crop.

3 The glasshouse structure restricts the ingress of hyperparasites, which

are therefore unlikely to cause trouble on short-term crops. They have not interfered with glasshouse trials so far, but are difficult to eliminate from long-term laboratory cultures.

4 Conversely, the natural enemy is confined with its host and therefore its effects are not diminished by emigration.

5 It is possible that natural enemies may overwinter in the glasshouse structure, and certainly can continue where year-round culture is practised.

6 Natural enemies are known to be distributed with propagation material and there are plans to exploit this fact as an appropriate means of dissemination.

7 Finally, although the development of insecticide resistance has not been demonstrated in the natural enemies so far investigated, they can be functionally resistant if they spend much of their life within the body of a resistant host. On the other hand, a large proportion of research time is being devoted to the screening of chemicals in the hope that they can be used selectively against one pest or disease while leaving the biological interaction of another pest unimpaired.

Biological control on cucumbers

The first pest to be investigated at the Glasshouse Crops Research Institute, with a view to its biological control, was the glasshouse red spider mite, *Tetranychus urticae*. This species is a major pest of cucumbers (and many other crops) and has the ability to diapause in the glasshouse structure during winter and reinfest the young crop in the spring. The appearance of resistant strains of the mite on cucumbers, and the consequent inability to maintain adequate control on British nurseries, coincided with reports from Holland of a promising predatory mite from Chile (Bravenboer & Theune 1960). This species, *Phytoseiulus persimilis*, was therefore imported and immediately gave encouraging results (Hussey *et al.* 1965). A method of rearing the predator on trays in an incubator was devised (Scopes 1968) and a precise programme for commercial use was developed and defined (Hussey & Bravenboer 1971).

The system consists basically of infesting each cucumber plant with 10 to 20 spider mites, 21 days after planting, and introducing 2 predatory mites to every other plant at 30 days. By 60 days from planting the pest should be overwhelmed by predators and decline in numbers. If the procedure is carried out correctly, the leaf damage at this stage will be well below the level that causes economic loss (Hussey & Parr 1963). Ten days later spider mites are completely eliminated, but since more mites are still emerging from diapause, the remaining predators are maintained for a considerable period.

Figure 1. Tray rearing of *Phytoseiulus persimilis* and its collection in gelatin capsules for commercial application.

It is advisable to continue introducing spider mites at intervals of three weeks when diapaused mites cease to appear, and this procedure forms the basis for a modified programme in subsequent years. This system has now been adopted on almost half the acreage of cucumbers in Britain. Growers are experiencing increases in crop estimated at between 10% and 30%, not, apparently, solely due to the elimination of the pest but also due to dispensing with chemical control and its attendant phytotoxicity. Added to this, the cost of biological control is only approximately one sixth of the cost of equivalent chemical control.

Once biological control is adopted in a glasshouse, chemical control must be relaxed. This introduces a major problem in any biological control system under glass, for other pests are able to establish. Control measures must be found for each new pest, preferably on a biological basis, but if on a chemical basis they must be suitably integrated into the system to avoid interference with the biological control components of the programme. Thus for each crop a comprehensive programme must be drawn up which permits the control of at least the most likely pests and diseases.

The second pest which will almost inevitably appear on an unsprayed cucumber crop is the glasshouse whitefly, *Trialeurodes vaporariorum*. Biological control of this pest had long been practised in Britain (Speyer 1927) by the distribution of the parasite *Encarsia formosa* from the Experimental and Research Station, Cheshunt. Production was ended in 1954 following the success of synthetic insecticides. With the renewed interest in biological methods, research was recommenced in 1965 to determine precise methods of introduction. A programme has now been devised, based on principles similar to those used for spider mite, namely the prior uniform introduction of the pest itself. Ten whitefly scales are placed on every fifth cucumber plant and, two weeks later, 100 black parasitized scales are introduced at the same sites. In both cases scales are introduced on portions of tobacco leaf vaselined to the cucumber leaves. After four weeks, 15% to 20% of the infestation should be parasitized, and thereafter whiteflies should decline.

This system has proved successful, but failures have occurred when the pest was established naturally before the artificial introductions were made. There is an inherent difficulty in that both pest and parasite tend to appear in distinct generations, so, unless these are synchronized, the parasites may emerge when the whiteflies are in an unsuitable stage for parasitization. Also the parasite and whitefly are differentially favoured by temperature. If localized outbreaks should occur, they can be spot-treated with malathion sprays without unduly hindering red spider predators.

The cotton aphid *Aphis gossypii* became an unexpected problem on biologically protected cucumbers. It can increase its numbers about 10 times each week and thereby overwhelm the plants in about a month. This high reproduction rate makes control by parasites difficult. The best parasite species investigated so far has been *Aphelinus* sp. aff. *flavipes* Kurdjumov, but this increases at only about 6 times a week. It can obviously never overtake the aphid until the pest population is so dense that it becomes self-limiting. The only practicable solution found is to introduce populations of aphids more than half of which are already parasitized. Much development of this approach has yet to be done. Meanwhile the Commonwealth Institute of Biological Control in Pakistan is seeking and testing further likely parasites, and have found that *Trioxys sinensis* has a potential increase similar to that of the aphid.

An alternative control is fortunately available, for the systemic aphicide, pirimicarb, gives a good kill of *A. gossypii* without affecting the biological programmes for spider mite and whitefly.

Another pest which gives intermittent trouble is *Thrips tabaci*. This tends to invade from outside in summer. It fortunately pupates in the soil, so control can be readily achieved by drenching the beds with 0·02% gamma-BHC or 0·04% diazinon, without affecting biological controls on the plants above. This may be regarded as a form of integrated control, separating biological and chemical treatments in space.

The so-called French fly, *Tyrophagus longior*, is a mite which emerges from the straw beds and attacks the leaves soon after planting. It can similarly be controlled by a 0·01% parathion spray as long as the biological control programme is delayed for at least a week. Thus biological and chemical treatments are separated in time.

Not only do controls for all pests have to be integrated into the total system, but also those for disease. Mildew is a major problem on cucumbers and must be controlled without adversely affecting the overall programme. Many fungicides (including vaporized sulphur) have an adverse effect on *P. persimilis*. Fortunately some of the recent systemic fungicides can be used. Dimethirimol has least effect on pests and natural enemies and is, therefore, to be preferred. Benomyl, at least when applied as a soil drench, impairs the reproduction of red spider, does little harm to the predator and partially controls aphids. These attributes may be regarded as advantages, but make the biological control programme less predictable. Triarimol, while not affecting mites, appears to be potent against *A. gossypii*. In view of present difficulties with aphid control, this could prove a useful 'dual purpose' compound.

Biological control on chrysanthemums

Interest in the use of biological control on chrysanthemums was first aroused through the appearance, in another pest, of strains resistant to many insecticides. This was *Myzus persicae*, an aphid not previously regarded as a primary pest of chrysanthemums. During experiments on the resistance of certain cultivars to aphid attack, the parasite *Aphidius matricariae* was accidentally introduced and completely eliminated the aphid (Wyatt 1970). Subsequent experiments and theoretical considerations have shown the importance of understanding the threefold relationship between plant, pest and parasite. The aphid breeds so fast, on a susceptible cultivar such as Tuneful, that the parasite is unable to catch up until the aphid's increase becomes self-limiting, with resultant spoilage of the crop. On cultivars of intermediate susceptibility, such as the Princess Anne sports, aphids increase fairly rapidly but can colonize only the upper leaves. Their population density soon becomes limiting (but at a level and earliness which causes no economic loss) and parasites are able to take control. On the more resistant varieties, such as Dawn Star or Portrait, aphids breed more slowly than the parasite and are readily controlled. They would, in any case, never reach damaging densities.

A. matricariae occurs naturally on chrysanthemum nurseries and at cutting producers', despite heavy spray programmes. It is consequently distributed with cuttings, along with resistant *M. persicae*. If its effects could be fostered by the cultivation of the more resistant varieties and restricted use of chemicals, efficient biological control would probably result. Scopes (1970) suggested the deliberate introduction of highly parasitized aphid populations into boxes of rooted cuttings for distribution throughout the industry. Even the most susceptible varieties could be protected in this way. Partial control of the aphid by nicotine smokes, followed by mass release of the parasite, has also proved effective in commercial trials.

The glasshouse red spider mite also presents problems on chrysanthemums, particularly when insecticidal treatments are relaxed. Here again, several approaches are possible using *P. persimilis* as a biological agent. This predator can readily be distributed in boxes of cuttings as with *A. matricariae*. Commercial trials involving the distribution of, first, the spider mite and, later, the predator (based on the cucumber programme) have shown promise on a commercial scale. Finally, the resistance of certain cultivars to the pest could be exploited, to foster the predator. Unfortunately, resistance to the spider mite is not linked with resistance to *M. persicae*, in fact the cultivar Portrait, though most resistant to the aphid, is extremely susceptible to the mite.

The chrysanthemum leaf miner, *Phytomyza syngenesiae*, is another major pest liable to cause trouble in biologically-based control systems. Several parasites are available, and work is in progress to determine practical methods for their use. Control is readily achieved in the summer, but difficulties arise in winter conditions. Even crude methods of parasite management have given outstanding commercial control, namely the collection of mined wild *Sonchus* leaves, and the rearing of parasites in cages and subsequent release in the glasshouse.

Two other aphids are likely to be troublesome: *Brachycaudus helichrysi* and *Macrosiphoniella sanborni*. The former can cause severe distortion of the growing tip even at low densities. Fortunately a parasite, as yet unidentified, has been found which seems promising as a biological agent. No suitable parasite has yet been found for *M. sanborni*, a species which forms crowded colonies on the upper stems. However, it is frequently attacked by a fungus (*Cephalosporium* sp.) which might be manipulated so as to give adequate control. The high humidities usually required for the success of pathogenic fungi could readily be obtained in chrysanthemum culture, since a polythene blackout system is always provided for year-round production. Research on this and other insect pathogens is now being undertaken by an Insect Pathology section recently established within the Entomology Department.

The last three pests (leaf miner and two aphids) may also be amenable to restricted chemical control. The adult leaf miner rests, feeds and oviposits on the upper leaves. Both aphids feed at the plant apex until forced lower by overpopulation. All three are still susceptible to a wide range of insecticides, so it may be possible to control them by superficial spraying while leaving the biological control of *M. persicae* and spider mite to proceed unhindered on the lower leaves. Research is in progress to determine suitable insecticides. Capsids, another sporadic pest causing severe damage at low densities, could probably be dealt with in the same way.

Finally, various moths invade chrysanthemum houses sporadically, particularly when artificial lighting is used in late summer. Parasites are unlikely to be effective controls, because of the unpredictable nature of the attacks. Insecticides are also difficult to use, as larvae hide in the dense stand, or, in the case of cutworms, burrow in the soil during the day. Here the use of pathogenic bacteria and viruses presents a promising line of attack. Work has been started in the new Insect Pathology section.

Conclusions

At the Glasshouse Crops Research Institute a practicable programme for the integrated control of cucumber pests has been devised, and its basic principles are already being widely adopted commercially. Despite the wide range of pests liable to attack chrysanthemums, satisfactory controls are apparently available for all. Their integration is now under investigation. Similar studies are being extended to tomato pests.

Biological control is probably cheaper than chemical control, and does not suffer from attendant phytotoxic and human hazards. However, its adoption by the glasshouse industry requires a much greater comprehension, by the grower, of the biology and even the population dynamics of each pest. No pest can be treated individually, but must be considered in the light of all the biological processes in action in the glasshouse at the time. The greatest problem then, comes, not in the control of the pest, but in persuading and educating the grower.

References

BRAVENBOER L. & D. THEUNE (1960) The new predator of red spider *Tetranychus urticae*. *Proefstation v.d. Groenten en Fruitteelt onder Glas te Naaldwijk*. 1960, p. 139.

DEMPSTER J.P. & T.H. COAKER (1974) Diversification of crop ecosystems as a means of controlling pests. In *Biology in Pest and Disease Control*, eds. D. Price Jones and M.E. Solomon, 106–14. Blackwell, Oxford.

HUSSEY N.W. & L. BRAVENBOER (1971) Control of pests in glasshouse culture by the introduction of natural enemies. In *Biological Control*, ed. C.B. Huffaker. Plenum Press, New York.

HUSSEY N.W. & W.J. PARR (1963) The effect of glasshouse red spider mite (*Tetranychus urticae* Koch) on the yield of cucumbers. *J. Hort. Sci.* 38, 255–63.

HUSSEY N.W., W.J. PARR & H.J. GOULD (1965) Observations on the control of *Tetranychus urticae* Koch on cucumbers by the predatory mite *Phytoseiulus riegeli* Dosse. *Entomologia exp. appl.* 8, 271–81.

SCOPES N.E.A. (1968) Mass rearing of *Phytoseiulus riegeli* Dosse for use in commercial horticulture. *Pl. Path.* 17, 123–6.

SCOPES N.E.A. (1970) Control of *Myzus persicae* on year-round chrysanthemums by introducing aphids parasitized by *Aphidius matricariae* into boxes of rooted cuttings. *Ann. appl. Biol.* 66, 232–7.

SPEYER E.R. (1927) An important parasite of the greenhouse whitefly (*Trialeurodes vaporariorum* Westwood). *Bull. ent. Res.* 17, 301–8.

WAY M.J. (1974) Integrated control in Britain. In *Biology in Pest and Disease Control*, eds. D. Price Jones and M.E. Solomon, 196–208. Blackwell, Oxford.

WYATT I.J. (1970) The distribution of *Myzus persicae* (Sulz.) on year-round chrysanthemums. II. Winter season: the effect of parasitism by *Aphidius matricariae* Hal. *Ann. appl. Biol.* 65, 31–41.

Control of forest insects: there is a porpoise close behind us

D. BEVAN *Forestry Commission, Alice Holt Lodge, Farnham, Surrey*

I imagine that, like other contributors, I went looking for definitions of the Symposium's declared theme of Integrated or New biological control before starting to write. Whilst searching I found a very recent paper (Stark 1971) which had so much of what I had in mind to say myself that I should like to recommend it as a general reference to enlightened control-philosophy in forestry. I shall also refer to specific passages in it from time to time in this paper. I cannot remember which of the definitions I found I liked best, but I did register that the message of them all was that it is extremely unwise to operate from a basis of ignorance! Quite a useful platitude to have handy when dealing with folk who do not know your subject.

The rationale of traditional forest insect control

Through no virtue of their own, foresters have tended always to think in terms of ecological or management control methods. This has been a simple result of the physical size, area-scale and low value per unit area of the crops they tend. The other relevant feature of forests, of course, is their long life, and from the protection point of view this has both economic and biological importance. Costs of control action, for example, cannot necessarily be equated with the immediate yield saved, but must be compounded according to the period elapsing between the treatment and subsequent harvesting; a corollary of this is that any expenditure in the early part of a rotation is more heavily frowned upon than it would be towards the end. Biologically, crop longevity allows a diverse and complex fauna to establish itself, different at different ages of crop, but qualitatively remarkably constant from one forest of a similar type or species to another—within certain geographical limits. There will be oddities of distribution, but broadly one can produce a similar list of species from one place to another. Their populations will, however, be continually changing—in the case of the pest species, like the porpoise, they are always close behind us and they can tread on our tails.

The forester in his training spends a good deal of time on the subject he calls 'Choice of species'. It is the process by which the environmental demands of the different tree species available to him for afforestation are tried for ecological fit against the range of sites he is liable to meet. He knows that once a species has been chosen for any particular site and planted the die has been cast for half a century or more. There is then little more he can do other than see that his future tending is of a high order and perhaps, in more modern times, climb into an aeroplane to apply fertilizers or insecticidal materials once or twice in the life of the crop. The picture I wish to transmit is of a very long-lived low-value crop in which all the main decisions which will affect its biological life have been taken when the plants enter the ground.

It seems to be a fact of life that if all the right decisions have been made and a vigorous fast-growing crop—well up in the yield tables—results, then it is seldom troubled much by primary pests. A defoliator may from time to time rise in population enough to produce some visible signs of its activity, but there is likely to be little or no sign of any effect upon tree growth. The forester in charge of such a forest probably does not even know the name of the entomology staff! I would emphasize I do not include introduced pests in this picture—they may do anything and one only has to look over the Atlantic to see the depredation of gypsy moth, and the indirect tree mortality caused by *Scolytus multistriatus*, the vector of Dutch elm disease, to see what can happen through an interesting variation on the theme of diversification!

So all is well in our fast-growing forest. Stark (1971), referring to natural forests, says 'It seems to me that the tremendous economic growth of forestry in the past has blinded us to the fact that prior to our exploitation of the forests, forest pest problems were much less. In many of those areas that are still relatively undisturbed problems are usually minimal.' He goes on to say 'The majority of our pests are man made.' I have to say that I was a little surprised when I first read these statements since one understands that the great spruce budworm plagues, for instance, are more or less a function of old age and lack of vigour in ancient untended virgin forest. But one also knows the great sensitivity of many defoliators to relatively small changes in stocking density, etc. resulting from normal silvicultural operation, so I respect this statement from this voice emanating from the depths of the natural forest. Certainly Stark is right when it comes to secondary insects such as the *Scolytidae* or bark beetles, and the stump-infesting *Hylobius* weevils—man here is undoubtedly a very potent part of the ecosystem.

Now the *Scolytidae* are an interesting group. In their native territory hey are normally thought of as secondary—but some are more secondary than others! They vary in habit from a species like the ambrosia *Trypodendron lineatum*, which requires logs for breeding felled three or four months prior to its attack period, to species like the European *Dendroctonus micans* which requires a still standing slightly debilitated tree, and will not touch felled material however fresh. There are all shades of host-plant material condition which one or other species of scolytid demands for successful breeding, and a knowledge of a species' requirements can be turned to good use when the management of its populations becomes important to the protection of the forest. *Scolytidae* very often loom large as secondary pests after so-called primary attack by defoliators. Dealing with such a double infestation can make heavy demands on available basic information. One such combination which we might take as an example is the defoliator *Bupalus piniarius* (pine looper) and *Tomicus piniperda* (pine shoot beetle), and I should like to tell you how we dealt with such a combined outbreak in 1953, when the first British infestation of pine looper occurred, and compare it with our approach to the last one in 1969.

The classical defoliator/bark beetle outbreak

In 1953 in Cannock Chase, Staffordshire, about 50 hectares of 32-year-old pine were completely defoliated by pine looper and a similar surrounding area was damaged to a rather lesser degree (Crooke 1959). Pupae in the soil were sampled over the whole forest during the following winter, and, borrowing a central European figure of six pupae per square metre as the 'critical' number for a control decision, we sprayed 1,500 hectares with 1 kg of DDT per hectare. The trees, although leafless or partly so, were not dead, and it was not therefore expected that *Tomicus piniperda* attack would be important. How wrong that was—it took more than two years to clear up the mess! Much of the resulting timber-produce was blue-stained and otherwise reduced in value. Management in this case had had to cope with a sudden outbreak of a defoliator about which they had little or no knowledge, and with a bark beetle which they knew well but of whose reaction to the particular circumstances they had no experience. We were ignorant indeed!

In 1969 at Wykeham forest in Yorkshire, another outbreak of *Bupalus* involving a similar area of 40-year-old pine occurred. In fact, the outbreak slipped past the countrywide early-warning pupal-sampling system which had been set up as an annual routine since the 1953 infestation. Sampling of pupae was carried out in the autumn and winter of 1969/70; analysis of the data, based on a locally derived critical pupal density of 20 pupae/m^2, indicated that some 700 or so hectares were at risk.

It was evident that unless we were very fortunate an insecticide was going to have to be used over the whole 700 hectares. Screening tests had been carried out in the intervening 16 years, but the field is a rapidly changing one and a number of new materials had immediately to be put under test. In fact, a half to one kilogramme per hectare of DDT had proved extremely effective in the past, but for obvious reasons we wished to get away from this choice. To cut a long story short, tetrachlorvinphos was chosen and given its maiden flight in Britain on the grounds of its generally good performance in trials and very low mammalian toxicity (Scott & Brown 1973). At the rate of 0·6 kg/ha, it was generally accepted that a kill as low as 80% to 90% might have to be accepted. On the face of it, this might seem to have been rather a risk to take—leaving 10% of a population running up to 150 or so female pupae per square metre, each with an egg-laying potential of 100-plus, might seem too much like gambling. In fact we knew, from previous infestations when DDT had been used, that any pupae appearing after the operation were normally quickly 'cleaned up' by the pupal parasite *Cratichneumon nigritarius*—a species normally of significance only at low population levels (Davies 1962). A similar effect was noted with *Lygaeonematus abietinus* (Bejer Peterson 1958). Of course, if treatment had proved only palliative and further infestation threatened, then there would have been no reason why the area could not have been treated again in a future year. In fact there was a general natural collapse of the population for reasons not well understood. We think one of the few larval parasites, *Campoplex oxyacanthae*, a species from which we have learnt to expect very little in the past, came into its own on this occasion and had a significant effect. Unfortunately, it came too late to affect the control decision.

One always hopes when preparing to take control measures that they will become unnecessary through natural collapse, and one tends to delay the decision to control to the very last moment. Therefore the egg assessments which are necessary for timing of correct application also supply a useful check on the success or otherwise of oviposition. This shows a further difference from the 1953 situation, when a firm decision to control was taken on pupal counts alone. It is unfortunate, however, that it has never been found possible to develop a method of counting *Bupalus* larvae sufficiently accurately on a forest scale—one would then be counting the actual culprit on the job. Late-stage larvae can be sampled with a good deal of difficulty in a research study-plot, but the technique is too cumbersome for forest-scale use. This is particularly sad in the case of *Bupalus* since it has been shown (Klomp 1966)—and our own work supports the observation to some extent—that it is during the first and second instar that critical mortality often occurs, at least at sub-outbreak densities. These early instars have defied efforts to assess them accurately even on a research scale, and so we know little more than the fact that quite high numbers can and do disappear from unknown causes. It will be seen from Fig. 1 not only that eggs were successfully deposited but also that hatching had reached about 90% by mid-August. Spraying, as I have already indicated, was carried out and good control obtained. The changes in approach to control from 1953 to 1969 were therefore that we could take our control decision on egg assessments rather than pupae, that we could limit the area to come under control by knowing that

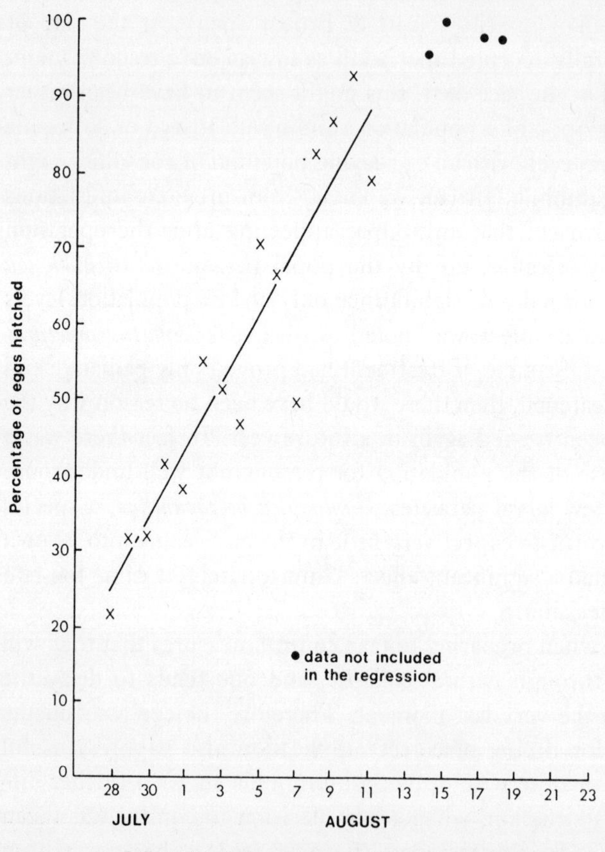

Figure 1. The progress of egg hatch during an outbreak of *Bupalus piniarius*; × indicates mean values used in the regression for timing treatment, ● post-treatment values (Scott & Brown 1973).

a far greater number of pupae per square metre were tolerable in British forests than in central European ones, that we could take advantage of a known parasite's presence to augment artificial control and, perhaps most important of all, we could use modern alternatives to DDT to provide practical control under our conditions.

The other pest species to be dealt with is the scolytid, *Tomicus piniperda*

(Bevan 1962). The life cycle of this insect is relatively simple, and all one really needs to know is that attacks for breeding purposes take place during March, April, May and beginning of June (see Fig. 2) (Davies, personal communication). There is a traditional method of controlling this pest by what is called the 'Six weeks rule'. This recommends that all material suitable for breeding in must be either removed or treated within six weeks of

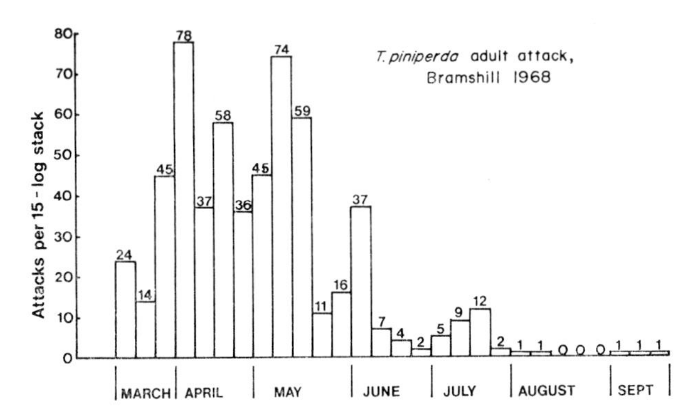

Figure 2. The pattern of attack upon logs by *Tomicus piniperda* for breeding purposes (Davies, personal communication).

felling during the whole of the spring and summer period. We can now reduce this period therefore to include only the months of March–June. If the rule is not followed, the beetle will breed in the logs and proceed to attack the crowns of the trees for maturation feeding, and then may well make further attacks for breeding purposes in the stems of weakened trees. The precise condition of material selected for breeding is critical. If stems are felled at varying dates throughout the year and then made available to the adult during its flight period (Davies, personal communication), a pattern of preference appears. Fellings made in July and August produce material with greatly reduced susceptibility (Fig. 3)—and apparently immune from all attack in Holland (Doom & Luitjes 1971). Knowledge of such a period can be very useful when fellings are carried out, for instance in nature reserves, where removal of small amounts of produce can be a practical embarrassment and application of insecticides undesirable.

So management of this beetle's populations in a normal forest, where fellings and removal are well ordered, will not normally present much trouble. A sudden arrival of 50 hectares of highly susceptible bald Scots pine is an entirely different matter, however. I think it tends to surprise even an entomological audience that there could be an endemic population—a porpoise close behind us—in an apparently healthy forest, in high enough numbers to account for the 50,000 odd trees on an area of this size. If those beetles were allowed to breed up unchecked they would then turn their attentions to the less defoliated categories and even healthy trees.

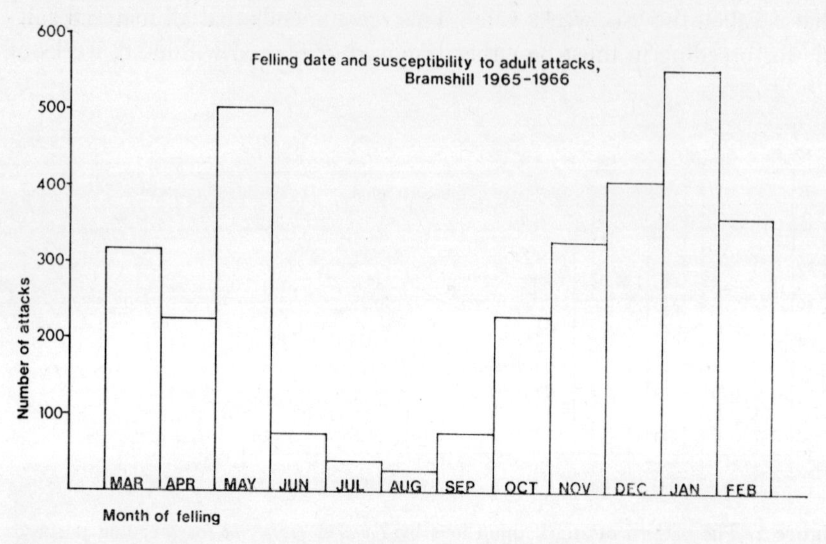

Figure 3. The relative susceptibility to breeding attacks by *Tomicus piniperda* of material felled in different months of the year (Davies, personal communication).

Tomicus is a most interesting beetle in its choice of breeding site. In the United Kingdom it is not the frightening pest it is sometimes claimed to be elsewhere, except in particular circumstances. Normally-growing healthy pine trees are well able to resist attack. One can see attempted entry into stems, and often the attacking beetle incarcerated in a blob of dried-up resin. Heavy infestations (resulting from large-scale breeding in log-stacks left in breach of the six week rule) may cause severe shoot-pruning through maturation feeding, but this does not often result in weakening of the host plant to the point of allowing breeding to take place on the stems. Let the same thing happen to a crop growing, say, in sand-dunes and you may well be in trouble. A crop completely defoliated by pine looper is even more susceptible. However, the freshly-felled log appears to be number one choice for breeding, and far more attractive even than a sick or leafless tree.

The attack of pine looper at Wykeham had thus left a legacy of trees destined to certain attack by *T. piniperda*. Flight and attack of beetles could be expected in the following April to June, and fellings were therefore started as soon as the snow permitted in February/March. It was most important to ensure that fellings were in progress throughout the attack period of the adults, and, particularly at the peak of flight, that there were plenty of fresh-felled logs lying on the ground to keep the beetles away from the damaged standing trees. Fellings were thus used as trap-logs. The ploy worked, and the surrounding forest of semi-defoliated trees not only escaped attack to any extent, but was also given a year's grace to recover and reclothe its crown more fully.

Control by management

This is the story of one infestation, but it is not complete. We are back with a choice of species, because the ground is now bare again. Thirty-two years ago the forester had no means of ameliorating these hard-pan, indurated calcareous grits of the Yorkshire dissected plateaus. Now he has the machines to break the root-impenetrable layers. In such a place, when tools were only available to scratch the surface, a pine was a reasonable choice. We may now very well turn to another, more demanding but more productive species altogether. I am not certain whether complete removal of the host plant constitutes integrated or some other kind of control!

I hope I have succeeded in indicating how, over a period of 16 years, research information was gathered and used to provide a more enlightened approach to an insect problem. It is obvious that such situations are best tackled from a basis of knowledge rather than ignorance—again the platitude —and that information on the behaviour of the pest species and its dependants, on the site and the soil, and on the crop itself, are all essential to this end. I am only sorry that knowledge of the population ecology of pine looper is so incomplete and so difficult to acquire.

Apart from any inherent instability in the character of monocultures, it is to be expected that those of plantation forestry will always be particularly vulnerable to insect plague, owing to their dependence, from inception to final felling, upon so many subjective decisions by man. Schindler (1971), referring to a projected scheme for the afforestation of two million hectares of old agricultural land in Germany, says that 'Under these conditions, nothing much can be hoped for from "natural factors" because they do not exist in farm land and we must recognize that the only alternative lies with insecticides.' Taken out of context, this sounds rather a blunt statement, but the author makes it quite clear that by the lack of natural factors he refers to the total absence of a forest condition or ecosystem—and his is a realistic statement. In the *Bupalus* example above, we are fortunate in that in the second time round we have an alternative tree species to plant, but this is not always the case. Schindler sees that when a political decision to blanket a huge area with trees has been taken, then the consequences may very well include the necessity of using insecticides. It is very often with the setting of objectives that the trouble lies, and it might always by borne in mind by those responsible for land use in the broad sense that the call for recreational areas, car parks and even housing estates could be integrated with a broader policy on, among other things, insect control. It is well known in forestry that successive infestations of certain insects tend to take place in certain susceptible areas, and these could very well be turned to other use.

The need for basic information

If I may again quote Stark (1971): 'Forest entomologists have to a considerable degree, been guilty of parochialism in their approach to the problem. They have sought explanation of the eruption (of populations) in entomological terms rather than ecological.' This is perhaps a little hard, but nevertheless has much truth in it, but I do not think this particular trouble is peculiar to forest entomologists. Certainly, because of their long life, tree crops do provide relatively untouched populations of insects convenient for entomologists and ecologists to study population dynamics—in fact forests can become playgrounds for biologists with strong mathematical leanings. Sometimes one feels that we too often become slaves to our own models and, having failed to make them work, content ourselves with polishing them! We as forest entomologists need these models, and in working order. If they can be predictive, all the better, but if they can only point to where further search might reveal the underlying cause, then that is good enough. It is probably true to say that most specialists tend to be parochial, or perhaps it is that the baton of knowledge is not always easy to pass over from one discipline to another. It is true though that, when a pest problem arises and it is decided to investigate it thoroughly, too often the next step is the finding of a place where populations are assured and infestation is as frequent as possible. In the very best cases, after 10 years a set of life-tables and a perfect working model is the product—applicable only to the n square metres chosen for study. It does not include any information on what makes that n square metres susceptible and another n square metres quite pest-free. This is a pity, because many pest populations are so evidently influenced by host-plant condition that to ignore it is to be parochial indeed. Now the host plant is something we can manipulate a good deal more easily, intelligently and permanently than the insects that feed on them. And with the help of plant physiologists and biochemists we can learn to do so better. The host plant after all is the reason for our employment!

Conclusion

The choice of a control measure to fit particular circumstances depends on the knowledge available at the time. Some information on the biology of well-known pest species often exists already or is relatively easy to acquire from short-term studies, and can be made the basis upon which many control decisions are taken. Such information indeed may be adequate for dealing with many of the insect problems that arise from management action or other identifiable primary causes. Sometimes, however, outbreaks occur of species whose epidemiology is more obscure, and a more intensive research effort is then needed. Such occasions often call for immediate palliative control measures to be taken, but the long-term solution lies in an understanding of underlying causes.

Forest areas which once suffer an outbreak of an apparently primary pest can be considered to be for all time susceptible to that species. They can be expected always to have populations of the pest species in question, to experience fluctuations in these populations, and to suffer periodic outbreak at times dependent on the character of the species and the site. In such forests it is possible to monitor population changes within a single generation and between a series of successive ones. The construction of a set of insect pest life-tables is the normal result of such work. Quite simple analysis of the tables can indicate topics—not necessarily entomological ones—which need investigation. A by-product of the tables may be a first attempt at the construction of a model representing the fluctuations in the pest populations. It is probable, however, that such a model would need a deal of feed-back from further field experience, before it could perform usefully as a predictive or diagnostic tool over a sufficiently wide range of site variables.

This is the kind of work essential to the development of more enlightened methods of insect control in ecosystems known to be sensitive and reactive to interference. Of all environments, perhaps forests arouse the greatest concern over matters of conservation, partly because a proportionately large area of the world's surface is occupied by forests, and partly because man dislikes to see changes in a crop he regards both as a source of enjoyment and as a very nearly permanent background to the landscape. Unfortunately forests produce a low financial return. Traditional economic calculations discourage the expenditure of large sums on the kinds of ecological investigations needed to aid in conservation and to ensure a responsible attitude towards the environment. This type of study needs many kinds of expertise, not least those of the specialist forester, since the questions posed are practical ones, and the answers are required for direct application in the field.

References

BEJER PETERSEN B. (1958) The Sawfly *Lygaeonematus abietinus* Christ. 2. *Forst. ForsVaes* 25, 49–61.

BEVAN D. (1962) Pine shoot beetles. *Leafl. For. Commn* 3.

CROOKE M. (1959) Insecticide control of the pine looper in Great Britain I. Aerial spraying. *Forestry* 32, 166–96.

DAVIES J.M. (1962) The pine looper moth, *Bupalus piniarius*, at Cannock Chase in 1960. *Rep. Forest. Res. for 1961*, pp. 176–82.

DOOM D. & J. LUITJES (1971) The influence of felled Scots pine on the population density of the pine shoot beetle. *Ned. BoschbTijdschr.* 43 (9), 180–91.

KLOMP H. (1966) The dynamics of a field population of the pine looper, *Bupalus piniarius*. *Advances in Ecological Research* 3, 207–305.

SCHINDLER U. (1971) Changes in the use of choice of insecticides against forest insects in Central Europe. *Proc. 6th Br. Insectic. Fungic. Conf. (1971)* 2, 463–6.

SCOTT T.M. & R.M. BROWN (1973) Insecticidal control of the pine looper in Great Britain III. Arial spraying with Tetrachlorvinphos. *Forestry* 46, 81–93.

STARK R.W. (1971) Integrated control, pest management, or protective management? *Proc. 3rd A. NEast. For. Ins. Work. Conf. 1970. U.S.D.A. For. Serv. Res. Pap. NE-194*, 111–29.

Section 4
The future of biological control in Britain

The future of biological control in Britain: a grower's view of the short term

R. GAIR *Ministry of Agriculture, Fisheries and Food, Agricultural Development and Advisory Service, Cambridge*

Possession of a few hundred square yards of back-garden does not give me the status of a *bona fide* grower. However, as an adviser in daily contact with farmers and growers, I believe that I am qualified to give a fairly accurate idea of how they—the farmers and growers—regard biological control. Even so, my own views will undoubtedly intrude into the following arguments.

A recent farming conference here at Oxford took as its theme the pressures under which British farmers are operating. As these pressures play an important part in determining the farmers' attitude towards pest control in general and biological control in particular, I think we must look at them in a little more detail.

Pressures on the farmer

The latest available figures for land values (Anon. 1955a, 1970a) (Table 1) show that, over the period 1954–69, land values have increased almost threefold over England and Wales as a whole, and by more than that in the arable area of Eastern England. The increases have no doubt continued into 1972.

Farm rents have increased by a similar factor over the same period (Anon. 1970b). I have included the figure for 1870 to show how steeply rents have climbed in the last few years.

The total numbers of registered and part-time workers engaged in farm work in 1954/55 and 1968/69 (Anon. 1959, 1969) are also shown in Table 1. The figures illustrate strikingly the declining labour force available to British agriculture.

Net farm income (NFI) is defined as the return to the farmer and his wife for their own manual labour and their management and interest on all farming capital, excluding land and buildings. The NFI in 1954/55 was £15·8 per ha (Anon. 1955b) and in 1968/69 £26·7 per ha (Anon. 1969b). Despite the apparent increase, due mainly to inflation, the NFI in absolute terms has hardly changed.

Table 1. Land values, farm rents and farm labour force in England and Wales

	1870	1954/55	1968/69
Land values (vacant possession) (£ per acre)			
England and Wales	–	80	204
Eastern Area	–	–	283
Farm rents (£ per acre)			
England and Wales	1·33	1·76	4·84
Eastern Area		2·20	7·20
Farm labour force			
Total registered workers		478,669	249,812
Total part-time workers		79,184	42,223

The rise in importance of the food-processing industry is having far-reaching effects on British agriculture. The prepacker, canner and freezer require uniform produce completely free from pest damage. Such freedom from insect and other pest damage to apples, pears, carrots, peas, Brussels sprouts and other vegetables is rarely, if ever, obtained in the produce delivered to the factory, so that a meticulous inspection process has to be installed there at great expense. In practice, the tolerance levels vary according to the supply/demand situation for each commodity. In time of glut, crops with only a small amount of pest blemish can be returned to the grower, to his financial loss. For pests of British crops, 80% control is achievable, even of such intractable pests as soil cyst nematodes. Complete control is simply not a practicable proposition.

So far, the crops affected by such stringent quality standards include only potatoes, fruit and a few vegetables. Some 38% of the 29,400 acres (11,900 ha) of maincrop carrots grown in England and Wales in 1971 were for processing, and the percentage figure is even higher for peas, French and broad beans and red beet. In contrast, only 13% of the Brussels sprout crop (totalling 42,300 acres (17,000 ha) in 1971) were grown for processing. While only a proportion of the above crops eventually find their way into the processor's hands, virtually the whole of the acreage of these crops is grown to the exacting standards of pest control demanded by the canner and freezer, and we have already seen how very high these standards can be.

The demand for high-quality produce is not limited to horticultural crops. Potatoes destined for the increasing canning, freezing, prepack and other processing markets must similarly be free of holes and lesions caused by wireworms, slugs, chafer grubs and other soil pests. In 1968/69 11% (540,000 tons) of the total home production of potatoes went for processing; by 1974/75, it is likely that 20% (950,000 tons) of the total home production will be processed. Standards set for cereals, field beans, sugar beet and other crops are getting higher—a trend which will intensify when Britain joins the European Economic Community.

Much attention has recently been focused on the need to preserve beneficial and general wildlife and to prevent environmental contamination. The British farmer is therefore asked to limit the size of his fields to a maximum of about 40 acres, to plant suitable copses, to leave plenty of flowering plants in his hedgerow and to eschew the use of such persistent organochlorine insecticides as DDT, aldrin and dieldrin. There is a price to be paid if the conservationists' demands are to be met, and right now the farmer and grower are footing the bill. The consequences of withdrawing organochlorine insecticides have recently been reviewed (Gair 1971). Apart from having to pay more for pesticides, the farmer is obliged to use materials that may be more toxic to operators and to crop plants and that may give less effective control than the organochlorines. These pressures are all economic or (as in conservation) have economic consequences for the farmer or grower. He reacts to them by lowering his unit costs of production and by supplying his market outlets with the high quality needed. The need for pest control is itself a pressure on the farming community, but, in our temperate maritime climate, one which is extremely variable. Catastrophic losses from pest attacks are much less frequent in Britain than in many other parts of the world, a point to which we will return later.

Immediate prospects for biological control

What are the prospects for biological control in such a situation? Let us first consider biological control in the somewhat narrow sense of manipulation of parasites and predators. The recent developments in biological control of glasshouse pests have been enthusiastically acclaimed by the industry, for Dr. Hussey and his colleagues at the Glasshouse Crops Research Institute have shown not only that predator control of red spider mite is cheaper than traditional chemical methods, but also that it leads to increased yields; and, as our chairman once remarked (Kennedy 1968), in doing so they 'exploded the popular idea that biological control is incompatible with unblemished high-quality produce'.

So far so good. But biological control as practised at the Glasshouse Crops Research Institute and that practised in commercial holdings are two different things, and it is already clear that some growers do not possess the necessary managerial ability to make a success of the Hussey blueprint. And there are snags to the blueprint (e.g. whitefly control) as Dr. Hussey will no doubt admit.

Even in the controlled environment of the British glasshouse, prospects for biological control are therefore still somewhat uncertain. Let us now go outside to consider pest control in British field crops. What are the prospects for biological control there? Advisory entomologists have on two occasions, in 1967 and 1971, reviewed the possibilities of using biological methods of pest control. They concluded that farmers and growers were alive to the possibilities of using such methods in integrated programmes, especially for pests such as nematodes which are often difficult to kill by chemical means alone. Advisers and their farming clients are anxious not to use chemicals indiscriminately, preferring to use selective pesticides whenever possible. Thus, if a farmer is offered the choice of two equally priced aphicides, one of which is appreciably less toxic to beneficial insects than the other, he can be relied upon to choose the former. Even if the preferred aphicide costs rather more, most farmers will still use it. But if there is a large difference in cost or effectiveness between the two aphicides, I feel that many farmers will go for the cheaper material or the one which is more effective against aphids, and ignore the greater deleterious effects which it may have on beneficial fauna. This brings us back to my question: who is going to pay the price of conservation?

The sporadic nature of pest attacks on outdoor crops militates against effective biological control in this country. Only a few pests of major economic status are with us in appreciable numbers year after year. Cyst nematodes, wheat bulb fly, carrot fly and cabbage root fly come easily to mind. Slugs, leatherjackets, cereal aphids, cutworms and caterpillars are notoriously prone to surge into prominence, then sink into comparative oblivion for varying numbers of years. The prospect for a specific parasite or predator is not very bright, and our present inability to forecast the likelihood of attacks by all but a very few pests does not help if we intend introducing the biotic agents at the start of an epidemic.

If we now widen our definition of biological control to embrace the use of resistant or tolerant plant varieties, then the prospects are quite different. The plant breeder has done a great deal in providing cereal and potato varieties resistant to nematodes, lettuce varieties resistant to aphids, fruit rootstocks resistant to woolly aphids, and so on. Provided these varieties give satisfactory yields and meet any quality demands which the market may impose, they are accepted with alacrity by both advisers and farmers. I believe that tremendous possibilities exist for the use of resistant or tolerant plant varieties in British agriculture and horticulture, and that for the control of pests (as distinct from diseases) these possibilities have hardly begun to be exploited. But plant breeding is a long-term exercise in biological control, and for the 1972 season we can hardly expect the plant breeder to make a much more significant contribution to pest control than he has done up to now.

Conclusion

In conclusion, I would claim that the British farmer or grower is basically sympathetic to the principle of biological control, for he is a conservationist at heart and derives little or no pleasure from pouring chemicals over the land. But although farming may be a way of life, it is primarily a means of earning a living. Experience with synthetic pesticides from DDT onwards has shown how quickly and effectively pest control can be achieved by chemical means, and, although most farmers would welcome cheap non-chemical methods, these simply are not available at present and may well fail to give farmers the degree of pest control required in our high-quality markets. Increasing costs of production, coupled with the need for marketing high-quality produce, demand pest control methods that are cheap, effective, easily applied and quick-acting. Biological or integrated control measures need more careful management than the average grower can provide. As the possibilities for introducing such measures into the U.K. seem limited, chemical control is unlikely to be supplanted for some time.

Acknowledgements

I thank my colleagues T.W.D. Theophilus, A. Moore and J.J. North for providing extracts of data from 'Agricultural Statistics'.

References

ANON. (1955a) *Agricultural Statistics 1954/55*. H.M.S.O., London.
ANON. (1955b) *Farm Incomes, England and Wales, 1954/55*. H.M.S.O., London.
ANON. (1969a) *Agricultural Statistics 1968–69*. H.M.S.O., London.
ANON. (1969b) *Farm Incomes, England and Wales, 1968/69*. H.M.S.O., London.
ANON. (1970a) *Agricultural Land Prices in England and Wales*. Agricultural Land Service Technical Report 20/1. H.M.S.O., London.
ANON. (1970b) *Farm Rents*. Agricultural Land Service Technical Report 19/2. H.M.S.O., London.

GAIR R. (1971) Organochlorine alternatives—a review of the present position in the United Kingdom. *Proc. 6th Br. Insectic. Fungic. Conf.* 3, 765–70.

KENNEDY J.S. (1968) The motivation of integrated control. *In* Integrated Pest Control–papers presented at a Conference of Advisory Entomologists on 11–12 April 1967, ed. D.W. Empson. *J. appl. Ecol.* 5, 489–516.

The future of biological control in Britain: a grower's view of the long term

A. H. STRICKLAND *Ministry of Agriculture, Fisheries and Food, Plant Pathology Laboratory, Harpenden, Hertfordshire*

It is easy enough to talk about trends, but sometimes not so easy to show that they exist. So, to avoid misunderstandings and to provide a quantitative basis for subsequent discussion, it is worth illustrating the main trends in British farming before trying to project them into the future. It is only within about the last decade that the Ministry's Statistical and Economics Services have been collecting complex data in sufficient detail to enable certain trends to be clearly illustrated, and it is right to say that this paper could never have been written without the unstinted help of colleagues in these Services.

New methods of crop production

One might almost say new methods of mass-production are being forced on growers by the high cost of labour. The accent is on mechanization, leading to techniques like tape- or pellet-seeding to a stand of crops like vegetables and sugar beet. Direct drilling of cereals, and even of some vegetables, is now possible and chemical weeding is universal. Much work is being done on crop spacing, leading, for example, to flat-bed planting of potatoes, and the sowing of carrots at 25 mm square spacing; also on commercial production of strawberries in polythene tunnels, and on factory-farming of livestock. The move towards mechanization (Table 1) is particularly strong on general cropping farms and on horticultural enterprises, with emphasis on mechanical harvesting, and this in turn is leading to calls for crop cultivars which mature evenly and which can be harvested in one operation.

In the last few years, nearly a quarter of the regular farm labour has left the land, and the capital tied up in new machines, and in more versatile models of long-established machines, has risen proportionately on those farms growing a range of crops. The rise has not been quite as great on farms specializing in cereals, partly because mechanization programmes were more advanced by the mid 1960s. The next big decline in labour might well follow

Table 1. Decline in numbers of regular farm workers (all farms), and increase in average machinery valuations (general cropping, and mainly cereals, farms), England and Wales, 1965–70

Year	Decline in numbers of regular workers, March census (1965 = 100)	Index numbers of average machinery valuations at depreciated cost (1965 = 100)	
		General cropping	Mainly cereals
1966	− 5	+ 7	+ 7
1967	− 9	+ 13	+ 10
1968	− 15	+ 18	+ 10
1969	− 20	+ 22	+ 11
1970	− 24	+ 23	+ 12

in 1975, when the Equal Pay Act, 1970, becomes fully operative and there will be an even greater need than at present to use machines in horticulture.

Growth of specialized farming systems

Specialization has become almost an across-the-board phenomenon, partly because of wage pressures by highly skilled workers, and partly because farming is now so complex that many growers find it pays to master a few techniques and concentrate their production accordingly. Table 2 illustrates the extent to which specialization has increased in recent years, while the generalists have declined. There is a clear trend (supported by other data not shown here for reasons of space) towards specialization in livestock and cereal production, and a decline in horticulture which is not as big as implied by the table: in practice, modern pest, disease and weed control, and the extension of mechanical cultivation, have led to movement of many horticultural crops from specialist holdings to arable farm production; in 1969/70 U.K. horticultural production was 14,000 hectares greater than in 1960/64, mostly farm-grown peas, beans and carrots for the processing industry (Anon. 1971). Other horticultural enterprises have declined in a land-use sense, but the drop of about 4,000 hectares has not affected total production, because in nearly every case yields per unit area have increased from use of modern methods. Although there may be a further decline in certain horticultural activities if entry to the European Economic Community (EEC) leads to a lowering of tariff barriers, there is always likely to be a market for home-produced 'quality' fruit and vegetables, as well as demands from the food processing industry for field-fresh produce.

Table 2. Changes in numbers of full-time holdings by farm type, England and Wales

Type of holding		Number ('000) in 1963	Change ('000) by 1970	Change in per cent of total SMD,* 1964–70
Specialist	Dairy	29·7	+ 0·2	+ 3·1
General	Mainly dairy	34·9	− 14·9	− 3·6
Specialist	Cattle rearing	2·9	+ 3·1	+ 0·4
General	Mixed livestock	14·2	− 1·8	
Specialist	Mainly cereals	6·3	+ 2·8	+ 3·0
General	General cropping	18·5	− 3·9	− 0·5
Specialist	Mainly fruit or vegetables	4·8	− 1·2	− 0·4
General	Mixed horticulture	11·2	− 1·7	− 0·7

* Standard man days. Accepted estimates of labour content of certain operations or enterprises.

Growth of large farm units

It is as easy to practise bad husbandry on a large farm as on a small one, and the mere expansion of large farms may reflect little else than a speculative interest in land values. Table 3 illustrates the extent of change in the past decade: in the smallest size group the numbers of holdings and the total land area used have declined; and, although the land area occupied by the next size group has declined, the proportion of holdings has increased. In the biggest size groups there has been substantial growth, and these now account for nearly 40% of our crops-plus-grass area, compared with 28% in 1960. But this is only part of the story: it is effective use of the enlarged holdings

Table 3. Changes in farm structure: growth of large 'crops + grass' units, England and Wales, 1960–70

Size group (hectares)	Per cent of area in 1960	Per cent change by 1970	Per cent of holdings in 1960	Per cent change by 1970
− 40	29·1	− 6·5	76·8	− 4·7
− 120	42·7	− 4·4	19·0	+ 2·2
− 280	20·5	+ 4·4	3·7	+ 1·8
280 +	7·7	+ 6·5	0·5	+ 0·7

that is important. Table 4 gives a representative selection of data on the growth that has taken place within the larger farm units, and relates to holdings where the indicated enterprise is the biggest on the holding in standard man-day (SMD) terms. For example, in 1960 in England and Wales about 17% of the dairy cows were on holdings with 50–99 cows, and 4% were on bigger holdings; but by 1970 nearly half the dairy cows were concentrated on these increasingly specialized holdings which were expanding at a rate of 5·6% per annum compound. Growth by the bigger cereal growers was

Table 4. Increases in size of major farm enterprises, England and Wales, 1960–70

Enterprise	Growth of larger units			Average annual increase in size of major enterprises (% per annum compound)
	Size group	Percentage of total enterprise in year: 1960	1970	
Dairy cows	50–99 cows	17·2	32·7	5·6
	100 and over	3·9	14·8	
Beef cattle	50–99	11·4	18·2	4·1
	100 and over	3·7	9·1	
Wheat	40–120 ha	21·3	37·2	7·4
	Over 120 ha	4·8	16·3	
Barley	40–120 ha	31·2	40·1	5·7
	Over 120 ha	9·4	19·2	
Maincrop potatoes	20–40 ha	13·3	18·0	6·0
	Over 40 ha	5·4	13·2	

similar for barley, where 60% of the crop is now grown on the larger holdings, but greater for wheat, where over half the crop is in the hands of relatively few big growers. But these compound growth rates obviously cannot continue indefinitely, and, even if cereal enterprises continue to expand at an average rate of 6·5% per annum in the 1970s, it does not follow that total production will expand at the same rate, because the smaller growers will be dropping out and their land is unlikely all to be taken by the big cereals men.

Marketing by forward contract

Growers are increasingly aware of the need to plan production with conscious reference to market requirements which, increasingly, demand a continuation of 'quality' produce. Forward contracting is a device to ensure a continuing supply of produce of the quantity and quality needed to satisfy consumer demand, and it also lessens the risks inherent in chance marketing, and tends to stabilize prices, and hence permits the grower to plan well in advance. It is equally liked by the processors and, more recently, by the big supermarket chains, who have large capital sums tied up in buildings and machinery to provide packaged goods for home consumption. It is almost true, in fact, to say that modern intensive livestock enterprises are little more than machines designed to convert plant products into animal protein at least cost: along with other kinds of processing plants, they require complex equipment and good transport facilities.

In the UK the British Sugar Corporation provides a classical example of forward contracting; the Corporation owns all sugar beet seed, which is supplied only to contract growers. In earlier years this led to a concentration of beet growing in the neighbourhood of sugar factories, and a big flare-up of beet eelworm, a problem that has been largely overcome by a crop rotation clause in all beet contracts. Soft fruit, particularly blackcurrants, is another example where forward contracting has long been the rule, but has tended to decline a little in recent years because of lessened popularity of currants. Also recently, there has been a big expansion of the frozen, and other, processed food industries, with expansion of factory and, in some cases, on-farm storage facilities. Table 5 shows the currently near-complete link-up of

Table 5. Estimated percentage of crop area grown under contract, England and Wales

Commodity	1960/61	1965/66	1970/71
Green peas	68	84	88
Brussels sprouts	2	9	20
Carrots (maincrop)	10	21	27
Potatoes (maincrop)	6	10	16
Raspberries	26	20	41
Blackcurrants	74	63	69

green pea growers to forward contracts from canning, quick-freeze, and dehydration processors, and the start of similar links between growers of Brussels sprouts, carrots, and potatoes, and their respective processors. It has been estimated (Hampson 1969) that about 25% of potato production may go to the processors by 1975, and a longer-term projection by the Potato Marketing Board suggested that the proportion might in due course rise to 50% (as in the U.S.A.).

With cereals, a third of the seed grain was grown under contract to merchants in 1967, and there is interest by both growers and merchants in extending forward contracts to cover the disposal of grain after harvest: in 1967 about a quarter of the grain was sold 'forward'; only a few merchants have sufficient storage capacity to hold grain for months to ensure an orderly flow onto the market; also, there is an incentive for growers to store on-farm for periodic local delivery (Britton 1969). In summary, it is continuity of supplies, whether of vegetables, roots, or grains, that is increasingly important in a situation where as many, if not more, people are employed in processing and distribution as in primary production.

Projection to the 1980s

Many of the recent projections for U.K. agriculture (Anon. 1971, 1971a, Britton 1969, Josling & Lucey 1971; Wallace 1967) have been made in the light of U.K. entry to the Common Market, and the consequential effects of the Common Agricultural Policy (CAP) and of various Directives which are mandatory in member States. The latest projection for agriculture (Josling & Lucey 1971) is based on analysis of current trends of the kind shown in Tables 1–5, and assumes continuation of (a) present U.K. trends in labour, machinery and specialization, and (b) the CAP in its present form. The next decade, if the projections are right, will lead to a considerable increase in meat production, especially of beef, pigmeat and poultry, within the European Economic Community (Table 6): the U.K. will expand grain

Table 6. Projection by Josling & Lucey (1971) for main commodities, 1968–80, assuming ten member European Economic Community and continuation of present Common Agricultural Policy ('000 tonnes per annum)

Commodity	U.K. production in 1968	U.K. projection for 1980	EEC-10 projection, total for 1980
Grains	13,363	22,839	120,826
Milk (butterfat equiv.)	510	486	4,577
Beef and veal	906	1,036	7,187
Mutton and lamb	247	264	(Not estimated)
Pig meat	826	1,122	8,630
Poultry meat	490	722	3,887
Eggs	900	1,028	4,316

production, mostly barley, because profitability in terms of gross margin (at present about £37 per acre = £91 per hectare) will be greatly increased. Another, independent, projection suggests that growers' net income from barley will improve by 13–15% in terms of the Unit £ by 1978/79. Thus instead of continuing to import about 8 million tonnes of grain per annum, we shall become a grain exporter to the Community, especially to Ireland and Denmark.

Josling and Lucey's grain projection is summarized in Table 7 along with three estimates of the land needed. The U.K.'s crops-plus-grass area is 12·32 million hectares, and cereals are currently grown on 3·81 million hectares annually. But the land needed to produce a given tonnage of a

Table 7. Projection of grain production in the U.K., 1980: estimates of land needed

Factor	Projected production in 1980	Increase over 1968
'000 tonnes of grain needed annually:	22,839 (Table 6)	9,476
Hectares ($\times 10^6$) needed, assuming yields:		
(a) Maintained at 1968 level (3·44 t/ha):	6·6	2·8
(b) Improved to 'healthy level' (5·02 t/ha):	4·5	0·7
(c) Improved half-way (4·23 t/ha):	5·4	1·6

given commodity depends on unit area productivity, and the estimates in Table 7 therefore assume (a) that grain yields are maintained at the 1968 level (there is recent evidence of a levelling-off in the big improvements in yield recorded in the 1950s and 1960s (Wallace 1967)); (b) that, by 1980, yields will have improved to the so-called 'healthy crop' level of 5 tonnes per hectare, and (c) that yields will be improved half-way, that is by around 20%. Though such an improvement may well be possible with existing techniques (see, for example, Strickland 1970), recent evidence suggests that an improvement of nearer 11% seems more realistic (Britton 1969).

Optimistically assuming that a 20% yield improvement will be general by the 1980s, it seems that an additional 1·6 million hectares of land will be needed for cereals in the next 10 years. Can this land be found? Perhaps more important, can we devise a system that will allow other than monocultural production?

To start with, there is ample evidence that cereals do best in our southern and eastern counties, and a little less well in the west, where it is wetter and diseases are more important in some years. So presumably the 'more land' hunt will start in the south and east, where there are currently 1·4 million hectares of land under temporary and permanent grass (South-Eastern, Eastern, and East Midlands Regions). It is perhaps unreasonable to postulate a complete plough-out of grass in these areas, but possibly reasonable to suggest considerable pressure on the land at present in grass, and that an appreciable proportion of it may well go to cereals.

A clear majority of the bigger cereal growers are already aware of the advantages of expanding production under the EEC (Britton 1969), especially in the Eastern and South-Eastern counties, and there is also evidence that about 15% of these expansionist growers hope to tie-in their increased production with more intensive use of grassland. The biggest current expansions in intensive livestock rearing are in the eastern half of the country, close to the sources of feed grains; but dairying remains more important in the west, and intensive pig production is increasingly important in Cornwall and in parts of the West Midlands. Thus, there may be some cereal expansion in Wales and the south-west (29% of growers replied affirmatively, compared with 17% negatively, in a recent survey), where there are over 2 million hectares of land under temporary and permanent grass; but undoubtedly the main pressure will be in the east, and inroads may well be made into current arable cropping practices.

The first things to go will surely be the less profitable crops: e.g. potatoes, which are currently over-produced, for which demand is tending to decline, and which are labour-intensive when grown on a small scale; one recent estimate suggests that the less efficient growers will be losing about £25 per hectare on potatoes by 1978/79. Then there is the very tight EEC Directive on sugar production which may lead to some decline in U.K. output because the EEC price for beet within the basic quota is lower than the persent U.K. guaranteed price (though there are some mitigating circumstances and, if our average costs can be lessened by better productivity, beet may remain profitable).

Finally, there are possible land savings in horticulture. In the past decade there has been a fall of about 4,000 hectares in land under crops like broad beans, cauliflowers, green onions, market peas, rhubarb, and some soft fruit, though in nearly every case yields have increased so that total productivity has more than balanced the land loss. So it seems that a combination of cutbacks on less profitable crops, land savings from enhanced productivity, less fallowing, and substantial ploughing-out of grass, could provide the land needed to satisfy the cereal projection, provided cereal yields can be increased by about 20% in the 1970s. The price will include a further simplification of cropping which, with cereals, will be close to a monoculture.

Prospects for biological control

If the less efficient growers are going to the wall, and those that remain are becoming precision growers, we can anticipate increased concern about pests, diseases and weeds, which lead to instabilities in planned production and may continue to affect profit margins. But these 'new men' will certainly probe the cost-effectiveness of any control measures which prevent them following near-monocultural production plans; in particular, they seem likely to object to techniques which further restrict their already limited freedom of action: for example, the adjustment of planting dates which may lead to cultivation or harvest bottle-necks. They may also object to labour-intensive techniques, such as the costly hand-grading of lightly infested or infected produce, or repeated crop inspections to determine the optimum time to apply controls. Thus it is important that control methods are properly costed so that growers can compare advantages within their chosen farming systems; and proper costing must include components for professional advice, ease or otherwise of harvesting, and possible loss of peak market price due to planting cultivars chosen for resistance to pests or disease. In what is undoubtedly a buyer's market, it is very much up to the biologists to sell their control methods in similar economic terms to those used by the chemists.

What, then, are the prospects for biological methods of control if the Josling and Lucey projection comes to pass, if about 44% of our crops-plus-grass land is under cereals, possibly 20% (compared with 33% at present) is under grass, and the rest is under intensive specialist cropping of one sort or another? Expenditure on machinery and power comes next to labour in farm out-goings, and the less the grower has to manipulate his land the happier he is. Minimal cultivation, and techniques like direct drilling, may provide acceptable opportunities for control of some pests by lessening the disturbance of predators which must occur when land is ploughed. Drainage can adversely affect pests like slugs, which are more than just a nuisance when they hollow-out winter cereal seed; and irrigation can affect both crop growth and the pests. When ploughing or soil cultivations are essential, there seem to be many ways in which harmful organisms can be affected: compacting the soil surface may lessen the ability of soil pests to migrate, and surface tilth can affect oviposition; some species will lay eggs on bare soil but not if the surface is covered with vegetation, and so on; there are obviously many ways in which land manipulation can be exploited to the detriment of pests.

Secondly, plant spacing can be manipulated to make the micro-environment more or less attractive for harmful organisms, and it is worth noting that the trend in a number of crops is now towards dense planting which nevertheless gives optimum spacing in terms of yield and quality per unit land area. Since pest attacks often seem to start on field headlands, it has recently been suggested that total damage may be less on large fields than on small ones, and it might be worth doing some trials to see whether mid-field infestations can be affected by planting the headlands more or less densely than the rest of the field (a variant of trap cropping without the disadvantage of having to handle two different types of crop).

Thirdly, and perhaps most promising, is the development of crop cultivars resistant to the worst pests and diseases, though there are still formidable problems to overcome in the form of resistance-breaking pathotypes of pests and diseases, and maybe also consumer resistance.

Finally, the future for so-called classical biological control with natural enemies does not seem very bright under our humid, insular conditions. This is not merely because high productivity farming demands almost complete pest control, but because pests, diseases and weeds interact to affect each other, and it seems inevitable that treatments which are not applied specifically to kill insects will interfere with pest/predator/parasite populations. But this is not to say that classical methods can be dismissed out-of-hand: there is always a chance, as the Glasshouse Crops Research Institute workers have shown, that partial control of a potential pest outbreak can lead to worth-while savings on the spray account.

Conclusions

Four trends in British farming are likely to continue throughout the 1970s: new methods of crop production will be developed; farming systems will become more specialized; the growth of large farm units will continue as the small, less efficient, growers sell up and leave the land; and there will be a new order in marketing, including much more forward contracting. These trends are projected into the 1980s and possible phytopathological problems are considered in the light of feasible controls.

When these trends are projected into the 1980s, they suggest that the growers who survive will probe the cost-effectiveness of all recommended control measures, tending to avoid those that are labour-intensive or impose intolerable constraints on farming operations. Cultivars resistant to pests and diseases will be acceptable provided they are satisfactory in other respects. The prospects for classical biological control do not appear to be bright.

Acknowledgements

Grateful thanks are due to colleagues in the Ministry's Economics and Statistics Divisions, and especially to C. R. Orton, J. M. C. Rollo and P. J. D. Smith for extracting and checking many of the data in this paper.

References

ANON. (1971) *Examination of the Horticultural Industry*, 1970. H.M.S.O., London, pp. vi + 114.

ANON. (1971a) *Farmers' and Growers' Guide to the EEC.* National Farmers' Union, London, pp. 90 + 7 tabulations.

BRITTON D.K. (1969) *Cereals in the United Kingdom: Production, Marketing and Utilisation.* Pergamon Press, Oxford, pp. xvi + 835.

HAMPSON C.P. (1969) Processing trends in potato production. *Agriculture* 76, 605–11.

JOSLING T. & D. LUCEY (1971) *The Market for Agricultural Goods in an Enlarged European Community.* Irish Agricultural Economics Society, Dublin, pp. 1–25.

STRICKLAND A.H. (1970) Some economic principles of pest management. In *Concepts of Pest Management*, eds. R.L. Rabb and F.E. Guthrie. University of North Carolina Press, Chapel Hill, pp. xi + 242.

WALLACE D.B. (1967) The economics of intensive cereal production. *Ann. appl. Biol.* **59**, 309–12.

The future of biological control in Britain: a research worker's view

N. W. HUSSEY *Glasshouse Crops Research Institute, Littlehampton, Sussex*

Since research workers concerned with integrated pest control do not question the future relevance of the concept, they may not be the best judge of its future development. Expansion of the research effort designed to achieve more rational pesticide usage should be politically inspired and the cropping systems within which such expansion is recommended should usually be chosen by those responsible for environmental quality.

On the other hand, further development of resistance to pesticides, both in terms of the species involved and the chemicals rendered useless, may, in certain restricted situations, create special problems that the reduced flow of alternative compounds is unable to solve.

I propose to discuss the future problems of the worker in integrated control. My views are naturally coloured by problems encountered within the environment in which my colleagues and I at the Glasshouse Crops Research Institute (GCRI) have attempted to develop practical systems of biological control. However, as no single pest can be controlled in isolation, I am really concerned with integrated control. Despite this bias, it is probably convenient to follow the sequence of steps adopted in our work on biological control under glass, for, although it is a restricted situation, many of the problems encountered are common to other cropping systems.

I believe that a thorough appreciation of the whole commercial scene surrounding the cultural systems under investigation is basic to research on integrated control. I, therefore, regard much of the international effort in this field as merely a contribution to scientific knowledge in that the scientists involved have been solely concerned with one or two of the component pest/natural enemy interactions and are largely unfamiliar with other factors limiting the cultural system within which the pests operate. Since integrated control embraces all the pests and diseases affecting a single crop, the scientist must understand the economic pressures affecting cultural techniques, must be sensitive to the effects of political interests from both within and outside the industry, and must have a wide knowledge of the current chemical controls for the whole range of pests and pathogens affecting the crop. To remain so fully informed about an agricultural ecosystem, the research worker must spend what many might regard as excessive time, with both the growers and those who service the crop in other ways.

Assessment of the problem

Before embarking on research to develop integrated control programmes, it is necessary to establish a justifiable level of expenditure. This may, of course, be almost infinite where, for political reasons, it is deemed expedient to reduce pesticide usage in order to avoid difficulties with environmental pollution. Currently, it is more likely that a biological solution will be considered so as to avoid resistance problems and, in these circumstances, the cost of conventional pesticide usage is usually set against the capital value of the crop to gauge an acceptable level of research investment. It is common practice to compare the direct costs of any biological solution with those involved in chemical control and then, according to one's leaning, ascribe advantages to one or other system. As integrated control is largely an exercise in compromise it will usually prove more expensive; however, any economic advantage attributable to biological control techniques cannot always be judged merely by direct pecuniary benefit to the grower, for reduction in intensive pesticide programmes may by itself lead to increased yields. Where natural enemies or pathogens have to be mass-produced or selective chemicals developed, we are apt to forget that the producer must see a profitable return on his capital outlay before he can be induced to service the technique, however great the advantage to the grower. It is obvious that, in business, economics have a considerable political element and so the research worker must be sensitive to human attitudes if he is to avoid providing a programme that no one wants to use.

A prerequisite to the initiation of work on integrated control must be the clear definition of an ultimate goal. This must not be stated in vague generalized terms. In the glasshouse the problem was simple—merely to provide reliable control of resistant spider-mites. Integration of the control of the other components of the pest and disease complex, which followed, was necessary to ensure effective biological control of the mites. Outdoors, the goals may be more difficult to define. In intensive cereal growing, for instance, acceptable levels of disturbance to game, degree and costs of weed control, economic yields from the cereal crop, and conservation requirements for field structure must all be considered.

Integrated control of pests demands a research team composed of strange bedfellows and cannot be achieved merely by some friendly co-operation between entomologists and pathologists. The agronomist should probably be the dominant partner.

Once it has been decided to go ahead it is necessary to select those pests within the ecosystem that are most likely to be successfully controlled by natural means. Likelihood of success is, however, only one parameter. Some pests (e.g. red spider mite and cabbage root fly) could be chosen because of their potential for developing resistance to chemicals. On the other hand, target pests could be chosen because they are so wide-spread that insecticidal control might lead to undesirable side-effects. Others, such as pine sawflies, might be selected because they occur in habitats where chemical control would be unduly expensive. Yet other targets could be identified by the need to avoid undesirable residues in the harvested produce. An interesting example is the tomato leaf-miner; this pest is present throughout the life of the crop, so that there is a conflict between the required frequency of insecticide application and the daily harvesting of the crop. Another reason for seeking natural controls may be phytotoxicity, such as that induced by the incorporation of insecticides into the growing media for control of mushroom flies, or the early production of tomatoes resulting in seedlings which, in the short winter day-length, cannot be treated with insecticides without phytotoxic hazards.

The target pests having been identified, their economic thresholds must be established. Then the most likely techniques for natural control must be surveyed. Where the pest is well-known and well documented, this analysis is relatively simple. Obviously, the cheapest and most direct method must be chosen. Sometimes, this could be a relatively simple cultural adjustment, as in the artificial manipulation of day-length to prevent the hibernation of glasshouse red spider mites. Too often, the agronomist insists on perpetuating cultural techniques that confront the pest control operator with impossible tasks. The massive and continuous monoculture created by year-round production under glass is a relevant example. Indeed, as integrated control demands a consideration of the whole ecosystem, it may, on occasion, dictate the cultural techniques. The aphid problem on year-round chrysanthemums under glass could be solved most economically by utilizing resistant varieties or by arranging for a break in the otherwise continuous system of culture. Pest control, although commonly accounting for only 1% to 2% of production costs, may become a factor limiting further cultural improvement. Before benomyl became available, an example was provided by the mushroom phorids. Although existing chemical controls have, in general, reduced populations to a level at which direct damage has been eliminated, the role of the flies as vectors of the fungus disease *Verticillium* has become critical, for disease losses are no longer acceptable in modern economic circumstances.

Difficulties in solving the problem

Having decided to introduce an element of biological control into an ecosystem, the remainder of the pest and disease complex (including hitherto minor components) must be identified and a theoretical programme devised around those control techniques that appear most suitable for harmonious integration. Normally, species that have already become resistant to chemicals, or are likely to do so, are strong candidates for natural control. We have, for instance, predicted that tolerance to the selective aphicidal carbamate, pirimicarb, could be selected in glasshouse populations both of *Aphis gossypii* on cucumbers and of *Myzus persicae* on chrysanthemums. It must be recognized that some pests are unlikely to be controlled by natural means. For instance, thrips have few specific natural enemies. A major effort, conducted by the Commonwealth Institute of Biological Control on our behalf, has shown that an undescribed pathogenic fungus, *Entomophthora* sp., was the only natural organism likely to limit populations.

This is an appropriate moment to consider the present unsatisfactory situation that commonly occurs where natural enemies have to be sought and then introduced into an ecosystem. With this situation at its simplest, it is only necessary to establish, from the literature, the identity of suitable natural enemies and then to obtain a nuclear stock from a research institute that already maintains cultures. More usually we have to rely on the Commonwealth Institute of Biological Control (CIBC) to locate suitable enemies in the field. Whatever the ultimate source, the short-list of suitable species may be further reduced when it is subjected to the exigencies of transport.

I regard as quite remarkable the variations in the efficiency of packing, etc., commonly encountered. Even the experienced CIBC has been involved in many unhappy and costly mishaps due to careless packing. Regrettably, delays caused by the customs all too often nullify the value of otherwise efficient transport and packing. So our original selection may be reduced to only one or two candidate species which must then pass a further hurdle, namely, our attempts to establish a culture. Very rarely do any of those items of information, so necessary in culturing insects, come with the consignment. Another accident, which can sometimes turn out to our advantage, is the inclusion of species that had not been sought. Our studies on coccinellid larvae included an example of such fortuitous success. After considering the available information, we chose and collected five species in the hope that one would perform effectively over a wide temperature range and variety of crops. In the event, *Cycloneda sanguinea* included, through the enthusiasm of an assistant, with a shipment of other species from Trinidad, turned out to be the most suitable species.

It may be necessary to select particular strains of parasites and predators to ensure that they are adapted to the environmental situation within which they are expected to operate. Often, only the crudest attempts are made to select such strains. It would normally be cheapest to select on the basis of tests in the country of origin. We have such a programme for parasites of *Aphis gossypii* in Pakistan, though shortage of funds limits its effectiveness.

As with pesticides, a successful screening programme to select natural enemies should operate at several levels. The first basic requirement is to establish that the natural enemy has a rate of population growth exceeding or approximating to that of its host. Preliminary determinations of this parameter can be made in small cabinets within which the environment is maintained at similar levels to the range expected in the ecosystem. It is, perhaps, ironic that it has become accepted that some biological disciplines, especially plant physiology, must be equipped with specially designed control environmental facilities, whereas others, especially in pathology, are often expected to do without. Such an apparently simple method of selecting natural enemies is beset with many hazards—an interesting example concerns our work on parasites of agromyzid leaf-miners. A common British parasite, *Diglyphus isaea*, performed excellently against chrysanthemum leaf-miner in our experimental cages when the host plants were artificially lit. However, when it was introduced on chrysanthemums in glasshouses in the late winter, the cultural conditions reduced the efficiency of the parasite; only later, as growing conditions improved, was effective control regained.

Having produced bionomic data for the construction of suitable mathematical models, it is useful to calculate the effects of different variables such as temperature, host/parasite ratio and time of introduction, etc., so as to predict those areas where further experiments are needed in a glasshouse. Mathematical models can save much experimental time and effort, but their construction demands more resources in scientific man-power and equipment than are normally available. We developed our models of the predator/red spider mite interaction after a biological control programme had come into commercial use, and it should not be overlooked that, however carefully such predictions are made, they may become irrelevant quite unexpectedly. One of the parasites, *Aphelinus flavipes*, which we have studied extensively for the control of *Aphis gossypii*, seems to be very sensitive to the method by which it is introduced. The problem is not yet fully understood, but it appears that adult parasites must emerge on to a natural surface rather than a glass tube or other piece of equipment chosen by the researcher.

Glasshouse experiments on biological control are demanding in space because each replicate must be separated from the next. In effect, this means a separate glasshouse for each. Even if the structures used are very cheap, they must contain suitable automatic control of light, ventilation and temperature, thus demanding more costly services than would compartments within a large single glasshouse. In our experience, the latter, even if pressurized, cannot provide the necessary separation. The air pressure puts considerable strain on the glass sealing, so that minute parasites are able to move between apparently isolated units.

Discovery of an effective natural enemy is only the first step in the development of an integrated control programme. The impact of other biotic factors, as well as the fungicides and insecticides used for the control of other components of the pest and disease complex, must also be evaluated.

Preliminary tests of selective action are usually made in the laboratory in three ways. Firstly, the effects of residual deposits are investigated by confining both pest and natural enemy on treated leaves. Secondly, the direct effects of the pesticides on natural enemies are studied. Should neither of these methods kill a predator then tests for 'food-chain' toxicity must be made. Such investigations are essential, following the demonstration of sterility induced in *Phytoseiulus* females after they have fed on red spider mites, each containing a sub-lethal dose of the fungicide benomyl. Unfortunately, in whatever detail, and however accurately such studies are made, they do not necessarily give valid information. This difficulty is caused by the failure of such tests to provide evidence of effects on arthropod behaviour; these effects can be unexpected. For example, if spider mite eggs are contaminated by dinocap, a material relatively innocuous to adult predators, they will not be consumed. Similar effects undoubtedly abound and will no doubt come to light as studies in this field expand. In practice, the only way to ensure that such effects are unimportant is to set up pest/natural enemy interactions to compare with similar interactions regularly sprayed. We usually make such comparisons in small glasshouses, and conclude that the candidate pesticides are suitable for integration if the biological control on untreated plants compares with that in the treated areas after about ten successive weeks of treatment.

Considering the sophistication with which the chemical industry screens pesticides, the current allocation of resources to the solution of the problems outlined above is obviously grossly unsatisfactory, though in line with the limited financial resources usually available in biological control.

The unenlightened application of quarantine regulations is another factor that could nullify the most efficient scientific approach. We have shown that red spider mite control by predators is most reliably achieved by introducing *Phytoseiulus* after the crop has been uniformly infested with small numbers of the pest. On cucumbers, these pest introductions must be made in late winter or early spring when the day-length is still sufficiently short to induce hibernation. As the mites introduced must be mass-reared under lights, they all too readily enter hibernation and, therefore, cease to reproduce when exposed on the plants. We have, however, imported several strains in which a non-diapause character is genetically fixed. Despite the obvious suitability of these strains for our purpose, with the added advantage that these mites could not survive beyond the season of their introduction, the authorities may find it difficult to permit their general release!

Assuming that we have developed an efficient system of biological control based on pathogen, natural enemy or sterile male introduction, it can only be successfully implemented on a commercial scale if the necessary organisms can be mass-reared and profitably sold by independent producers. This requirement puts a most important limitation upon research on rearing techniques. Usually the entomologist is happy to develop a convenient production system that maintains a viable culture. Profitable commercial production demands that the insects should reproduce at their maximum potential on the cheapest substrate and that every step in the technique should be accurately costed so as to ensure that the method is economically viable. Fractional cost advantages on an experimental scale may lead to significant profit margins on a large scale.

The selection of natural enemies, screening of their biological efficiency, the development of suitable breeding techniques, and the subsequent modification of integrated programmes when new pesticides have to be incorporated demand sizeable research investment. Further, the commercial demands imposed by the cultural system imply that research should be conducted rapidly. Long-term state research seldom satisfies the economic needs of industry. Studies of the complexity outlined call for a large team. Each facet within the project normally requires a scientific officer and assistants with appropriate experimental facilities. It is, therefore, unlikely that the research group could number less than twenty workers.

Teams of this size and larger must be managed effectively if the goal of commercially practicable control is to be attained within the desired time scale. Management must thus be both efficient and disciplined. I regard discipline as absolutely vital to the success of integrated control projects. The management of such mission-oriented research must be controlled through a matrix of research projects. Each must have clearly defined objectives, the progress towards which must be re-evaluated at frequent intervals. The objective will only be achieved if the scientists involved continually orientate their efforts to the final goal. Some may regard such limited objectives as too confined, but success is adequate reward to those who seek their inspiration from truly applied research.

Another important and, I think, underestimated problem in the development of integrated control programmes is the mechanism of obtaining official approval for minor pesticide usage. Not unnaturally, the pesticide industry cannot economically justify the expenses of residue determinations, etc. on crops which cover only limited acreages, especially since the work is designed to develop techniques of control which can be expected to reduce pesticide usage. Our techniques for thrips control provide an example. The Pesticides Safety Precaution Scheme regards soil drenches as foliar sprays. On cucumbers this means that fruits cannot be picked for 14 days after soil drenches of BHC or diazinon. Since cucumbers are normally harvested every two days, such rulings entirely prevent commercial exploitation of a biologically effective technique, despite the fact that residues determined on crops grown at the GCRI revealed such small residues that the interval between application and harvesting could probably be safely reduced to 2 or 3 days. Ironically it is the pesticide firm which must take the initiative, and provide extra residue data to alter its approved label merely to reduce its sales for the benefit of the grower or community. Similar problems are likely to arise in the production and use of biotic pesticides. It is high time that the mechanisms for clearance of insect diseases should be considered, if we are to avoid the development of control systems for which no safety clearance can be given.

In experimenting on integrated control in the future, the area required will be considerably larger than that traditionally demanded by pesticide trials. Where natural control is to be manipulated in agriculture, it is important to recognize that one is measuring the effects on a sizeable ecosystem. Thus, where reservoirs of natural enemies are involved, the study areas may need to contain large and diverse ecological niches to compare with similar areas without such reservoirs. Similarly, the side-effects of pesticides such as fungicides and weedkillers must be evaluated on a sufficiently large scale to avoid the complications that would arise if large numbers of organisms migrated into the study area. The use of larger experimental areas will naturally cost more for entomological work than the authorities have become accustomed to. Similarly, if large numbers of natural enemies, sterilized adults or biotic pesticides have to be produced and released for such trials then, again, costs will be high.

Any extension of our efforts in natural control will, therefore, inevitably lead to increased costs. These must be met if reduced pesticide usage is to be encouraged. Research on natural control must, in the final analysis, be prosecuted with the intense effort as that the chemical industry devotes to the development of new, efficient pesticides.

Future exploitation of biological control

Where integrated control demands the introduction of natural enemies or microbial agents, it may well be important for the research centre to have direct advisory control both of the production systems and of the sales policy adopted by the commercial rearers. Just as abuse of pesticides is controlled by approved labels, so biological control must be monitored by an effective code of practice. For many people, biological control has an innate fascination which can all too easily become a major draw-back. Quite limited experience seems to create a confidence that easily leads to variations of the recommended methods of application. These may, in turn, lead to failure and loss of confidence in the whole concept. Not unnaturally, there is considerable resistance to the introduction of glasshouse pests as a prelude to control. Growers fear the consequences, and erroneously believe that the costs of control can be reduced by omitting this apparently dangerous technique.

However, there is no doubt that this pre-establishment is a prerequisite for satisfactory control based upon a single introduction of pest and natural enemy. Interestingly, at least two growers who have practised variants of our biological methods for the past five or six years have now found that they must follow our precepts. There should, therefore, be some restrictions on the sale of natural enemies to ensure that the techniques are so applied that the greatest benefit accrues to the grower. The most satisfactory system would demand pest control under contract, with the contractor using biological methods both to reduce his costs and to overcome resistance to pesticides. With such objectives, the contractor would rapidly find that pest introduction led to more effective control.

Regrettably, some commercial rearers of natural enemies tend to encourage simpler techniques, usually amounting to short cuts. Further, they tend to regard their responsibilities as over when they have shipped natural enemies to the grower.

At present, the Agricultural Research Council believes that its responsibilities for pest control end at the perfection of a method at the research level. However, a technique perfected by the research worker has usually to be considerably modified in commerce, and this may create situations that could seriously endanger the whole future of the technique. The future of integrated control is bleak indeed, if sharp administrative demarcations prevent intelligent adaptation of new methods in commercial use. Such development could, of course, be supervised by the Agricultural Development and Advisory Service (ADAS), but, where advice can only be given and not enforced, exploitation of research results costing hundreds of thousands of pounds may be endangered. In integrated control the goal must be clearly seen before research is commenced, and the likely methods of exploiting successful techniques must be envisaged at a very early stage. This may involve some form of legal enforcement of the codes of practice involved. Under glass this dramatic course seems neither necessary nor even advisable, but the principle could well be important in other cultural situations.

For success, integrated control must usually operate over a wide area to avoid the migration of pests from non-co-operating farms which could upset the population balance upon which biological techniques depend. Similarly, successful integrated control depends on adherence to a predetermined pattern of pesticide usage. In this new world of pest control, complete freedom would no longer be possible, and all changes in techniques would have to be based on scientifically sound conclusions. Growers are unlikely to be the best judges of the impact of, for example, fungicides or changes in cultural methods on integrated control. Such matters must be controlled by adequate advice which, ultimately, should be legally binding.

Finally, I should stress the changed attitudes that must precede widespread exploitation of integrated control. Our administrators must grapple with such problems as those associated with the clearance both of pesticides for minor use and of microorganisms for field use. If successful integrated control programmes are to contribute to environmental enrichment then the administrators must recognize that, like noise abatement and the curbing of the motor car, they will cost the community money, and cannot be regarded merely in terms of profit or loss to the farmer.

Other contributions to this Symposium have drawn attention to the dangers inherent in the 'blanket' application of any technique whether biological or chemical. If we are to retain economic control of such biologically variable organisms as pests and pathogens, we must, even in Britain, with its smaller problems, make greater use of more complex control systems in the future.

Integrated control is an exercise in 'integration' by the community, and failure to recognize this fact will possibly be the most serious deterrent to the more rational use of pesticides.

The future of biological control in Britain: a manufacturer's view

W. F. JEPSON *Cyanamid International, Gosport, Hampshire*

I felt a certain trepidation when I was asked to express the 'manufacturer's' viewpoint, since this will vary widely according to the point which he may have reached in his scale of operations. So this paper must be regarded as an individual contribution to the very vexed problem of harmonizing ideas and interests involved in modern pest control for the benefit of all.

In the first place let us dispense with arguments as to the moral attitude that it is necessarily desirable to reduce the use of agricultural chemicals to an absolute minimum. I would go as far as to suggest that we should welcome the ever-increasing part played by chemicals, as an economic input resulting in the wider availability of good, pure and varied foodstuffs to sections of our population which never knew them in the past. In my own lifetime I have seen citrus, banana, and tomato become food for the masses, and these are now being joined by cucumbers, avocado pears, aubergines and a host of new delights. Perhaps this is a facet of the 'quality of life'. I am going to assume that it is common ground that this situation can only be extended and developed in modern cultivation and marketing conditions by the more intensive use of chemicals. Even the arch-enemy DDT has recently become the subject of a spirited defence by no less a person than Dr. Norman Borlaug in his address to the 16th Governing Conference of FAO in November 1971. Dr. Borlaug voiced alarm at what he described as the 'current crusades of the privileged environmentalists in the United States' to bring about a legislative ban on the use of DDT in the United States which he said would almost certainly lead to a movement for a world-wide ban. 'This must not be permitted to happen, until an even more effective and safer insecticide is available, for no chemical has ever done as much as DDT to improve the health, economic and social benefits of the people of the developing nations.' I have a feeling of unease that we in the northern affluent societies are facing a dilemma, which may even engender political trouble if we are compelled to advocate for the developing countries a substance which we condemn as a dangerous contaminant in our own.

Relations with the public

Having said that, I am at once aware that the use of chemical plant protection methods has to be justified on a multiplicity of grounds, most of them old, but a number of them recent, urgent and fashionable. No one appreciates better than the manufacturer, if only because it may appear in his balance sheet, that the age of complacency in the environment and pollution fields is firmly behind us. It is almost platitudinous, but also true, to say that the manufacturer is fully conscious of being a member of a community, operating within a framework set by the community, and of having public responsibilities. It is not generally realized that the evolution of the ever-increasing list of demands by the many regulatory agencies has actually been hastened by the constant improvement of both field sampling and of analytical techniques by the manufacturing companies themselves.

Thereafter, safety in the environment can only be secured by the education of the user, both on the technical and social planes, and in this task the manufacturer works in close co-operation with the extension and advisory services, though this co-operation must now move closer to the research institutes. The practical approach to these problems is beset with difficulties, but may have to begin, as far as the manufacturer is concerned, with a public relations exercise, designed to inform the general public on matters which the specialist has known for a very long time, namely the many hurdles that any new chemical has to surmount before it can be accepted for safe and effective use.

Decisions within the industry

The next step is decision on investment in product development. I think a non-economist can see from a Discounted Cash Flow diagram that a capital sum of money, once committed (allowing for an annual interest rate or discount of say 12%) will be subject to a strong negative cash flow for several years during development. Only after five and often six years is there a possibility that the pay-back line may be crossed, and then the product is cast on the troubled waters of competition, patent expiries and, now, environmental demands. How little choice a manufacturer really has if, in response to popular clamour, he confines his efforts to materials with, say, selective action, acute oral LD_{50}, rat, over 100 mg/kg, persistence in the environment of not more than three months, non-hydrocarbon, organophosphate, carbamate, etc. It is instructive to look at the realities of the present situation. This I have tried to do (Table 1) by taking a few figures from the latest edition of the Pesticide Manual (Martin 1971). These figures

Table 1. Mammalian toxicities of insecticides used in Great Britain*

Insecticide classes	Total no. of insecticides	No. of insecticides with acute oral LD_{50} less than 100 mg/kg
Organochlorines, inorganics and vegetable products	47	11
Organophosphorus	71	41
Carbamates	17	10

* The insecticides studied were those given in Martin (1971). Data on mammalian (mainly rat) toxicity from the same source.

may suggest that the laborious and expensive synthesis and screening systems employed during the past twenty years have not produced revolutions, but have laid their products approximately evenly around a toxicity level of 100 mg/kg. It is no easy task by taking thought to add inches to our stature. I do not want to labour the point that the constant demand for safe, selective, non-persistent insecticides is unlikely to be realized in the near future. We shall have to make better use of what we have.

We can perhaps summarize the process of integrated decision-making, which confronts the industrial biologist in the field, in Table 2.

Table 2. Factors involved in pesticide selection

1. Operator factors (safety and convenience of formulations)
2. Consumer factors (residues, toxicological 'reputation')
3. Biological factors
   - (a) Varietal changes
   - (b) Extension of monoculture
   - (c) Clean weeding
   - (d) Shift of planting date
   - (e) Destruction of parasites and predators
   - (f) New break crops
   - (g) Demands of processors, e.g. for perfect, uniform products
   - (h) Hedgerow orchard cultivation
   - (i) Monogerm beet seeding

Since those of us who work in the practical field of plant protection are for the most part brought up and trained in the biological sciences, we may be expected to include the first seven of these factors in our routine of making trial protocols. In fact most of us are not concerned with how much chemical can be sold per acre but with how economical our application can be. If we do not do this, someone else will. What we are most concerned with in practice is that our treatment will either leave an unacceptable degree of damage in the crop, or that it will create a sub-lethal dosage situation with danger of inducing resistance. Our instinct is, then, in the annual crop at any rate, to 'kill pests dead', and this will normally include those natural enemies present at the time in susceptible stages or in critical numbers.

Rapprochement with biological control

We would like to modify our schedules to avoid the destruction of any biological controlling factor of real significance acting within the production cycle of the crop. How can this be done? At present the communication between research on the biota and the practice of chemical control leaves much to be desired. Too frequently, the creation of pejorative clichés and of chemical and biological bandwaggons has tended to aggravate and perpetuate a state of apartheid between the proponents of biological and chemical solutions to pest and disease problems.

The publication of the Rothschild report may perhaps provide, both by irritation and stimulation, an opportunity for the creation of some machinery enabling both sides to sit down and take stock of our combined assets, with a view to deciding how best to allocate scarce resources. Targets could be set for each side: the manufacturer to do his DCF sums with some confidence in the future; the biologist to achieve continuity and defined objectives in his study programmes which give answers to questions on integration of all methods of control. At present both sides are tending to proceed on parallel lines.

I should like to feel that my chemical control operations were backed by a biological criterion or rating allocated to each class of problem. If one could go into a field problem forewarned with a kind of trinomial biological rating, for example 007 (a familiar cliché), meaning 0 for parasites, 0 for predators, and 7 for abiotic factors, this would alert the chemical control man to the probable ecological situation. More realistic is the case of citrus, almost an artificial crop, with highly uniform clones grown under strict irrigation regimes in a highly equable summer climate. Here our trinomial rating might be 352, which would express the relative weighting of biotic and abiotic factors. In this way it might be possible to build up a 'population' of applied problems which lend themselves to concomitant studies from both biological and chemical angles. Such studies, if their objective were to define in the greatest possible detail the optimal integration of chemical with biological measures, would doubtless attract the support of industry. Longer-range studies of population dynamics in both crop and non-crop situations are highly desirable as a basis for both biological and chemical intervention, but these programmes would seem to belong to university departments of pure science rather than to applied departments and institutes whose declared function is to investigate methods of economic pest control and to train practitioners. The business of synthesizing and testing new compounds will be left properly to the competitive forces within industry, but with the knowledge that the prizes for success and innovation are not unduly restricted by extremist ideologies.

To turn to a more specific aspect of the manufacturer's edifice, we might look at some of the developments in the efforts to improve the environmental safety as well as the economy of pesticides.

The introduction of methyl analogues of the more toxic organophosphorus compounds has proceeded steadily ever since the original Schrader discoveries. We have methyl parathion, fenitrothion, dimethoate, and azinphos-methyl as examples of developments which have proved beneficial as regards both toxicity and selectivity. The diversification of the vinyl phosphates and of the triazine weedkillers to meet selected insect, mite and weed problems provided further examples of the development of the potentialities of pesticidal molecules. If we consider the 140 or so insecticides listed in the 1971 Pesticide Manual, we must surely accept that we have a reasonably broad spectrum of activity with some reserve for resistance problems, which are still a minority interest when spread over the pest field as a whole. A plea for the retention of a wide range of plant protection chemicals was recently made by Professor Schumann, President of the 38th German Plant Protection Conference held in Berlin in October 1971. Professor Schumann pointed out that there are over 1,500 registered pesticide formulations in Federal Germany, and expressed the view that it would be risky to become dependent on an arbitrarily limited number of materials. But as we can see, with nearly 50% of these products classed as moderately to highly toxic in the environment, industry must devote much time and many resources to formulation and labelling for safe use. Our own Ministry has set an excellent example in framing recommendations for safe use.

Of particular interest at present is the development of pesticides in low-concentration granular form, for accurate placement in the soil, thus avoiding contamination by drift and the killing of mobile parasites and predators. However, granular systemic treatments are slow to progress against the more conventional spray application, especially when a well-timed spray, applied on a warning system, is available (as in the sugar beet crop in this country). It is, however, still questionable whether such sprays can always be timed accurately enough to avoid damage to the biotic factors in the environment. Concomitant studies over a period long enough to produce consistent results are unlikely to be achieved by a student working for a higher degree. Co-ordination of biological and chemical interests is needed in research programmes in which industry might be invited to participate at the outset. One might even look forward to the integration of disciplines in something resembling the German system of Plant Protection Institutes, a system which is also found in many other European countries.

A more recent development in application technique is that of ultra-low-volume spraying, which has formed the subject of a new book by Maas (1971). The idea of cutting out the transport of water in many parts of the world, and of improving the speed and precision of application by new equipment providing better droplet spectra, has been countered by allegations of increased contamination from the more concentrated material used. Here again there is scope for controlled investigations in the same area by biological and chemical interests in which the effects of ULV spraying on the biota are properly quantified and the foundation laid for a *modus vivendi* between chemical and biological controls.

A third example of opportunities for joint work is chemical growth regulation. I am aware that some plant breeders regard this operation as 'not playing the game', but I submit that it should be regarded as a valuable accessory tool in the better cultivation of better varieties. To reduce lodging and eyespot in wheat, to stimulate flower production and retention in top fruit and vines, to ensure even ripening and assist harvesting in citrus—these are some of the fields in which chemicals will be used to change the environment of the crop. The whole subject merits close integrated development of the chemical and biological approaches, but the scientific and technical difficulties put any intensive basic study beyond the reach of the manufacturer.

Conclusions

So, to sum up the manufacturer's viewpoint, I would say improvement is needed in the planning and priorities of research based on objectives; in the mutual communication of laboratory and field criteria which help to define the role of biotic, cultural and chemical interventions in terms of intensity, timing or even of mutual exclusion. Great progress has been made already in Britain on these lines on major problems such as sugar beet pests, cabbage and carrot flies and the red spider mites. Let us turn our backs on the flood of sterile recriminations for the past errors of society and make a new start from common ground already achieved to make pest control ever safer, more effective and more rational.

References

Maas W. (1971) *ULV Application and Formulation Techniques.* Pp. 164. Philips-Duphar, Crop Protection Division, Amsterdam.

Martin H. (ed.) (1971) *Pesticide Manual*, 2nd edn. Pp. 495. British Crop Protection Council, London.

The future of biological control in Britain: a conservationists' view

K. MELLANBY *The Nature Conservancy, Monks Wood Experimental Station, Abbots Ripton, Huntingdon*

The word 'conservation' can mean several things. My understanding of the term is first that it implies the safeguarding of our whole environment, and the provident use of all our resources. Secondly, it covers the specific requirements for wildlife conservation, that is for safeguarding our native flora and fauna. This includes the protection of all our indigenous species, and the prevention of any of them from becoming extinct. We must accept that we cannot prevent individuals of many species from being killed and from being eliminated from some areas—changes in numbers and distribution have always occurred—but we must strive to ensure that a viable population is maintained, and that characteristic areas of all types of habitat and of all plant and animal community are preserved. We must prevent damage, for instance by pollution, to these safeguarded areas. And we should try to make sure that as large a number as possible of all species remain, so that they may be seen, enjoyed or studied by as many people as possible. I shall deal in this paper mainly with the conservation of wildlife. I shall not use the term conservationist to include those who wish to go 'right back to nature' and eschew the use of any modern methods in agriculture or in life generally, nor shall I use it for those who wish always to put wildlife before human interests.

This symposium is concerned with the control of pests. Here again we have a word with many meanings. What is a pest to one person may be an interesting and valuable form of wildlife to another. Thus if lions are eating the cattle, or even the children of an African villager, he looks on them as pests, and if elephants are destroying his crops he may consider them to be 'vermin', yet wildlife conservationists are doing their best to safeguard both lions and elephants. In Britain few people, other than those who grow sugar beet, look on larks as pests, and most gardeners enjoy watching bullfinches except when their fruit trees are being attacked. What are weeds to the farmer may be pleasant wild flowers to the naturalist. A pest is thus any organism which is doing damage and it is only a pest when such damage is serious or at least when it is likely to be significant if the potential pest is not controlled.

Pesticides and wildlife

In recent years there have been many spectacular successes in the control of pests by chemicals. However, there is also widespread concern about the possibility of environmental damage from pesticides, and this whole symposium, by its title, presupposes that biological rather than chemical means of pest control are desirable. This suspicion of chemical pesticides has largely been engendered by conservationists, some of whom are often fanatically 'anti-chemical', and, as a result, willing to welcome any type of pest control which can possibly be called 'biological'.

Conservationists have good reason to be suspicious of chemical pesticides. In Britain we know that, in the late nineteen fifties and early nineteen sixties, very large numbers of seed-eating birds died as a result of eating grain that had been dressed with dieldrin and aldrin to protect it against attack from the wheat bulb fly. We know that many predators, including foxes and raptorial birds, died as a result of eating birds poisoned in this way. Populations of many of these species were severely affected, almost certainly because of this insecticidal poisoning. In many countries other organochlorine insecticides, particularly DDT, have killed many birds, and have had serious effects on populations of fish and other aquatic organisms. Residues of these pesticides are widespread, and occur in many organisms at levels which, though not lethal, have some pathological effects. I believe that in Britain we have restricted the use of organochlorines sufficiently early to avoid permanent harmful effects on wildlife, but serious and probably permanent damage has been done in other countries. DDT is so persistent, and so susceptible to biological concentration by living organisms, that, even when properly used, it can have harmful ecological effects. This is why its banning is advocated in areas which can afford to use less dangerous (if more expensive and possibly less effective) substitutes, and its gradual phasing out in areas where disease vector control by other means would, at present, be impracticable, has a good deal of responsible support. A total and immediate ban on DDT in all tropical countries would be a disaster of unparallelled magnitude, but stricter controls on its use should be introduced without delay.

Other pesticides have caused ecological damage, especially when misused. Aerial spraying is effective in covering crops and forests, but it is difficult to avoid contaminating non-target areas. Nevertheless when carefully and responsibly used, many herbicides and insecticides do little damage, and have results very similar (e.g. the elimination of weeds in cereal crops) to those obtained by labour-intensive cultivations before the pesticides had been put on the market. It is even true that conservationists may find it satisfactory to use some non-persistent herbicides to control scrub on scheduled nature reserves. Nevertheless the minimum use of these chemicals, in all areas, is desirable, not only because of their dangers, but also because unnecessary spraying is costly and wasteful.

Ecological implications of other control methods

Previous papers have dealt with many types of biological—or at least of non-chemical—control. I shall now consider these from the point of view of wildlife conservation.

Several speakers have described the progress that has been made by plant breeders in producing cultivars resistant to pests and diseases. This approach is entirely acceptable. Pests usually reach epidemic numbers because man has produced conditions (e.g. large areas of monoculture) which encourage this to happen. If the crop plant does not support the pest, then little or no damage is done to wildlife; in fact damage may be avoided, when there is no large reservoir of, for instance, a polyphagous pest insect, to spill over into uncultivated areas.

Many agricultural practices discourage the build up of pest populations. Thus the rotation of crops had this as one of its objects. Crop rotation is often thought of as 'ecologically desirable', and it may well be so, but it is even more 'unnatural' than repeatedly growing the same crop on the same area of ground. Various speakers have discussed cultural means which discourage pests and encourage their natural enemies. This approach is again entirely acceptable to the conservationist.

The 'classical' examples of biological control include the introduction of potential enemies, often parasitic or predatory insects, into a new area in which they are not indigenous. The spectacular successes of this method have occurred in many cases where the pest itself was not native to the area. I think we should be very careful before using these methods in Britain. Today most ecologists are not at all happy about deliberately introducing exotic animals or plants into Britain, nor can we be sure that, once introduced, their effects will be solely beneficial. Control of pests in the artificial micro-climate of the glass house is perhaps the least to be criticized, particularly if the beneficial organisms used already occur in this country.

Another form of biological control which is becoming increasingly important is the use of pathogens. In Britain we have seen how effective this was, in the reduction of the rabbit population by the myxoma virus (myxomatosis). It is perhaps ironical that this undoubtedly successful example of biological control gave rise to such protest from many of those most critical of chemical pesticides. This was partly because of our ambivalent attitude to the rabbit. Had the species entered this country before the land connecting it to the continent of Europe was submerged, instead of having been introduced by the Norman invaders, it would have been accepted as a desirable, indigenous species. Even with this doubtful pedigree, rabbits have an emotional appeal, and their grazing saved some conservationists from the effort of scrub control. For whatever reasons, we rejected the deliberate use of myxomatosis to control an animal which, in many years, was an undoubted pest.

Many other pathogens, viruses, fungi and bacteria, are used to control pests. The conservationist should view all these with some caution. It is impossible to be completely sure that such a pathogen will not mutate into a form which attacks species other than the target pest. It may thus be more dangerous ecologically than even a persistent organochlorine insecticide, which will, inevitably if slowly, be eliminated from the environment, and which will at least not multiply in amount after its application.

Occasionally the conservationist may be inclined to agree to an introduction, as in the case of the grass carp to control water weed by grazing. This fish will not breed in Britain, and so could be said to act rather like a persistent herbicide, unlikely to spread outside the desired area.

Vectors of human diseases have seldom been controlled except by chemicals, or by the elimination of a habitat (i.e. draining a marsh, which also affects other, more desirable, wildlife). Progress is reported in genetic engineering, with the introduction of genes into a population and the production of sterility. Except that another mosquito may occupy the niche vacated by one so controlled, with unforetellable ecological consequences, these techniques are generally welcomed by conservationists. Otherwise large amounts of persistent insecticides may be liberated, and, as human life may be at stake, this may have to be accepted.

Today there is much enthusiasm for what is called 'integrated control', which means the sensible use of pesticides in such a way as to allow the natural effects of beneficial insects and other organisms to operate also. I assume that this also means that pesticides will only be used when they are needed to avoid economic damage. To be effective, these practices depend on thorough ecological studies of all crops. Where such information exists, integrated control is obviously good sense as well as good conservation. Thus the conservationist must accept *some* chemical control, and he must reject some non-chemical, 'biological' methods.

Wildlife conservation in the future

In the future in Britain wildlife conservation will have to be an active process. It is obvious that changes in the pattern of agriculture, as well as the spread of towns because of the increase of the population and of its affluence, will have very serious effects. Recently there has been an excellent dialogue between some farmers and some conservationists, and we have found that many forms of wildlife, particularly birds, can be encouraged to remain in large numbers on a modern farm if very slight changes in its planning (e.g. planting with trees and shrubs in odd corners to replace the uneconomic hedges which are removed) are made. This dialogue may have caused conservationists to be too optimistic. Birds are indeed an important part of our wildlife, particularly important because of the way they may serve as an 'early warning system' for new forms of environmental pollution, but plants and insects are equally important. Modern agriculture, e.g. the ploughing of old grassland, or its 'improvement' with fertilizer, completely and permanently eliminates most of the interesting plant species and the insects which depend upon them. If these species and communities are to be preserved, then the areas in which they live must be totally preserved, and managed for conservation and not for productive agriculture. This means that we need more nature reserves, and the need for their establishment is a growing one. Pesticide control is important for conservation, but pesticides are probably not, today, the greatest danger. Some forms of 'biological control' are obviously even more unacceptable than most chemicals. But it is the whole system of modern agriculture and industry with its pressure on all types of land which most endangers our wildlife.

Appendix: miscellaneous contributions

Introduction to a session on approaches to control

D. L. GUNN *Taylors Hill, Chilham, Canterbury, Kent; formerly Agricultural Research Council, London*

It is the business of forward-looking scientists to watch all the indications and to discover (and advise how to meet) any threatened disaster. How soon and how severe a coming storm is likely to be is a matter of opinion. Some people think that predictions of doomsday alert the stupid; they may irritate the already wary and bring the subject of environmental conservation into disrepute. Thus a great acreage of pesticide use seems to be entirely successful and harmless, skilfully managed and ecologically sound; but enough secondary, unintended and unwanted effects have occurred to stimulate the active movement, of which this Symposium is a part, towards a return to control of pests using pesticides less, if at all.

This afternoon, we are to hear from people whose main devotion is the protection of human health and human food from insect pests; they have not neglected the environment, but they have put our survival and health first.

Introduction to a session on the application of biological control

M. B. GREEN *Imperial Chemical Industries Limited, Mond Division, Runcorn, Cheshire*

The Society of Chemical Industry is very pleased to be one of the sponsors of this Symposium. The impression is given by the popular press—and sometimes also by reputable scientists—that, on the one hand, there is the pesticides industry concerned only with profit and wantonly and aggressively marketing chemical pesticides with unspecified hazards to the consumer and to the environment, while, on the other hand, there are the pure biologists and ecologists who offer a biological control which is 'natural' and free from risk. The opinion is also expressed that the pesticides industry has no interest in biological control 'because there is no money in it'.

I want here and now to scotch these ideas. We all have the same objective, namely, to develop methods of crop protection and pest control which are effective and economical, which protect crops, increase yields and safeguard essential supplies of food, and which offer maximum safety to the consumer and to the environment. I hope that the fact that this Symposium is organized jointly by the three Societies will bear witness that we are all united in this common aim.

I believe that a crop protection technology will evolve in which both chemical and biological methods have their rightful place, complementary and not antagonistic to each other. This can happen only on the basis of a unified science of crop protection and pest control which brings together all the relevant disciplines. Progress has been hindered and mistakes made because each discipline has worked within its own framework with insufficient intercommunication. We must breed men capable of comprehending all the complexities of a crop protection problem.

We know already from this Symposium that biological control presents its own particular problems and its own inherent risks. All of us here—and everyone concerned with the subject—have a grave responsibility to ensure that these problems and risks are fully appreciated and that biological control is used wisely and well. We must not let our own or public concern for the environment lead us to jump from the frying-pan into the fire.

The present group of contributions moves from the principles of biological control to its applications. These present a number of new problems both to the grower and to the manufacturer—not the least of which are economic problems. The pesticides industry is interested in biological control because its business is the whole field of crop protection and pest control, and it will play its full part in solving those problems which are its proper concern.

The development of biological control in the West Palaearctic Regional Section under the influence of OILB/IOBC

L. BRADER *General Secretary, IOBC West Palaearctic Regional Section, Instituut voor Plantenziektenkundig Onderzoek, Wageningen, the Netherlands*

The development of biological control (using the term in the broad sense adopted by both the IOBC and this conference) in the West Palaearctic Regional Section was strongly influenced by the original Organisation Internationale de la Lutte Biologique. This same influence is now being exerted by its successor, the International Organisation for Biological Control, represented in this region by the West Palaearctic Regional Section (WPRS), with its own Council, Executive Committee and hierarchy of Commissions and Working Groups, as explained below. IOBC is a non-commercial organization: it has no source of funds other than its membership subscriptions and, consequently, much of its work is dependent on voluntary participation. Members and co-operators are working towards the same objective, namely, the better use of the results of research activities in the control of noxious animals and plants.

The idea of an international organization for biological control was first mooted at a Symposium of the International Union of Biological Sciences in Stockholm in 1948, and the decision was taken to create the Commission Internationale de Lutte Biologique (CILB), the name being changed to OILB in 1965. Initially, the organization, concentrating on the use of parasites and predators, stimulated the idea of biological control among the authorities in different Western European countries. Contacts were established with FAO for the preparation of biological control projects in the developing countries. For example, a Working Group on biological control in cereals in the Middle East was directed by officers of OILB. This Working Group established the feasibility of biological control of *Eurygaster integriceps* at costs considerably below those of chemical control.

The Organization can certainly claim some credit for the creation in the 1950s of the laboratories for insect pathology in Madrid and Zurich, the establishment of a laboratory for biological control at Zemun in Yugoslavia, the transformation of the Research Institute at Darmstadt into an Institute for Biological Control and the foundation of the Laboratory for Microbial Control at La Minière in France. At the same time, the laboratories at Alès and Antibes in France oriented their research activities progressively towards biological control.

The real start of co-operative research activities came after the creation in 1956/57 of Commissions and Working Groups on specific problems such as the Commissions on Identification of Entomophagous Insects, and on Insect Pathology, and the Groups on San José Scale, *Dacus oleae*, and *Ceratitis capitata*, *Hyphantria cunea* (Fall Webworm), Defoliators in Mediterranean Forests, Ants of the *Formica rufa* Group, and *Leptinotarsa decemlineata*.

In all these groups, classical biological control formed the scientific basis. In the Commission on Integrated Control, where initially the pest control problems in orchards received most attention, a newer approach was adopted: an attempt was made to relate the action of natural enemies of pest species to the use of other methods of control, including chemical control. The activities of these Working Groups have certainly contributed to our knowledge of the advantages and shortcomings of biological control.

The Commission on Identification of Entomophagous Insects has rendered great service to many research workers through its Centre at Geneva.

Apart from its many other activities, much credit is due to the Commission for Insect Pathology for the establishment, at the Institute Pasteur, of a Centre for the Identification of Strains of *Bacillus thuringiensis*. The importance of this Centre has been clearly demonstrated in recent years, these entomopathogenic bacilli having attracted much attention.

The Working Group on the San José scale based its activities on the work done by Klett at Stuttgart and was strongly supported by workers in Germany, Switzerland and France. It finally confirmed the feasibility of controlling San José scale with *Prospaltella perniciosi*. The results obtained in the field with this parasite quickly vindicated earlier expectations, and led to a rapid consummation of the work of the Group. In 1969, it being considered that the aims had been achieved, the Group was dissolved.

The Group working on *Dacus oleae* started with the study of the implications of the widespread application of chemical control and came to the conclusion that an extremely unstable situation had been created; this was characterized by a reduction of the numbers of *Dacus oleae* but at the cost of aggravating attack by *Saissetia oleae*. For this reason, the Group concentrated its activities on biological control, its efforts resulting in the creation of a combined United Nations Special Fund-FAO project on the control of olive pests.

The activities of the Group on *Hyphantrea cunea* were centred around the laboratory at Zemun in Yugoslavia. With the help of OILB funds, three missions to North America collected parasites and predators, and helped at the same time to establish contacts with American workers. The limitations of biological control of this insect were clearly demonstrated and the Group has since shifted its attention to the microbial control of *Lymantria dispar*. Defoliators in Mediterranean forests proved to be a very complex problem. Although encouraging results were obtained with virus control of *Thaumatopoea pityocampa*, much more detailed ecological studies will be necessary before permanent solutions can be obtained.

The combined efforts of different workers on ants of the *Formica rufa* group led to a better understanding of the significance of these species in the regulation of noxious insects in forest trees. The conclusion was drawn that the protection of these ants would be of the greatest importance in forestry. Consequently the Group submitted a proposal to the Committee for the Conservation of Nature of the European Parliament to legislate for this. A law was enacted in 1964.

Workers in ten different countries collaborated in the early 60s in a combined effort to solve the problems created by the massive attacks of *Leptinotarsa decemlineata* on potato crops. The possibility of introducing a predator, *Perillus bioculatus*, was investigated. This predator was produced in eight different laboratories for two release programmes in Hungary in the years 1961 and 1962. Unfortunately, the introductions proved unsuccessful, probably because the predator dispersed too rapidly.

This summary of the activities of the first Working Groups of the OILB, although giving only a few of the results obtained in the field of biological control, reveals the potential for collaboration within Europe. Even so, it would be unfair to judge the value of the Organization on this information alone: these results must be viewed in context. In 1961, Grison came to the conclusion that, in France, investment in biological control amounted to only 0·1% of that in chemical control; in a similar study in Germany, Franz obtained a figure of 0·26%. In recent years this figure has changed in favour of biological control but still leaves by far the greater investment in chemical control. It is therefore still not possible to make a fair comparison between the two kinds of investment in respect of the benefits received.

Outside Europe, modern biological control techniques have produced some outstanding results. These aroused over-optimistic expectations in some quarters but it was soon found that success was the exception rather than the rule. Experience taught that managing insect populations was an extraordinarily complicated task; in most cases, success could be expected only after a long series of research experiments. The International Organisation for Biological Control of Noxious Animals and Plants considers that biological control offers great possibilities, but feels that more effective co-operation between workers is essential.

Reverting to the objectives of IOBC, it could be said that Phase I, namely awakening the different authorities to the need for the study and application of this more sophisticated approach to pest management, has made considerable progress. But Phase II, the transformation of the approach into practical techniques, has still to be achieved in the great majority of pest situations. The West Palaearctic Regional Section of IOBC wants to provide facilities for workers scattered throughout its region to collaborate in their specific fields. To this end, the system of Working Groups has been created and, in most cases, functions very satisfactorily. Currently, 18 of these Groups are operating. Some of the long-established Groups have, in recent times, given a good demonstration of the value of the system. Thus, the Working Group on Integrated Control in Orchards has produced valuable criteria for the application of integrated control in these situations. I am confident that effective solutions for pest control problems in apple orchards will be available in the very near future.

Wilson, during the Commonwealth Entomological Conference in 1960, expressed the opinion that ultimately the special need for collaboration in this field of biological control must lead to the formation of a world centre to serve the needs of national biological control institutions. The existence of strong regional organizations in the British Commonwealth, the Americas, Western Europe and the Soviet block should facilitate the formation of such a world organization. The recent administrative changes within IOBC meet these requirements. We may rest assured that those labouring in this extremely difficult field will justify the confidence with which we now accept the general philosophy of biological control.

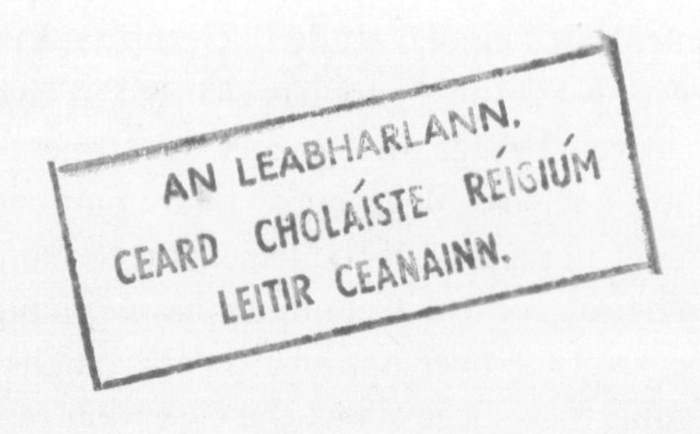

Author Index

Figures in italics refer to pages where full references appear

Adams D.J. 142 *145*
Adams T.G. 78 *83*
Alam M.M. 62 *71*
Alcock M.B. 32, *39*
Alexander T.T. 259 *267*
Allen G.P. 46 *53*
Allen J.A. 222 *229*
Alston F.H. x 73–86 *83* 284 *292*
Anderson M.M. 78 *83*
Andrewartha H.G. 166 *173* 205 *206*
Anon. 135 *143* 197 199 *206* 280 *291* 315 316 *319* 322 326 *330 331*
Anslow R.C. 32 *39*
Antonovics J. 28 39 *40*
Arant F.S. 131 132 *144*
Armour C.J. 215 *229*
Averre C.W. 240 *247*

Bailey V.A. 18 *26*
Baker K.F. 239 241 *247*
Baker R.H. 172 *176*
Balachowsky A. 152 *158*
Balch R.E. 121 122 123 *127*
Balis C. 241 *247*
Banks C.J. 290 *292*
Banks W.A. 132 133 *144 145*
Baranyovits F.L. 183 *195*
Barrer P.M. 140 *143*
Bartlett B.R. 63 *71*
Bartlett F.J. 132 133 *143 144 145*
Bate P.G. 47 *53*
Baumeister G. 77 *83*
Bawden F.C. 179 *194*
Beakbane A.B. 74 (unpublished)
Beard R.L. 183 *194*
Beatson S.H. 131 *143*
Beck S.D. 140 *143*
Becker H. 74 81 *83*
Bejer Petersen 305 *312*
Bell E.A. 7 *14*
Belton P 130 *143*
Bennett F.D. 62 *71*
Berkoff C.E. 155 156 *158*
Beroza M. 149 152 *160*
Berridge M.J. 157 *158*
Betts M.M. 214 *230*
Bevan D. x 302–14 *312*
Bhatnagar-Thomas P.L. 152 *161*
Biesterfeldt R.C. 141 *144*
Biever K.D. 123 *128*
Biffen R.H. 89 *95*
Bing A. 130 *145*
Bingham J. 92 *96* 101 *104*
Birch L.C. 205 *206*
Bird F.T. 120 122 123 124 *127*
Bishara I. 152 *158*
Bishop P.M. 130 132 140 *144 145*
Bjerke J.S. 148 *160*
Black M. 36 *40*
Black W. 88 89 *95*
Blackburn F. 263 *267*
Blake C.D. 251 260 *266*
Blanche D. 131 135 141 *143*
Blencowe J.W. 270 *276*
Bliss D.E. 243 *247*
Bohn G.W. 284 *293*
Borlang N.E. 90–5 *95*
Börner C. 74 *83*
Boston M.D. 168 *176*
Boswell A.L. 142 *145*
Boubals D. 74 *83*
Boudreau G.W. 221 *229*
Bowers W.S. 149 150 152 *159 161*
Boyce H.R. 251 *268*
Boys P.A. 79 *83*
Brader L. x, 359–62
Bransby-Williams W.R. 168 *173*
Bravenboer L. 198 *207* 296 *301*
Breider H. 74 *83*
Bremond J.-C. 221 *230*
Brenchley W.E. 36 *40*
Brian P.W. 94 *96*
Briggs J.B. 75 76 79 81 *84*
Britton D.K. 326 327 *331*

Broadbent L. 198 *206*
Brook J.M. 35 *40*
Brough T. 221 *230*
Brown R.M. 306 *312*
Brown W.L. 131 132 *143 144 146*
Brues C.T. 52 *53*
Bryan J.H. 163 170 *173 174*
Bryant P.J. 158 *161*
Buckner C.H. 66 *71*
Buczacki S.T. 242 *247*
Burdette W.J. 156 *159*
Burges H.D. 115 *127*
Burke A.W. Jr. 157 *159*
Bushland R.C. 165 *174*
Busnel R.-G. 221 *230*
Butenandt A. 154 *159*
Butterworth J.H. 156 *159*

Cameron E. 50 *53*
Campion D.G. 150 158 *159*
Canerday D. 120 *127*
Canfield R.H. 28 *40*
Carl K.P. 62 *71*
Carlisle D.B. 157 *159*
Carls G.A. 135 *145*
Carman R.M. 140 *145*
Carpenter J.M. 261 *267*
Carson R. 73 132 *143*
Carter A.L. 283 *292*
Carvalho J.C. 135 *143*
Casida J.E. 157 *159*
Cavers P.B. 52 *53*
Champness S.S. 36 *40*
Chancellor R.J. 43 44 46 47 *54*
Chang I. 240 *247*
Channon A.G. 284 285 *291*
Chant D.A. 68 *71* 106 *114*
Chapman R.K. 286 *291 292*
Chapple P.J. 216 *230*
Charles A.H. 45 46 *53*
Chater E.H. 52 *53*
Chaudhary K.D. 156 *160*
Cherrett J.M. x 130–46 *143 144* 197 *206*
Child R.D. 81 *84*
Chippindale H.G. 36 *40*
Chitwood B.G. 259 *266*
Christensen L.D. 166 *176*
Christie J.R. 249 *266*
Clark E.C. 121 *127*
Clark L.R. 16 *26*
Clarke A.J. 258 *266 268*
Clarke B. 222 *229*
Clarke S.P. 220 *231*
Clements R.F. 277 279 281 *291*
Cleveland L.R. 157 *159*
Clifford B.C. 95 *96*
Clothier S.E. 121 124 *127*
Coaker T.H. x 106–14 *114* 198 201 *206*
207 284 287 288 289 *291 293* 294 *301*
Cochran D.G. 158 *159*
Cockbain A.J. 271 *276*
Coffee Board of Kenya 184 *194*
Cohen C.F. 155 *160*
Cohen D. 36 37 *40*
Cole L.C. 202 *206*
Collyer E. 107 *113*
Colussa B. 165 169 *174*
Coluzzi M. 169 *174*
Concienne E.J. 132 *144*
Conference of Advisory Entomologists
289 *291*
Conway G.R. 165 *174*
Cooke D.A. 254 255 264 *266 268*
Cooke R. 263 *266*
Cooke R.C. 262 263 *266*
Cotten J. 261 *266*
Cox J. 165 169 *174*
Craig C.H. 130 140 *144*
Craig G.B. 163 171 *174 175*
Cramer H.H. 134 *144*
Crane M.B. 75 *84*
Critchley B.R. 150 158 *159*
Crooke M. 304 *312*
Cross D.G. 235 237 *238*
Cruickshank P.A. 149 151 *159*
Cuéllar C.B. 163 165 169 *174*
Cunningham R.T. 166 *176*
Curtis C.F. 171 172 *174*
Cussans G.W. x 49 *53* 64 *71* 97–105 *104*

Dale W.T. 283 *292*
Dame D.A. 168 *174*
Daubeny H.A. 75 *84*
Daulton R.A.C. 249 *266*
Davich T.B. 169 *174*
David 121 124 *127*
Davidson x 162–77 *174* 197 *206*
Davies J.M. 305 307 308 *312*
Davis D.E. 218 *230*
Dawkins P.A. 47 *55*
Dawkins R. 222 *230*
Day M.F. 75 *84*
DeBach P. 23 25 *26* 45 51 *53* 61 62 63
66 *71* 97 *104* 202 *206*
Dempster J.P. x 106–14 *113 114* 201 *206*
213 *230* 294 *301*

Dennis 290
Dethier V.G. 130 *144*
D'Herde J. 259 *266*
Dibley G.C. 281 *292*
Dimond A.E. 124 *127*
Doane C. 121 *127*
Doane J.F. 286 *291*
Dodd A.P. 44 *53* 102 *104*
Doggett H. 76 *84*
Donnelly J. 166 167 *174 175*
Doom D. 307 *312*
Doutt R.L. 61 *71*
Downing F.S. 151 *159*
Drechsler C. 262 263 *266*
Du Charme E.P. 253 *268*
Duclos L.A. 261 *267*
Duddington C.L. 262 263 *266*
Duncan C.J. 221 *230*
Dunn J.A. 284 286 290 *291 293*
Dunning R.A. 107 *114* 254 *268* 270 *276*
Dyte C.E. 150 158 *159*

Earle N.W. 155 156 *159*
Eastop V.F. 75 *84*
Echols H.W. 135 141 *144*
Edwards R.W. 233 *237*
Ehrlich A.H. 132 134 *144*
Ehrlich P.R. 132 134 *144*
Eisner T. 131 *144*
Elder W.H. 218 *230*
Elgee D.E. 120 *128*
Elliott J.G. 42 44 47 *53*
Ellis P. 151 155 156 *159*
Ellis P.E. 157 *159*
Ellis P.R. 287 *291*
Elton C.A. 202 *206*
Elton S.S. 106 *114* 212 213 *230*
Empson D.W. 257 *266 267*
Ensor H.L. 289 *291*
Entwhistle x 66 *72* 115–29 *127*
Epps J.M. 261 *267*
Eren J. 263 *267*
Esser R.P. 262 *267*
Evans K. *267*
Evans S.A. 43 46 47 49 *53 54* 178 *194*

Falcon L.A. 120 126 *127* 201 202 *206*
Feldmesser J. 259 *266*
Fennah R.G. 179
Fenner F. 214 215 216 *230*
Fernald H.T. 130 *144*
Fieser L.F. 156 *159*
Fieser M. 156 *159*
Finch S. 198 *206* 288 *291*
Fleschner C.A. 103 *104*
Fletcher J.T. 90 *96*
Forbes S.A. 60 *71*
Ford H.R. 168 *176*
Foster G.G. 172 *174*
Foster J. 50 *53*
Fox R.A. 243 *247*
Franca A. 80 *84*
Franklin M.T. 43 *54* 261 *267*
Franz J. 123 *127*
Friedman L. 149 *160*
Frings H. 221 *230*
Frings M. 221 *230*
Fryer J.D. 43 44 46 47 49 *54* 178 *194* 233 *237*
Fujimoto M.S. 166 *176*
Fujioka S. 155 *160*
Fye R.L. 168 *175*

Gadgil M.D. 39
Gahan J.B. 168 *175*
Gair R. x 190 *195* 277 278 279 288 *291* 315–20 *320*
Gajic D. 100 *104*
Galley D.J. 198 *208*
Gardiner B.O.C. 121 124 *127* 157 *160*
Garrett S.D. 65 *71* 239 241 243 *247*
Gassner G. 89 *96*
Gates L.F. *276*
Geier P.W. 16 *26* 165 *174*
George R.A.T. 284 *292*
Giban J. 221 *230*
Gibb J.A. 213 214 *230*
Gilbert B. 153 *160*
Gilbert E.G. 81 *83*
Gill J.S. 156 *159*
Gill R. 172 *174*
Gilmore S.K. 262 267
Glancey B.M. 130 140 *144*
Glynne M.D. 88 *96*
Godfrey B.E.S. 262 *266*
Goeden R.D. 103 *104*
Gonzales D. 201 202 *206*
Goodfellow A. 131 *145*
Goodwin T.W. 154 *161*
Gouck H.K. 168 *175*
Gould H.J. 78 *85* 296 *301*
Grace T.D.C. 121 *127*
Gradwell G.R. 16 17 18 25 *27* 204 *208* 212 *232*
Gramet P.M. 221 *230*
Grayson J.M. 158 *159*

Green C.D. 258 *267*
Green H.B. 132 *144*
Green J.O. 32 *39*
Green M.B. 357–8
Greenhall A.M. 211 *230*
Greenslade R.M. 75 *84* 191 *195*
Greet D.N. 258 *267*
Grist D.H. 179 *194*
Grogan R.G. 284 289 *293*
Grosdaude C. 242 *247*
Gross H.R. Jr. 132 *144*
Guarniera D. 168 *175*
Gudauskas R.T. 120 *127*
Guerrieri G. 166 *176*
Gunn D.L. 357

Hagen K.S. 190 *195* 196 201 202 206 207 *208* 211 *231*
Hall D.W. 131 *145*
Hamlyn B. 270 *276*
Hammerton J.L. 46 *54*
Hampson C.P. 325 *331*
Hanawalt R.B. 100 *104*
Hansen C.J. 76 81 *84 86*
Hardman J.A. 281 287 288 *291*
Hare W.W. 261 *267*
Harper J.L. 28 36 39 *40* 44 45 46 47 49 50 51 52 *54* 100 *104*
Harris E.J. 166 *176*
Harris P. 44 *56*
Harris V.T. 213 218 *231*
Hart W.G. 166 *176*
Hartley G.S. 191 *195*
Hassell M.P. 18 24 25 *26*
Hawkes C. 288 *292*
Hawkes G.R. 259 *267*
Hawkins J.H. 90 *96*
Hayes J.D. 95 *96* 261 *266*
Hayes W.A. 263 *267*
Hays S. 131 132 *144*
Heathcote G.D. 43 *54* 271 *276* 281 *292 293*
Heimpel A.M. 117 118 *127*
Hemming C.F. 130 *144*
Hensley S.D. 132 *144*
Hesse C.O. 76 *86*
Hickling C.F. 234 *237*
Higginson Y. 284 *291*
Hill A.R. 78 *84*
Hill S.O. 142 *145*
Hinde R.A. 221 *230*
Hinse C. 156 *160*
Hočevar S. 241 *247*
Holcomb R.W. 285 *292*
Holloway J.K. 102 *104*
Holm L. 100 *104*
Hooker A.L. 91 *96*
Hooper D.J. 261 *267*
Horsfall J.G. 124 *127*
Hostetter D.L. 123 *128*
Howard H.W. 76 *84* 88 *96* 261 *267*
Howard W.E. 213 221 *230 231*
Huber G.A. 75 *86*
Huffaker C.B. 43 47 52 *54* 98 102 *104*
Hughes R.D. 16 *26* 110 *114* 198 *207*
Hull R. x 197 *207* 255 264 266 269–76 *276*
Hussey N.W.,x 115 *127* 198 *207* 296 *301* 317 318 332–41
Hutchins R.F.N. 135 *145*

Ilan J. 153 *159*
Imai S. 155 *160*
Imms A.D. 62 *71*
Impekoven M. 222 *230*
Isaacson A.J. 221 *231*
Ishihara T. 179 *195*
Iszard R.E. 235 237 *238*

Jacob M.C. 221 *231*
Jaggard K.W. 273 *276*
James P.J. 257 *266 267*
Janzen D.H. ix 3–14 *14*
Jaques R.P. 121 *128*
Jenkins D. 213 *230*
Jenly 139
Jepson W.F. x 342–8
Johnson G.V. 130 *145*
Johnson O.A. 172 *176*
Johnson R. 91 93 94 *96*
Johnson R.T. 259 *267*
Jones D. Price x 60 *71* 178–95 *195* 202 *207*
Jones F.G.W. x 201 *207* 249–68 *266 267* 270 *276*
Jones I.T. 95 *96*
Jones M.G. 45 48 52 *54*
Joshi G.C. 167 *175*
Josling T. 326 327 *331*
Jost E. 172 *175*
Joyner S.C. 155 156 *160*
Jumber J. 221 *230*

Kahn R.P. 130 *145*
Kamakahi D.C. 166 *175*

Kaplanis J.N. 155 156 *160 161*
Karlson P. 154 157 *159*
Kasasian L. 179 *195*
Keep E. 76 77 79 *84 85*
Kelman A. 240 *247*
Kempster R.H. 130 *143*
Kempton D.P.H. 286 *291*
Kennard C.P. 135 *144*
Kennedy J.A. 67 *71*
Kennedy J.S. 75 *84* 202 *207* 317 *320*
Kester D.E. 81 *84*
Kilama W.L. 171 *175*
Kilpatrick J.W. 167 *176*
King H.E. 75 *85*
Kitzmiller J.B. 165 169 *174*
Klomp H. 214 *232* 305 *312*
Knight R.L. x 73–86 *84 85* 284 *292*
Knipling, E.F. 164 *175*
Kochba Y. 82 *85*
Kommedahl T. 240 *247*
Kovachich W.G. 279 280 *292*
Krafsur E.S. 172 *175*
Krieg A. 123 *127*
Krishnakumaran A. 148 149 157 158 *159 160 161*
Krishnamurthy B.S. 167 *175*
Kropàč Z. 36 *40*
Krstič M. 241 *247*
Krumholz L.A. 214 *231*
Krupauer V. 234 235 *238*
Kulkarni V.G. 149 *160*
Kust C.A. 36 *40*

LaBrecque G.C. 168 *175* 218 *232*
Lamb K.P. 152 *161*
Landa V. 150 153 *160*
Langenbuch R. 123 *127*
Larbey D.W. 250 251 *267*
Lashley R. 51 *54*
Laven H. 170 172 *175*
Law J.H. 148 *160*
Leach R. 242 *247*
Leaphart 242 *247*
Legowski T.J. 78 *85* 201 *207*
Leigh T.F. 201 202 *206*
Le Pelley R.H. 179 184 *195*
Le Roux P.M. 91 *96* 196 *207*
Letchworth P.E. 149 151 154 *160*
Lever R.J.A.W. 179 *194*
Lewis C.T. 156 *159*
Lewis T. x 107 *114* 130–46 *144* 197 *206* 281 *292*
Lin S.Y. 234 *238*
Lindley G. 74 *85*
Lindquist A.W. 173 *175*
Link K.P. 88 *96*
Linley J.R. 120 121 *128*
Linn J.D. 190 *195*
Liu G. 212 *231*
Lloyd H.G. 216 *281*
Lofgren C.S. 132 133 142 *143 144 145* 168 *176*
Lofts B. 218 220 *231*
Logan D.M. 166 *176*
Lombard P.B. 81 *86*
Long D.B. 198 *207*
Long W.H. 132 *144*
Lopez R.E.D. 135 *145*
Lovett J.V. 32 *39*
Lownsberry B.F. 76 81 *84 86*
Lucey D. 326 327 *331*
Luitjes J. 307 *312*
Lupien P.J. 156 *160*
Lupton F.G.H. x 87–96 *96* 284 *292*

Maas W. 347 *348*
MacArthur R.H. 106 *114* 202 *207*
MacConnell J.G. 135 140 *145*
MacDonald W.W. 171 *175*
Macer R.C.F. 88 93 *96*
Macfadyen A. 46 *54*
Machin D. 32 *39*
MacLeod J. 166 *175*
Madel W. 46 *54*
Maddrell S.H.P. 157 *159 160*
Maguadda P.L. 168 *175*
Mai W.F. 179 *195*
Makepeace R.J. 233 *237 238*
Malo S.E. 82 *85*
Mamet R. 214 *231*
Mankau R. 263 *268*
Mariconi F.A.M. 135 *145*
Markin G.P. 142 *145*
Marler P. 221 *231*
Marshall I.D. 216 230
Martignioni M.E. 122 124 *128*
Martin B.B. 130 140 *144*
Martin H. 178 195 344 *348*
Martin J.S. 135 *145*
Martin M.M. 135 140 *145*
Maskell F.E. 190 *195* 201 *207*
Masner P. 150 153 *160*
Mason G.F. 169 *174*
Massee A.M. 75 *84* 181 *195* 196 198 *207*
Mastenbroek C. 89 *95*
Mauffray C.J. 142 *145*
McCormick W.J. 132 *144*
McCray 167 *176*

N

McDonald P.T. 172 *176*
McGovern T.P. 149 152 *160*
McLaughlin R.E. 116 *128*
Mead-Briggs A.R. 216 *231*
Meerovitch E. 153 *161*
Meifert D.W. 168 *175*
Mellado L. 166 *175*
Mellanby K. x 69 71 349–53
Menn J.J. 149 151 154 *160*
Merrett M.R. 135 141 *144*
Mesnil L. 152 *158*
Meyer H. 172 *175*
Mian A. 172 *176*
Miaullis J.B. 149 151 154 *160*
Micks D.W. 169 *174*
Millard W.A. 241 *247*
Miller G.R. 213 *230*
Mills W.R. 89 *95*
Milstead J.E. 122 *128*
Milton W.E.J. 36 *40*
Mitchell R. 240 *247*
Mitchell W.C. 166 *176*
Monro J. 166 *173*
Montiero H.J. 153 *160*
Moore F.J. 43 46 *54*
Moorhouse J.E. 157 *159*
Mordue W. 158 *160*
Morgan E.D. 151 155 156 *159*
Morgan P.B. 169 *175*
Moriarty F. 263 *266*
Morlan H.B. 167 *176*
Morris K. 36 *40*
Morris R.F. 16 *26* 197 205 *207*
Mors W.B. 153 *160*
Moser J.C. 135 *145*
Mountain W.B. 82 *85* 251 *268*
Moutia L.A. 214 *231*
Mowat D.J. 198 *207*
Muftic M. 157 *160*
Mulder A. 265 *268*
Muller H.P. 288 *292*
Munakata K. 130 *145*
Munroe E.G. 63 66 *71*
Murdie G. 198 *208*
Murton R.K. x 211–32 *231*
Mustafa A.H.I. 140 *145*
Mykytowycz R. 216 *231*

Nadel D.J. 166 *176*
Neilsen H.T. 120 121 *128*
Neilson M.M. 120 *128*
Newton 166 *176*
Nicholson A.J. 17 18 23 24 *26*
Nickel C.A. 172 *176*
Nollen H.M. *265 268*
Norris R.F. 16 *26*

Odetoyinbo J.A. 165 *169 174*
Ødum S. 36 *40*
Ohinata K. 166 *176*
Oostenbrink M. 258 *268*
Orr L.D. 235 *238*
Osborne D.J. 157 *159*
Ota A.K. 130 *145*

Padovani I. 155 156 *159*
Paillot A. 122 *128*
Pallos F.M. 149 151 154 *160*
Palmblad I.G. 50 *54*
Palmere R.M. 149 151 *159*
Parker J.H. 76 79 *85*
Parnell F.R. 75 *85*
Parr W.J. 296 *301*
Parrott D.M. 201 *207* 250 251 254 256 261 *267*
Paterson H.E. 169 *174*
Patrick Z.A. 82 *85*
Patterson R.S. 168 *176*
Patton W.P. 169 *176*
Peacock A.D. 131 *145*
Pederson P.N. 93 *96*
Pelham J. 90 *96*
Pellegrino J. 153 *160*
Pentelow F.T.K. 235 *238*
Penzes B. 235 *238*
Perry V.G. 81 *86*
Peters N.C.B. 47 *53*
Petersen L.C. 89 *95*
Phillips D.V. 239 *247*
Pickett A.D. 196 *207*
Pilcher D.E.M. 157 *160*
Pimentel D. 63 *71* 108 *114* 202 *207* 214 *231*
Plant Protection Ltd. 190 *195*
Pollard E. 107 *114*
Potter C. 200 *208*
Potts G.R. 46 *55* 201 *207*
Pramer D. 263 *267*
Proctor J.H. 183 *195*
Prout T. 172 *174*
Proverbs M.D. 162 166 *176*
Punja N. 151 *159*
Putman W.L. 120 126 *128* 130 *145* 196 *207*
Putwain P.D. 50 *55*

Rabotnov T.A. 28 *40*
Radcliffe E.B. 286 287 *292*
Rai K.S. 172 *176*
Rapoport E.H. 253 *268*
Ratcliffe F.N. 215 216 230
Rautapää J. 78 *85*
Ray S.N. 167 *175*
Redfern R.E. 149 *152 160 161*
Reed B.C. 169 *176*
Rees H.H. 154 *161*
Reierson D.A. 131 *146*
Richardson N.L. 166 *173*
Ricker D.W. 103 *104*
Ricklis S. 153 *159*
Riddiford L.M. 150 *160*
Ridpath M.G. 221 *230*
Riley C.V. 74 *85*
Ripper 191 *195* 196 *207*
Rishbeth J. x 239–48 *247 248*
Rivers C.F. 122 126 *128*
Rivosecchi L. 166 *176*
Robbins W.E. 155 156 *160 161*
Roberts H.A. 36 *40* 47 51 *55* 283 *292*
Robinson D.W. 43 *55*
Robinson E.J. 36 *40*
Robinson W.B. 213 218 *231*
Robson T.O. 233 235 236 237 *238*
Roddy L.R. 132 *146*
Rogers D.J. 25 *26*
Rolfe S.W.H. 109 *114* 289 *292*
Roller H 148 152 *160*
Romanuk M. 148 150 151 *160 161*
Ross G.J.S. 201 *207* 261 *267 268*
Royama T. 25 *26* 224 *231*
Rudd R.L. 132 *145* 202 *207*
Ruscoe C.N.E. x 147–61, *159 160* 197 *207*
Rush G.E. 259 *267*
Russell E.J. 253 *268*
Russell G.E. 93 *96*
Ruston D.F. 75 *85*

Sacca G. 168 *175 176*
Sacher R.M. 153 *160*
Sadd J. 75 *84*
Sagar G.R. ix 28 *40* 42–56 *55* 102 *105* 205 *207*
Sakai M. 155 *160* 172 *176*
Salisbury E.G. 100 *105*
Salter D.D. 110 *114*
Sanchez-Riviello L.M. 169 *176*
Sarukhán J. ix 28–41 *40* 47 51 *55*
Sato Y. 155 *160*
Saunders J.H. 75 *85*
Sawall E.F. 202 *207*
Sawyer B. 163 *174*
Sayer H.J. 190 *195*
Sayre R.M. 262 *268*
Schilder F.A. 74 *83*
Schindler U. 309 *312*
Schmid P. 124 *128*
Schmidt C.H. 167 168 *174 176*
Schmidt C.T. 249 *268*
Schneiderman H.A. 148 149 157 158 *159 160 161*
Schwartze C.D. 75 *86*
Schwarz M. 149 *161*
Scopes N.E. 296 299 *301*
Scott P.R. 93 *96* 241 *248*
Scott T.M. 305 306 *312*
Seabrook E.L. 167 *176*
Seaforth C.E. 140 *144*
Seay R.S. 123 *128*
Seecof R. 121 *128*
Sehnal F. 158 *161*
Seinhorst J.W. 258 *268*
Selinger R. 172 *175*
Selman M. 43 *55*
Shanta C.S. 153 *161*
Sharpe R.H. 76 81 *86*
Shaw J.G. 169 *176*
Sheail J. 216 *231*
Shepherd A.M. 258 263 *268*
Shorey H.H. 130 *145*
Shortino T.J. 155 156 *160*
Sidor C. 118 *128*
Simmonds F.J. 97 *105*
Simpson C.J. 283 *292*
Sims B.G. 134 *144*
Skaff V. 153 *161*
Skinner G. 288 *291*
Slama K. 148 149 150 151 153 155 158 *161*
Sly J.M.A. 279 *292*
Smallman B.N. 106 *114*
Smirnoff W.A. 122 *128*
Smith C.N. 168 *175*
Smith F.F. 130 142 *145*
Smith H.S. 60 *72*
Smith I.C. 131 *145*
Smith J.G. 108 113 *114*
Smith K.M. 118 122 *128*
Smith P.M. 200 *208*
Smith R.F. 181 190 *195* 196 201 *208* 211 *231*
Smith R.H. 171 *176*
Snyder W.C. 239 240 *247 248*
Sobers E.K. 262 *267*

Solomon M.E.,60 *71* 205 *207*
Sonnett P.E. 149 *161*
Soo Hoo C.F. 123 *128*
Sorm F. 148 149 150 *161*
Southern H.N. 216 *231*
Southwood T.R.E. 106 *114* 199 203 205 *208*
Speyer E.R. 297 *301*
Spielman A. 153 *161*
Spink W.T. 132 *144*
Spishnakoff L.M. 169 *176*
Sprock C.M. 221 *231*
Stairs G.R. 124 *128*
Stapley J.H. 131 *145*
Stark R.W. 302 303 310 *312*
Staub A. 80 *86*
Steiner L.F. 166 *176*
Steiner O.D. 135 *145*
Steinhaus E.A. 120 124 *129*
Stella E. 168 *176*
Stephenson J.W. 130 *144*
Stern V.M. 191 *195* 196 201 202 *206* *208* 211 *231*
Stoczéń S. 213 *232*
Stokes F.G. 283 *292*
Stone A.R. 261 *267*
Stonehouse B. 214 *232*
Stott B. x 214 *232* 233–8 *238*
Straib W. 89 *96*
Strickland A.H. x 198 199 *208* 279 281 *292* 321–31 *331*
Stringer C.E. 130 132 133 140 *144* *145*
Stroganov N.S. 237 *238*
Sturhan D. 261 *268*
Sua V.I. 81 *86*
Suchy M. 149 *161*
Suit R.F. 253 *268*
Svoboda J.A. 155 156 *161*
Sweetman H.L. 212 *232*
Swift M.J. 242 *248*

Tamm C.O. 28 *40*
Tanada Y. 119 *129*
Tapley R.G. 184 *195*
Tassan R.L. 202 *207*
Taylor J.B. 240 *248*
Taylor T.H.C. 63 *72*
Taylor W.C. 163 164 *177*
Teschapek M. 253 *268*
Theane R.J.P. 218 221 *231*
Theune D. 296 *301*
Thomas P.J. 152 *161*
Thompson C.G. 124 *129*
Thompson H.V. 215 *229*
Thompson J. 218 *231*
Thompson M.J. 155 156 *160*
Thompson W.R. 23 25 *26* 64 *72*
Thurston J.M. 43 47 49 *54* 100 *105*
Tinbergen L. 214 222 *232*
Tinsley Ț.W. x 66 *72* 115–29
Tolg I. 235 *238*
Tomlinson J.A. 283 289 *292*
Townshend J.L. 250 *268*
Travis B.V. 131 132 *145*
Tribe H.T. 263
Trudgill D.L. 261 *267* *268*
Turnbull A.T. 106 *114*
Tydeman H.M. 75 80 *84*
Tyler P.S. 131 *145*

U.S. Department of Agriculture 181 182 *195*

Vallega J. 90 *96*
Valverde J.A. 214 *232*
van den Bosch R. 63 *71* 181 190 *195* 196 201 202 *206* *208* 211 *231*
van den Brande 259 *266*
van der Laan P.A. 263 *268*
van der Plank J.E. 91 *96*
van Dobben W.H. 213 *232*
van Emden H.F. 43 *53* 82 *84* 107 *114* *276*
Varley G.C. ix x 15–27 *27* 28 *41* 204 *208* 212
Vial P.V. 123 *128*
Vickerman G.P. 201 *207*
Villa R.B. 211 *282*
Vinson J.W. 150 *161*
von Steudel W. 275 *276*
Voûte A.D. 106 *114* 212 *232*

Wagner R.E. 131 *146*
Wagoner D.E. 172 *176*
Wain R.L. 193 *195*
Wakabayashi N. 149 *161*
Walker J.C. 88 *96*
Walker W.F. 149 150 152 *161*
Walkey D.G. 284 *292*
Wallace D.B. 326 327 *331*
Wallace H.R. 250 *268*
Wallbank B.E. 288 *292*

Wallis G.W. 243 *248*
Walton K.C. 216 *231*
Walton P.D. 151 *161*
Wapshere A.J. 44 51 *55*
Ward C.P. 213 *230*
Ward G.D. 213 *230*
Ward R.A. 171 *176*
Wareing P.F. 36 *41*
Warrington K. 36 *40*
Washino R.K. 190 *195*
Watson A. 213 *230*
Watson D.J. 275 *276*
Watson M. 270 *276*
Watson M.A. 275 *276*
Watt K.E.F. 25 *27* 106 *114* 202 *208* 212 221 *232*
Watts L.E. 284 *292*
Way J.M. 43 46 *55*
Way M.J. x 69 *72* 196–208 *207* *208* 281 290 *293* 294 301
Webber L.R. 250 *268*
Webley D. 131 *146*
Weidhaas C.S. 167 168 *176* 218 *232*
Welch J.E. 284 289 *293*
Wells G.J. 101 *105*
Wesson G. 36 *41*
Westigard P.H. 81 *86*
Westwood M.N. 81 *86*
Westwood N.J. 220 221 *231*
Wheatley G.A. x 198 *208* 277–93 *293*
Whitaker T.W. 284 *293*
White D.F. 152 *161*
White G.B. 167 *176*
Whitehead A.G. 254 260 *268*
Whitehead F.H. 201 *208*
Whitesell K.G. 190 *195*
Whitten M.J. 163 164 171 172 *174* *176* *177*
Wicker E.F. 242 *247*
Wiehe P.O. 242 *248*
Wier S.B. 154 *161*
Wigglesworth V.B. 148 149 150 *161*
Wildbolz T. 80 *86*
Wilhelm J. 240 *248*
Williams C.M. 147 148 150 151 153 158 *160* *161*
Williams D.A. *113*
Williams E.D. 49 *55*
Williams J.R. 179 *195* 263 *268*
Williams O.B. 28 *41*
Williams T.D. 258 *266* *268*
Willig A. 154 *161*
Wilson B.J. 47 *55*
Wilson C.L. 65 70 *72*
Wilson D.E. 5 *14*
Wilson E.O. 132 140 *146*
Wilson F. ix x 59–72 *72*
Winkler A.G. 74 *86*
Winteringham F.P.W. 185 193 *195*
Witcomb D.M. 233 *238*
Wolfe M.S. 93 94 *96*
Womeldorf D.J. 190 *195*
Wood B.J. 22 *27*
Woodbridge A.P. 151 155 156 *159*
Woodford E.K. 43 *56*
Worthington E.B. 59 *72*
Wright D.W. 290 *291*
Wright E.N. 214 221 *230* *232*
Wright J.E. 155 *161*
Wyatt I.J. x 198 *208* 294–301 *301*

Yde-Andersen A. 244 *248*
Yuan C. 148 *160*

Zaoral M. 149 151 *161*
Zárate L.L. 135 *146*
Zdarek J. 150 153 *160*
Zink F.W. 289 *293*
Zolotova Z.K. 235 *238*
Zwölfer H. 44 45 *56* 102 *105* 106 *114*

Subject Index

Abax and winter moth 16
Abramis brama 236–7
Abraxas grossulariata 126
Acacia host of bruchids 6
Acanthoscelides on *Mimosa* 10
Accipiter gentilis 213
Aceria spp. 78
Achromobacter nematophilus 115
Acridotheres tristis 214
Acromyrmex 134–42
Actinolaimus 262
Acrythosiphon pisi 188
ADAS (Agricultural Development and Advisory Service) 199 201 290 340
advisory entomologists, see 'ADAS' and 'technical services'
Aedes translocations 172
Aedes aegypti 152–3 163 167 172 184
Aedes albopictus 172
Aedes scutellarius 170
Aedes taeniorhynchus virus 120 121 128
aerial spraying 132–3 135–7 142 189 350
Africa 168–9 184 190 242 260
Agonum dorsale 112
Agricultural Development and Advisory Service, see 'ADAS'
agricultural experiment stations in tropics 13
Agricultural Research Council 340
agricultural revolution based on weed control 199
agriculture (see also 'barley', 'cereals', 'cover crops', 'economics', 'rotation', 'sowing to a stand', 'sugar beet', 'tape-seeding to a stand', 'technical services', 'vegetable production', 'wheat')
  arable: crops mostly annuals 269; differences from other systems 269; monoculture on Canadian prairies 203–4; shelter around fields as source of pests and predators 275; types of environment 269
agriculture—*cont.*
  attitude of 'new men' to control, grading, inspection 328–30
  catastrophic pest losses uncommon in UK 317
  changes in pattern affect wildlife 353
  consequences of withdrawing organochlorines 317
  crops in irrigated desert, California 202
  cultural pest control ecologically acceptable 351
  decline in labour, increase in machinery 321–2
  ecology of monoculture and feasibility of pest control measures 203–4
  expansion of on-farm storage 325 326
  factory-farming of livestock 321
  farmer's aims, and criteria of success 269
  four trends likely to continue in UK 330
  growth of large farm units 323–4 330
  growth of specialized farming systems 322–3
  how more land for cereals may be gained, UK 327–8
  husbandry 45–6
  immediate prospects for biological control 317–19
  improvement of old grassland: ecological losses 353
  industry demands complete freedom from pests 316
  influence of food-processing industry 316–17
  intensive culture, uniform environment 269
  intensive, needs skilful control of pests and diseases 269
  land use in Trinidad 135
  large farm units avoid diversity 202
  less land under some vegetables, fruit, but more produced 328
  livestock enterprises need equipment, transport 325

agriculture—*cont.*
man-power in processing and distribution 326
market requires flow of 'quality' produce 324 326
marketing, forward contract 324–6 330
mechanical harvesting demands evenly maturing crop 321
mechanization 321–2
movement of horticultural crops to arable production, increased output 322
need for costing of control programmes 329 330
new methods of crop production 321–2
plant spacing 321; influence on pest damage 329
potential of resistant cultivars 329–30
projection to the 1980s 326–8
projections in light of UK entry to EEC 326
prospects for biological control 328–30
regional predominance of various products, UK 327–8
rotation, see 'rotation of crops'
size of fields: demands of conservation 317
treatment of headlands, influence on pest damage 329
treatment of land, and influence on pests 329
tropical, ecological features 11
Agrilus ruficollis 78
Agriotes spp. 190 280 316–17
agroecosystems (see also 'agriculture' and 'ecosystems') 42 68 106 202–5 212–13
Agropyron gigantea 100–1
Agropyron repens 100–1 103–4
Agrostemma githago 50 100
aldicarb 258
aldrin 131 133 135 190 317 350
Aleochara spp. 111–12
alfalfa (see also 'lucerne') 201–2 204
alfalfa aphid (*Therioaphis trifolii*) 202–3
alfalfa caterpillar 122 124
algae as weeds 179
Allium vineale 99
allylisothiocyanate 288–9
almond 81
Amblycerus 6
America (see also 'Canada', 'Costa Rica', 'United States') 65 75–6 89 91 215
America, Central 4 5 6 10
American bollworm 184 202
4-aminopyridine 222
ammonium sulphamate stump treatment 244–6
amphibia for biological control 66
Amphorophora agathonica 75
Amphorophora rubi 76–7
Anastrepha ludens 166 169
Anatonchus, predatory nematode 262
Anopheles, hybrids 163 169
Anopheles atroparvus, myxoma vector 216
Anopheles farauti 170
Anopheles gambiae 165 184
Anopheles gambiae complex 169–70
Anopheles gambiae species A 163
Anopheles gambiae species A and B 172
Anopheles melas 169
Anopheles merus 169
Anopheles punctulatus complex 170
Anopheles quadrimaculatus 167
Antheraea eucalypti, virus in 121
Anthonomus 155
Anthonomus grandis 169 184
Anthriscus sylvestris 290
Antibes, France, laboratory 358
antibiotics and plant diseases 240
antibiosis (see also 'host-plant resistance') 287
antifeedants 130 155–6
ants 130–43 212 358 359
Apanteles flavipes 62
Aphelenchoides ritzemabosi 79 255
Aphelinus flavipes 336
Aphelinus sp. aff. *flavipes* 298
Aphidius matricariae 299
aphids 63 67 74–7 79 80 81 85 107 109–113 130 152 172 183 188 189 191 197 198 200–3 269–76 281–90 297–300 318 335–6
Aphis fabae 188 200–1 281–90
Aphis gossypii 297–8 335–6
Aphis grossulariae 79
Aphis idaei 77 85
apholate, chemosterilant 167–9
Aporcelaimus sp., Pl. opp. 262
apple 74–5 79–80 181–3 198 316
apple-grass aphid 80
apple mildew 80
apple sawfly 79 80
apple scab 80
apple sucker 80
apricots 81–2
Aquila chrysaëtos 213
arachnids 157
Arctostaphylos spp. 100
Armillaria mellea 242–3 246

arrestants 140
Arthrobotrys oligospora, Pl. opp. 262
Arvicola terrestris 213
Asolcus basalis 61
Aspergillus flavus 118
Aspergillus sp. 116 117
Aspidiotus perniciosus 82
Atomaria linearis 270
Atta 134–42
attractants, see under 'control'
Australia 28 61 63–6 98 216
Australian sheep blowfly 163 164
Austria 234
Avena fatua 47 49–50 100–1 103
Avena ludoviciana 100
azadirachtin 155–6
azinphos-methyl 152 346

Bacillus lentimorbus 116
Bacillus popilliae 116
Bacillus subtilis 240
Bacillus thuringiensis 116 118 183 358
bacteria 66 115–18 183 192 239–40 281 358
baits, toxic 130–43
Barbados 62
bark beetles 303–4 306–9
barley 93–5 101 324
bats 211
Bauhinia, host of bruchids 6
beans 200–1 229 281 282 316 317 322 328
beans, broad 328
beans, dwarf 282
beans, field 200–1 317
beans, French 316
Beauveria bassiana 118
Beauveria sp. 116 117
beet cyst nematode 259 265 270
beet eelworm 325
Beet Eelworm Orders (UK) 260 270
beet, red 271 316
beet, seakale 271
beet, sugar 93–4 107 182 189 197 198 269–76 281 317 325 328 346 347 349
Bellis perennis 50–1
Bembidion spp. 109 111 112
benomyl 298 334 337
γ-BHC 190 298 339
Biggleswade, Beds. 280
Bilharzia 184
biological control (see also 'natural enemies', 'parasites', 'pathogens for control', 'predators', 'sterile insect release', entries under 'weeds', and names or groups of pests, pathogens and weeds)
- advantages, demands on grower 301
- advantages and successes 66–7
- advantages of control by contract 340
- advisory entomologists' reviews 318
- and use of biological methods 59–71
- attitude of farmers and growers 318–19
- basis of OILB Working Groups 358
- biological forms of control: definition 60; examples 67
- compatible with unblemished produce 317
- deficiencies in transport of natural enemies 335
- development in West Palaearctic region 357–60
- difficulty, and need for better co-operation 359–60
- earliest use and meaning of term 60
- expenditure, and that on pesticide research and use 67
- FAO-OILB projects in developing countries 357
- future exploitation 339–41
- future in Britain 330 332–41 342–7
- growers' misgivings about planned introduction of glasshouse pests 340
- immediate prospects 317–19
- in an artificial environment 202
- in glasshouses 294–301 332–41
- in glasshouses, conditions influencing pests and natural enemies 294–6
- in glasshouses, experiments require space, equipment 336–7
- in tropics 11–14
- interest of pesticides industry 356
- limited funds 337
- limited potential in Britain 330
- management of, beyond the average grower 319
- models of 23–6 278 336
- need for advisory control 339–40
- need for caution in introducing natural enemies 44 351
- need for commercial producers of predators, etc. 338
- of chrysanthemum pests 299–300
- of citrus pests with ants 212
- of *Dacus oleae*: OILB Working Group, and UNO project 358

biological control—*cont.*
of *Eurygaster integriceps*, cheaper than chemical control 357
of glasshouse pests, difficulties of applying in industry 317–19
of glasshouse red spider mite, better and cheaper than chemical 297 317
of glasshouse red spider mite with *Phytoseiulus* 296–7
of glasshouse whitefly 297
of *Heliothis* upset by insecticides used against *Lygus* 202
of *Hyphantria cunea*, limitation of 359
of mosquito larvae with fish 214
of *Myzus persicae* on glasshouse chrysanthemums 299
of *Nezara viridula* with *Asolcus* 61
of plant pathogens 65–6 239–46
of rats with cats 213–14
of red locust with mynah birds 214
of San José scale with *Prospaltella* 358
of vertebrates 65 211–29 351–2
of water voles with weasels and ermines 213
of water weeds with grass carp 214 352
of weeds 64–5
organizational work of IOBC 70 357–60
partial control may be economically worth-while 330
preadaptation of natural enemies 68
problems and risks 356
procurement of natural enemies 335
prospects for 328–30
quarantine regulations an obstacle 337–338
regional organizations, need for world centre 360
research at laboratories in Europe 358–359
scope of 62–4
selection of appropriate (strains of) natural enemies 335–6
sporadic nature of pest attacks in Britain a hindrance 318
testing rate of increase of natural enemies 336
viewpoint of farmer and grower 315–19
with ants (*Formica rufa*) against forest pests 359
with bacteria 66 115–18 183 358
with birds 66 214
with fish 66 98 214 231 234–8 352
with fungi 66 116–18 244–6
with mammals 66 213–14
with mites 66 296–7 299 337
with nematodes 66 115
with parasites: properties for success 23 25–6
with pathogens, see 'pathogens for control'
with Protozoa 116
with Rickettsiae 116 117
with viruses, see 'pathogens for control'
Biological Control Subcommittee of British National Committee for Biology 59
birds 66 76 123 190 211 213–15 217–29 273 349 350 353
black bean aphis 188 200–1 281–90
blackberry 78
blackberry mite 78
blackcurrant gall mite 75 78
blackcurrant leaf midge 79
blackcurrant reversion virus 78–9
blackcurrants 78–9 325
'blanket' spraying of countryside 189
Blattella 155
blowflies 163 164 166 167
Bombyx mori 120
boric acid as chemical stressor 122
Botrytis cinerea 244
bracken 100
Bracon hebetor, venom 183
Brachycaudus helichrysi 81 300
brassica aphids, control 198
brassicas 108–13 270 279 280–2 286–7 289–90 316 325 328
bream 236–7
Bremia lactucae 281 284–5
Brevicoryne brassicae 107 109–13 281 286 289
British National Committee for Biology 59
British Sugar Corporation 270 325
brown rust of barley 94
bruchid beetles eating seeds 6 8 9 10 13
Brussels sprouts 108–13 280–2 286–7 289 316 325
Bryobia pratensis 82
Bulgaria 234
bullfinch 81 349
Bupalus piniarius 126 304–9
Burma (Rangoon) 170
Buteo buteo 213
Butlerius, predatory nematode 262
buttercup, see '*Ranunculus*'
buzzard 213
Byturus tomentosus 78

cabbage 286 287
cabbage aphid 107 109–13 281 286 289
cabbage caterpillars (*Pieris*) 63 108–10 112–13 121–2 124 213
cabbage looper 123 126
cabbage root fly 109–10 113–14 198 281 286–7 288–90 318 334 347
Cactoblastis cactorum 61 97 99–100 102
cadmium chloride as chemosterilant (birds) 218
calabrese 282
California 63 67 166 197 201–2 204 271–2 284 285
Cambridge Plant Breeding Institute 87 93
Campoplex oxyacanthae 305
Canada 25 61 63 122 126 166 196 203–4
canavine in seeds 7
Cannock Chase, Staffs. 304
Capnodis tenebrionis 81
Capri 166
capsid bugs 275 300
captan 240
carabid beetles 16 109 111–13
carbamate pesticides 191 344
carbon disulphide, soil fumigant 243
Carduus nutans 51
Carex pseudocyperus 235
carp, common 234 237
carp, grass 98 214 234–8 352
carp, silver 237
Carpocapsa pomonella 16 80 126 130 152 166 184
carrot fly 281 288–90 318 347
carrots 279–81 288–9 316 321–2 325
Caryedon on *Bauhinia* 6
carrier in toxic baits 131
catechol and onion smudge 88
Catenaria anguillulae, Pl. opp. 262
cats 213–14
cattle rearing 323–4
cauliflowers 110–12 280 282 286–7 289 328
Cavariella aegopodii 281
Cecidophyopsis ribis 75 78
cedar waxwing 123
celery 279 280
Cephalosporium sp. 239 300
Ceratitis capitata 81 166 358
Ceratophyllum demersum, a food of grass carp 235
Cercosporella herpotrichoides, see 'eyespot of wheat'
cereal cyst-nematode 254 258 261 264 265
cereals 88–95 98–101 189–90 198 201 240 272–3 285 317–18 323–8 347 350
Chaetomium globosum 240
chafer grubs (Scarabaeidae) 316–17
chemical (pesticidal) control (see also 'fungicides', 'herbicides', 'insecticides', 'pesticides', and names of chemicals, pests, weeds, pathogens)
  biological principles involved 282–3
  can be achieved quickly and effectively 319
  chemical weeding universal 321
  contribution to increased fruits, etc. 342
  contributions biology can make to 278
  effect of plant density 281
  farmers and advisers against indiscriminate use 318
  lack of adequate alternatives 319
  many spectacular successes 350
  numerous and varied insecticides available 346
  pesticides essential in agriculture 274
  rejection of moral desirability of reduction 342
  repercussions: see entries under 'fungicides', 'herbicides', 'insecticides', 'pesticides'
  research problems beyond scope of postgraduate students 346
  success, and unwelcome aspects 355
  suspicion largely engendered by conservationists 350
  ultra low volume (ULV) spraying, topic for research 347
  unlikely to be supplanted for some time 319
  well-timed sprays may be preferred to granular systemic treatments 346
chemosterilants against birds 217–18
chemosterilants against insects 150–5 167–9 288
cherries 81–2
Cheshunt Experimental and Research Station 297
chestnut blight 241 247
chickweed, see '*Stellaria media*'
China 234
chlorfenvinphos 110–13 198
chlorpropham (CIPC) 283
Chondrilla juncea 98 101
chrysanthemum aphids 334
chrysanthemum leaf and bud eelworm 79 255
chrysanthemum leaf miner 299–300 336

chrysanthemums 79 295 299–301 334 336
Chrysolina hyperici 102
Chrysolina quadrigemina 97 102
Chrysopa perla 118
Cicadetta montana 78
CILB (Commission Internationale de Lutte Biologique) 357
Circus cyaneus 213
citrus 134–6 182 345 347
climatic stability and pest control 203–4 294–5
clover as ground cover 108–13
club-root 279 286
coal tit 214
Coccoidea, selective control 183
Coccinellidae 60 180 275 335 359
cochineal insects (*Dactylopius*) 97–103
Cochliomyia hominivorax (screw-worm) 163 165–6
cockroaches, control 131
cocoa in Trinidad 134–6
coconut moth 61
codling moth 16 80 126 130 152 166 184
coevolution 6–9
coffee 179 184
Coleoptera of stored products, juvenile hormone insecticides 152
Coleoptera, selective control 183
Colias philodoce eurytheme 122 124
collar rot of apple 80
Colletotrichum circinans 88
Colorado beetle 151 358 359
Columba livia 218
Columba palumbus, see 'wood-pigeon'
Commision Internationale de Lutte Biologique (CILB) 357
Committee for the Conservation of Nature (European Parliament) 359
Common Agricultural Policy, and projections of UK and EEC production 326
Common Market, see 'European Economic Community (EEC)'
Commonwealth Entomological Conference (1960) 360
Commonwealth Institute of Biological Control 69 298 335
competition: effects of other plants on cauliflowers 110–12; effects of crop on weeds 100–1; in *Ranunculus repens*, density and mortality 32–3; in regulation of weed populations 47–9; influence of other species on *Bellis perennis* 50–1; of *Peniophora* against *Fomes* 244–6; of *Polyporus* and *Polystictus* against *Armillaria* 246; of saprophages against *Endothia* 241 247; species competition for control 173
competitive displacement 3 9
compounds disrupting insect growth 158
conditional lethal genes 171–3
Conference of Advisory Entomologists (1957) 289
confusants, sex pheromones as 130
conservation 311 317 349–53 355
conservationists 134 349–53
control (see also 'aerial spraying', 'herbicides', 'hormones', 'host-plant resistance', 'insecticides', 'natural enemies', 'parasites', 'pathogens for control', 'pesticides', 'pheromones', 'predators', 'sterile insect release', entries under 'weeds', and names or groups of pests, pathogens and weeds)
- achievable freedom from pest damage 316
- adjustment of planting dates may be unwelcome 329
- antibiotics against plant diseases 240
- apartheid between proponents of biological and chemical control 345
- attractants in toxic baits 131–2 135 140–1 166 288–9
- attractants: pheromones 130 140–1
- behavioural methods: insects 130–43 202
- behavioural methods, vertebrates (esp. birds) 221–9
- biological control (with parasites, predators, pathogens), see 'biological control'
- biological forms of: definition 60; examples 67
- biological (*sensu lato*) of pests and diseases of sugar beet 269
- biologically-based methods, danger of rushing, and need for mathematical models 278
- 'blanket' application of any single technique dangerous 341
- by diversification of crop ecosystem 106–13
- by predators, hindered by crop rotation 270
- by sterilization, see 'sterilization—chemosterilants'
- chemical and biological methods should be complementary 356

control—*cont.*
chemical control, see 'chemical (pesticidal) control'
chemicals for disruption of insect growth: screening 158
chemosterilants 150–5 167–9 217–18 288
combinations of methods (see also 'integrated control') 46 49 153 166–8 197 243 346–7
confusants: pheromones 130
consequences of withdrawing organochlorines 317
conservationists' viewpoint 349–53
cost-benefit 269
'critical' pest density for control decision 304
cultural control measures ecologically acceptable 351
cultural techniques may prevent pest control 334
direct and indirect (of weeds) 49
ecological implications of various methods 351–2
effective, harmless control as aim 355–6
environmental features and feasibility of methods 203–5
genetic, see 'genetic control'
habitat control against birds 214–15; and against disease vectors 352
hormones for control 147–58 (see also 'hormones, insect'): advantages 147; insects suitable for control with juvenile hormone insecticides 152–3; juvenile hormone and mimics 147–154; not yet practical 147
host-plant resistance, see 'host-plant resistance'
host-plant resistance as a form of biological control 87
human and environmental concern 355
in orchards: IOBC (WPRS) Working Group on Integrated Control in Orchards 360
in orchards: OILB Commission on Integrated Control 358
instability of ecosystem may assist 205
interplanting as an aid 202 204
isolation of crop 204 271
medical pests and juvenile hormone mimics 153
microbial, see 'pathogens for control'
models of control by insecticides and by sterile release 164–5

control—*cont.*
natural control: conditions favouring or hindering 203
need for cheap, effective methods 319
need for costing of control measures 329 330 333
need for early warning system 201
new start needed, not recriminations 347
of birds, see 'birds, control'
of forest insects 302–11: an ecological problem 310; need for life-table studies 311; role of models 310; *via* manipulation of host plant 310; with pathogens 359
of mammals 65 213–17 221 351–2
of migratory and windborne pests 130
of olive pests: UNO project 358
of pests: cost small, but may limit cultural development 334
of pests: research worker hindered from development work 340
of pests, subsidiary to weed control in Britain 199
of pests and diseases: grower's attention to relevant factors 269
of pests and diseases in intensive agriculture 269
of plant nematodes 249–66
of vegetable pests: no one method succeeds 278
of vertebrates 65 211–29 351–2
of weeds, see 'weeds, control'
pesticides, see 'chemical control', 'fungicides', 'herbicides', 'insecticides', 'pesticides'
planning research based on objectives 347
relative investments in biological and chemical control (France, Germany) 359
repellent sounds and chemicals against vertebrates 221–2
repellent sounds: simulated bat cries 130
reproduction inhibitors, see under 'sterilization'
research requirements 347
sterile insect release (q.v.)
'third generation pesticide' 147
through endocrine system 147–58
through feeding behaviour 130–43
too much reliance on pesticides: ecological approach needed 59

control—*cont.*
toxic baits 130–43
trap crops 130 243
'venereal disease' effect 153
veterinary pests and juvenile hormone mimics 153
Cordia, host of bruchids 6
cormorant feeding on roach 213
corn (see also 'maize'), numbers of pests 182
Costa Rica, studies in 5, 6, 10
Cotswold area, sugar beet 271
cotton 75 182 201–2 204
cotton aphid 297–8 335–6
cotton boll weevil 169 184
cotton bollworm, control with virus 126
cotton stainer bug 149
cottony cushion scale controlled by *Rodolia* 60
couch grass 100–1 103–4
cover crops 243 270–3
Cratichneumon and winter moth 18
C. nigritarius and pine looper 305
creosote treatment of stumps 244
'critical' density of pests 304
Crofton weed, biological control 63
Cronartium ribicola controlled by fungus 242
crop density, effects of 100–1 272 281 329; effects on infestation 101 281 329; effects on weeds 100–1
Crustacea and β-ecdysone 157
crustecdysone 154
Cryptomyzus korschelti and *C. ribis* 79
Ctenopharyngodon idella 98 214 234–8 352
cucumber mosaic virus 281 283 286
cucumbers in glasshouses 300 337–9 396–398
Culex, chromosome translocations 172
Culex pipiens 170 172
Culex pipiens complex 170
Culex pipiens fatigans 163 167–8
Culex tritaeniorhynchus 172
cultivation, effect on weeds 45 99
cultural practices 45–6 197 199 201–2 205 336
Curaçao island 165–6
Curculionidae (weevils) eating seeds 6 8 9
curly top of sugar beet 273
currant clearwing moth 79
cutworms (Noctuidae) 126 131 280 290 300 318
cyasterone, structural formula 155
Cycloneda sanguinea 335
Cyperus esculentus 184
Cyprinus carpio 234 237
cytoplasmic incompatibility 170
Cyzenis, parasite of winter moth 18
Czechoslovakia 234

Dactylopius spp. 97 103
Dacus cacurbitae 166
Dacus dorsalis 166
Dacus oleae 358
Dacus tryoni 166
dairy holdings, England and Wales 323–4
dairying in UK 328
damsons, pest resistance 81–2
Daphnia sp. 234
Darmstadt, Institute for Biological Control 357–8
Dasyneura pyri 81
Dasyneura tetensi 79
DCF, 'Discounted Cash Flow' 343 345
DD (dichloropropane-dichloropropane) 265
DDT 132 189 276 304–5 317 342–3 350
DEF 148
Deladenus siricidicola 66
demeton 191 202
demeton-methyl 274
Dendroctonus micans 304
Denmark 244
density-dependence 4 8 9 17–26 29–39 212 214
Desmodus rotundus 211
development work: research workers, ADAS 340
Diatraea saccharalis 62 71
diazinon 298 339
22,25-diazocholesterol dihydrochloride 218
Didymella disease, tomatoes 286
dieldrin 132 133 189 190 198 317 350
Diglyphus isaea 336
dimethirimol 298
dimethoate 346
dinocap 337
Dioclea megacarpa toxic seeds 7
Diplogaster, predatory nematode 262
Diprion hercyniae, see '*Gilpinia hercyniae*'
Diptera, selective control 183
Dipterocarpaceae 5 7
direct drilling 321
Directives of EEC, and projections of production 326 328

Discolaimus, predatory nematode 262
Discounted Cash Flow (DCF) 343 345
disease transmission, genetic mechanisms 171
disodium octaborate 244
dispersal of fauna from propagation houses to glasshouses 295 299
Ditylenchus dipsaci 255 259 260
diversity 3 106–13 202–5 212 269 294–6 303 317 339
DMF 148 152–3
dodecyl methyl ether 149
Dorylaimus predatory nematode 262
downy mildew of lettuce 281 284–5
downy mildew of sugar beet 93–4 271–3
drilling to a stand 199
Drosophila melanogaster 121 172–3
dryberry mite 78
Duboscquia penetrans 263
'Dursban', mosquito larvicide 153
Dutch elm disease 303
Dysaphis devecta 79–80 83
Dysaphis plantaginea 79–80
Dysaphis pyri 81
Dysdercus fasciatus 188

earwigs in houses 132
East Malling Research Station 73–82
α-ecdysone 154–5
β-ecdysone 154–7
economic injury levels 289
economic thresholds 334
economics (see also 'agriculture', 'fruit', 'vegetable production', etc.) 315–17 321–8 343 345
ecosystems
  agroecosystems and biological control 68 212–13
  agroecosystems: diversification for pest control 106–13 212–13
  longevity enhances diversity 302
  simplicity, complexity and stability 42–43 106 202–5 212–13
eggs, UK and EEC production 326
Egretta garzetta 214
Eichornia crassipes 184
Egypt, *Spodoptera* 152
Elateridae, 'wireworms' 190 280 316–17
Elodea canadensis 235
Empira tridens 78
Encarsia formosa 297
Endothia parasitica 241 247
England and Wales, agricultural statistics 279–80 315–16 322–8
Entomophthora 117 273 335
environment (see also 'conservation(ists)')
  capacity of 34–5
  damage to, hazards to, see entries under 'aerial spraying', 'insecticides', 'pesticides'
environment-crop-pest interactions on sugar beet 273
Equal Pay Act (1970) 322
eradication: by sterile release 164–6 168 170; campaign condemned 132; genetic control, replacement preferable 173; of weeds 43–4 49
Erioischia brassicae 109–10 113–14 198 281 286–7 288–90 318 334 347
Eriophyes pyri 81
Eriosoma lanigerum, host resistance 74–5 79 80 318
Eriosoma pyricola, host resistance 81
Erysiphe 89 101 271
Erysiphe graminis 89 101
ethirimol 95
Eublemma, predatory noctuid 180
Euonymus europaeus 290
Eupatorium adenophorum 63
Europe, Western 70 89 357–60
European corn borer 130
European Economic Community (EEC) 317 322 326
Eurygaster integriceps 357
Euzophera semifuneralis 82
extinction of populations 4 24 (see also 'eradication')
eyespot of wheat 88 347

fall webworm 358 359
FAO-OILB biological control projects 357
farmers (see also 'growers') 201 315–19 353
farnesol 148 153
fenitrothion 346
Feronia 16 112
Fiji 61 64
fire ants 131–4 140–1
fish 66 98 213 214 234–8 350 352
flea beetles 275
flies, chemical control 131
Florida 165–8
foliage eaters 4 7 10–11

Fomes annosus 243–8
Fomes lignosus 243
forests 3–11 203–4 302–11 358 359
Formica rufa group 358 359
foxes 350
France 234 358 359
'French fly' (*Tyrophagus longior*) 298
fruit (see also 'orchards', and fruits by name) 6 8 73–83 316 318 322–3 325 328 347 349
fruit flies (various species) 81 166
fruit tree red spider mite 80 82 107 120
fungi (see also 'pathogens, for control', and plant pathogens by name) 66 88–9 91 93 116–18 263–4, and Pl. opp. 262
fungicides (see also compounds by name) 50 94–5 183 277
fungitoxins in sugar beet 94
Fusarium oxysporium f. *lycopersici* 239
Fusarium roseum 240
Fusarium solani f. *phaseoli* 24
Fusarium wilt of peas 286

Galerucella cavicollis 81
Gambusia affinis 214
gamma-BHC 190 298 339
Gelechia demissae 82
genetic control: cytoplasmic incompatibility 170; meiotic drive 173; needs background studies and field trials 173; population replacement preferable to eradication 173; sex distorters 173; species competition 173; sterile insect technique 162–170; translocations 171–3; welcomed by conservationists 352
genetics, see above, and 'host-plant resistance'
Geneva 358
German Plant Protection Conference (Berlin 1971) 346
German Plant Protection Institutes 347
Germany 234 309 346 357–8 359
Gilpinia hercyniae 120–2 124 126–7
Glasshouse Crops Research Institute 294–301 317 330 332
glasshouse red spider mite 296–7 299 317 334 336 337–8
glasshouse whitefly 63 297 298 318
glasshouses 198 294–301 351
Glossina austeni, chromosome translocation 172
Glossina morsitans, sterile release 168–9
Glossina spp., trypanosomiases 163
golden eagle 213
gooseberry 75 78–9
gooseberry sawfly, common 79
goshawk preying on wood-pigeon 213
Gossypium, various spp., resistance to jassids 75
Governing Conference of FAO (Nov. 1970) 342
grape vines (*Vitis*) 74 182 347
Grapholitha molesta 61 82
grass carp 98 214 234–8 352
grasshoppers, chemical control 131
grassland 29–39 42 203–4 327–8
groundsel transmitting virus 283
grouse, raptors and 213
growers (see also 'farmers') 301 315–19 321–30 333 340
growth inhibitors, see 'hormones, insect'
gypsy moth, N. American forests 303

halo blight of beans 281
Harpalus rufipes 109 111 113
harvest spider 109 111
Hawaii 63 64 99–100 249
hedge parsley and cabbage root fly 290
hedges, copses, etc., near crops
- a source of pests and natural enemies 275
- affect local distribution of pests and pathogens 281
- capsid bugs and mice invade crops from shelter 275
- reservoirs of natural enemies a factor in control experiments 339
- treatment of headlands, influence on pest damage 329
- trees and shrubs to replace hedges on farms 353

Helicobasidium purpureum 279
Heliothis spp. 126 156 184 202
Heliothis zea, 'American bollworm' 184 202
Hemicycliophora spp. (sheath nematodes) 259
hempa, chemosterilant 167–8
hen harrier and grouse 213
hepialids, control 126
heptachlor 132 133 190
herbicide 2,4,5-T and habitat 215

herbicides: advantage of inactivation in soil 187; and plant population dynamics 48; 'blanket' spraying for scrub control 189; chemical weeding universal 321; effects on weed population 47; fear of 46; for control of water weeds 233; impact on environment 46 215; key to modern vegetable production 280; leading to pest upsurge on sugar beet 107; non-persistent, to control scrub 351; often preferable to biological methods 97–9; phenoxybutyrics, selective 193; phenoxyacetics, systemic, selective 191; pre-emergence 283; rarely make environmental problems 183; selectivity of 183; 'side effects', should be evaluated 339; success in broad-leaved weed control 183; triazines, diversification of 346
Herpestes, mongoose, Jamaica 214
Hessian fly, resistance to 285
Heterodera avenae 254 258 261 264 265
Heterodera glycines 261
Heterodera rostochiensis 252 254–9 261–3 264–6
Heterodera schactii 259 265 270
heterosis (hybrid vigour) 169
Holland, see 'Netherlands'
Holy Island (England) 166–7
hop aphid 81 189
Hoplocampa flava, host resistance 82
Hoplocampa testudinea 79 80
hops, spraying against aphid 189
Hordeum murinum, weed 180
Hordeum sativum 93–5 101 324
hormones, insect, for control
  codling moth control with juvenile hormone 152
  DEF, juvenile hormone mimic 148
  diuretic and salivary hormone mimics as insecticides 157
  DMF, juvenile hormone mimic, structural formula 148
  dodecyl methyl ether, juvenile hormone mimic 149
  ecdysone mimics: insecticidal effects and use 155–6; and selective systemic action 156; toxicology 156–7
  ecdysones and mimics 154–8: environmental hazards 157; inhibit growth, cause sterility 154–5
hormones, insect, for control—*cont.*
  expense of producing mimics 156–8
  farnesol, juvenile hormone mimic 148
  instability a factor in possibilities for control 157–8
  juvabione, juvenile hormone mimic ('paper factor') 148–9 151; structural formula 149
  juvenile hormone, structural formula 148
  juvenile hormone and mimics 147–54: induce supernumerary nymphs 151; insecticidal effects 150; instability, and more stable compounds 154
  juvenile hormone mimics: against aphids 152; against medical and veterinary pests 153; and insecticides 150; and sterile male release 153; block mosquito development 152–3; chemical action 149–50; hybrid compounds 149; insecticidal use 151–4; limitations 153–4; mammalian toxicology 151; structural formulae 149 150
  methyl (dimethylhexyl) benzoate, juvenile hormone mimic, structural formula 149
  passage of sterilants in copulation 150 153
  possible development of resistance to 158
  possibilities for use in control 147–58
  sesoxane, juvenile hormone mimic, structural formula 149
  terpenoid ethers, juvenile hormone mimics 149
horsetails as weeds 179
horticulture, see 'agriculture', 'chrysanthemums', 'fruit', 'glasshouse(s)', 'orchards', 'vegetable production', and vegetables and fruits by name
host-plant resistance
  an ecologically acceptable control 351
  aphid-resistant lettuce varieties bred 318
  aphid-resistant strains of alfalfa 202
  appearance of new races 88–90
  breeding for: principles, methods 73–4 82–3 87–95; successes against disease 88
  chrysanthemum cultivars and aphids 299
  crucifer cultivars and egg-laying cabbage root fly 287

host-plant resistance—*cont.*
  development of spinach resistant to
    cucumber mosaic virus 286
  field resistance 91
  genetic strategy in breeding 90–3
  genetic variability, and breeding 87
  fungitoxins in sugar beet 94
  healthy trees and pine shoot beetle 308
  host selection and antibiosis 286–7
  in integrated control of disease 94–5
  increased interest in 73
  insecticides preventing selection for 286
  in vegetables 284–5
  less exploited against pests than against
    pathogens 318–19
  need for combined resistance to pests
    and diseases 284–5
  nematode-resistant varieties of potato
    bred 318
  (non-)race-specific 91–5
  of apple to fungal pathogens 80
  of apple to various pests 74–5 79–80
  of blackberry to various pests 78
  of blackcurrant to various pests 78–9
  of cotton to jassids 75
  of crops to diseases 87–95
  of fruit plants to pests 73–83
  of gooseberry: to gall mite 75 78–9;
    to various pests 78–9
  of peach to various pests 75–6 81–2
  of pear to various pests 80–1
  of potato to nematodes 75–6
  of *Prunus* spp. to various pests 81–2
  of raspberry: to aphids 75–8; to
    various pests 78
  of sorghum to *Quelea* (weaver bird) 76
  of sugar beet to downy mildew 93–4
    271–3
  of various crops to nematodes 260–2
  of vines to *Phylloxera* 74
  potential of resistant cultivars for control
    329–30
  premature use of resistant varieties 340
  relative permanence of 74–6
  selection of potatoes for resistance to
    wart disease 286
  selection of tomatoes for resistance to
    Didymella 286
  successes, acceptability, possibilities
    318–19
  to onion smudge, and catechol and pro-
    tocatechuic acid 88
  'vertical' and 'horizontal' hypersensitive
    reaction: of potato to *Synchytrium* 88;
    of potato to viruses 88; of
    sugar beet to *Peronospora* 93
host-plant tolerance, of *Peronospora* by
    sugar beet 94
housefly 166 168 172 188 218
Hungary 234 237 359
Hyalopterus amygdali and *H. pruni*, resis-
    tance of *Prunus* spp. 81
Hydrocharis morsus-ranae eaten by grass
    carp 235
hygiene in control of virus 197 270–1
Hylobius weevils in stumps 303
Hypericum perforatum, control by *Chryso-
    lina* 97 102
Hyperomyzus lactucae and *H. pallidus*,
    resistance of *Ribes* 79
Hyphantria cunea 358 359
Hypophthalmichthys molitrix 237

Icerya purchasi 60
immigration (see also 'invasion') 4 10 13
    43–4 48 270 298 339 341
Imperial College 201
India 75 167
industry: commercial rearing of predators,
    etc., for biological control 338
industry, food-processing 316–17 325
industry, pesticides
  attitude and aims 355–6
  concern for economical but effective use
    344–5
  co-operation with advisory and research
    services 343
  development of new products: no profit
    in first five to six years 343
  factors in selection of new pesticides 344
  formulation and labelling for safe use
    346
  improving pesticides by molecular
    diversification 346
  new products subject to competition,
    patent expiries, environmental
    demands 343
  providing extra residue data to reduce
    sales 339
  public relations tasks 343
  responsible attitude to environmental
    hazards 343
  restraints on choice of materials 343–4
  rewards of developing new compounds
    should not be too restricted 346

industry, pesticides—*cont.*
should participate in research embracing biological aspects 347
view of future of biological control in Britain 342–7
Insect Pathology Section, Glasshouse Crops Research Institute 300
insecticide resistance: brief references 59 73 171 199; simple genetic basis 171
insecticides (see also 'chemical control', 'control', 'fire ants', 'industry, pesticides', 'pesticides', and names of chemicals)
against *Dacus oleae* aggravating attack by *Saissetia* 358
'blanket' spraying of countryside 189
breakdown products should be innocuous 186–7
broad-spectrum, persistent, to control caterpillars and beetles 181
deaths of birds (and their predators 350) eating treated seed 190 350
deaths of poultry, livestock, wildlife 132
demands for safety, selectivity, non-persistence not met 344
dosages of azinphos-methyl or juvenile hormone to control codling moth 152
effects of ecdysones on plants 157
environmental damage (see also 'aerial spraying') 132 162 190 350
environmental hazards 131 134 181 186–7 189 193 199 350
harmful to mushroom crop 334
high volatility; advantages and hazards 187
hormones as 147–58
indispensable 196 204
killing fish, and birds and their predators 350
killing natural enemies of pests 22 107 198 202
less than ideal method 162
limited persistence for selective action 186 193
main means of pest control on vegetable crops 277
methyl analogues of organophorus compounds 346
microbial, scope for mass production 71
numbers used in Britain, mammalian toxicity 344
organochlorine: permanent harm to wildlife, but not in Britain 350

insecticides—*cont.*
over-reliance on, and failure 59
pest increases induced by 22 59 107 132 181 202 276 358
Pesticide Manual 344 346
population effects, theory 15 18–22 26 164–5
pressure against use of persistent organochlorines 317
protecting weeds 50
random screening 147
reducing diversity and inducing pest outbreaks 107
resistance affects only a fraction of the field 346
resistance to 131 152 153 158 184 189 198 199 295 299 334 335 346
restricting application by row placement, seed treatment, spot treatment 189–90
Schumann's plea for retention of wide range of compounds 346
selection of compounds for screening 192
selective chemicals preferred by farmers 318
selectivity: advantage, and difficulty, of low application rates 188–9; a function of the ecosystem 185; and ultra-low-volume (ULV) spraying 190; demeton at reduced rates against aphids 202; demeton spares aphid predators 191; granular preparations 346; group-selectivity 183 194; importance of size of market 185; intrinsic selectivity 185; likely to be achieved empirically 193; menazon selectivity related to spray concentration 188; need for selective control of caterpillars 181–182 194, Coleoptera, Coccoidea and Diptera 183; new ideas on its achievement 196; of malathion in glasshouses 297; of organophosphates and carbamates 191; of pirimicarb against cotton aphid 298; principles of exploitation of 187–92; properties required for 185–7 194; rôle of chemist, biochemist and biologist in development of 192–4; rôle of controlled dosage 188, of controlled release 191–2, of effective placement 189–90, of systemics 191, of ULV spraying 190; search for selective insecticides

insecticides—*cont.*
against cabbage root fly 198; selectives for use in glasshouses 296 298; significance and enhancement 178–194; social influences 192; species-specificity not feasible 179 180 183 194, except for some major pests 184 194; testing for integrated control 337; timing 200–1
should not undergo concentration in food chains 187
significance of adsorption on soil 187
'side effects' 15 22 134
systemic: and selectivity 191; gross feeders, e.g. caterpillars, tend to be immune 191
ultra-low-volume (ULV) spraying 190
insects: see Contents List for Chapters concerned with insects, and numerous topics and names of species indexed
inspection at factory 316
Institute for Biological Control, Darmstadt 357–8
Institute Pasteur, Centre for Identification of Strains of *Bacillus thuringiensis* 358
integrated control
adopted only if insecticide problems unsolved 198–9
advisory entomologists' reviews of possibilities 318
an exercise involving the whole community 341
and agronomy 199
and conservation 199 201 352
as conceptual basis of field research on control 205
attitude of farmers and growers 318 319
based primarily on natural enemies 196
community, not only farmers, must pay 341
co-operation of varied specialists 333–4
definition of research goal 333 340
demeton supplementing natural enemies of alfalfa aphid 202
descriptive studies unproductive 197
desirable even where no immediate benefits 205
diversity a factor in experiments on 339
early consideration of ways of applying techniques 340
economic constraints 333
economic thresholds must be established 334
integrated control—*cont.*
enforce codes of practice? 340–341
environmental features affecting feasibility 204
evaluating impact of biota and pesticides 337
experiments will cost more in future 339
FAO definition 197
grower needs knowledge of crop and pests 273
harmonizing controls of various pests and pathogens 335
hindrance from restrictions on minor pesticide usage and on pathogens 338 341
host-plant resistance in 94–5
important, but uneconomic? 67
in an artificial environment 202
in Britain 196–206 269–76 332–41
inadequate resources for testing effects of pesticides 337
integration of biological and chemical control: by spatial separation 298; by timing 298; of *Myzus persicae* on chrysanthemums 299; of tree fungi 243 246
integration of biological and chemical studies 345–7
integration of chemical with other forms of control desirable 60
integration of control of various pests and diseases in glasshouses 297–8
integration of host resistance and fungicide treatment 94–5
integration of mildew control and pest control on cucumbers 298
IOBC (WPRS) Working Group on Integrated Control in Orchards 360
justifiable cost of research must be determined 333
less successful than chemical control: examples 198
limited contribution of any one discipline 200–1 275
management of, beyond the average grower 319
manipulation of environment 202
may be increasingly useful as British agriculture changes 206
may dictate cultural methods 334
may include *ad hoc* combinations of methods 197
meaning of x 196–7

integrated control—*cont.*
might include increases in diversity 107
must usually operate over wide area 341
need for collaborative research 206
need for extensive, complex experimental areas 339
need for large research groups 338
of alfalfa and cotton pests, California 201–2
of aphids on sugar beet: examination, spray warning 275; insecticides need backing by other measures 275
of chrysanthemum pests under glass 299–300
of glasshouse pests 198 332–41
of plant nematodes 264–6
of sugar beet pests and diseases 269–276; International Congress on 269–70
OILB Commission on 358
organizational work of IOBC 70 357–360
persuading and educating growers 301
required on multidisciplinary level 199
requires adherence to a set plan of pesticide usage 341
research in Britain 198 332–41
research management requirements 338
research strategy 197–8
research worker should understand economic-political background of crop 332–3
rôle of ADAS 199
rôle of research institutes and universities 199–200
secondary parasites assist weed control 102
selection of target pests 334
short-term and long-term efforts of industry 199
strip cutting of alfalfa 201–2
success through experimental method 197
testing selectivity of pesticides 337
trinomial biological rating 345
wide use, and attitude of administrators 341 within- and between- crop procedures 201–2
International Biological Programme, co-operation 69
International Organization for Biological Control (IOBC) 59 70 357–60
International Union of Biological Sciences 59 357
interplanting alfalfa in cotton fields to aid pest control 202 204
intrinsic rate of increase 28
invasion 28 132 166 275 298 300 303
inverse density-dependence 35
IOBC (International Organization for Biological Control) 59 70 357–60; constitution and aims 357
irrigation affects infestation of vegetables 281
Isaria, entomophagous fungus 116
isolation of crops 204 271
Italy, sterile release of *Musca* 166
Itersonilia pastinacae, parsnip canker 285

Jamaica, mongoose introduced 214
Japan, grass carp 234
Japanese beetle 81 116 189
jassids, resistance of cotton 75
Juncus effusus eaten by grass carp 235
juvabione ('paper factor') 148–9 151

Kenya 168 184
'Kepone' and fire ant control 132 133
key-factor analysis 17 212

Lake Kariba, release of sterile *Glossina* 168–9
Lagopus lagopus, grouse 213
La Minière, France: Laboratory for Microbial Control 358
larch, control of *Fomes* on 243–6
larks, pests of sugar beet 273 349
Laspeyresia funebrana 82
Laspeyresia nigricana 126 290
Laspeyresia pomonella 16 80 126 130 152 166 184
late blight of potatoes 89
L-dopa in seeds 7
lead arsenate in baits 131
leaf-cutting ants 134–43
leaf-miners 191
leaf spot fungi of sugar beet 272–3
leatherjackets (Tipulidae) 280 318
Lemna spp. eaten by grass carp 235
Lepidoptera on chrysanthemums 300

Leptohylemyia coarctata, see 'wheat bulb fly'
lettuce 279 281 282 283 284–5 289 318
lettuce mosaic virus 289
lettuce root aphid 284–5
Levuana iridescens 61
life expectancy of *Ranunculus* 32–3
life tables 16–17 311
Ligula intestinalis in roach 213
linden bug and juvenile hormone mimics 149
Litomastix spp., parasites in *Thera* 214
Littleport, Isle of Ely 280
livestock rearing expansion (UK) 328
locusts, tracking of swarms 130
Lonchocarpus, bruchids on 6
Longidorus attenuatus and soil texture 254
Longidorus elongatus, resistance of *Rubus* 78
Longidorus spp. (needle nematodes) 254 255 259 264–5
loose smut of barley 93
lucerne (see also 'alfalfa') 101
Lucilia cuprina 163 164
Lucilia sericata 166 167
Lygeonematus abietinus 305
Lygus hesperus, exploitation of host preference 202
Lymantria dispar 121 359
Lymantria monacha (nun moth) 123

Macrobiotus vichtersii Pl. opp. 262
Macrocentrus ancylivorus 61
Macrosiphoniella sanborni 300
Madrid, laboratory for insect pathology 357
magpie moth, microbial control 126
maize (see also 'corn') 91 240
maize seedling blight 240
Malacosoma disstria 124
Malacosoma fragile 121
malathion 190 191 297
male polygamy and female monogamy, insects 163
Malus spp., resistance to pests 79
Mamestra brassicae, microbiological control 126
mammals 7 16 65 66 169 211 213–14 215–17 220 255 275 351–2
management, effect on weeds 45
Manduca and moulting hormone mimics 155
mangolds 271
mangroves, pure stands 5
manzanita (*Arctostaphylos*) 100
Mariana Islands 166
'matrone' mating inhibitor 163
meat 192 326
Mediterranean fruit fly 166
Mediterranean region, OILB Working Group on Defoliators in Forests of 358 359
Meloidogyne (root-knot nematodes) 76 81–2 260
M. arenaria and *M. hapla* in groundnuts 260
M. incognita and *M. javanica* 76 81–2 260
melon fly (*Dacus cucurbitae*) 166
menazon, and various pests 188
Metarrhizium anisopliae, toxin 118
Metarrhizium spp., entomophagous fungi 116
metepa, chemosterilant 167–8
methyl (dimethylhexyl) benzoate 149
methyl parathion 346
Mexican fruit fly 166 169
Mexico, screw-worm flies 166
mice invading crops 275
Middle East, biological control in cereals 357
mildew of cucumbers, integrated control 298
mildew of sugar beet 271–2
mildew on Brussels sprouts 281
mildews, selective fungicides 183
milk: UK production 326
Mimosa pigra, bruchids on 10
Mimosestes, bruchid on *Acacia* 6
Ministry of Agriculture, Fisheries and Food, safe use of pesticides 346
'Mirex' and ant control 132–5 141–2
Mission, Texas, screw-worm rearing 166
mites 66 75 78 79 80 81 82 101 107 120 188 191 296–7 299 317 334 336 337–8
  as biological control agents 66 296–7 299 337–8
Mobile, Alabama, fire ants 131–2
moles (*Talpa europaea*) 16 213
MON 0585, structural formula 153
mongoose introduced to Jamaica 214
Monk's Wood Experimental Station, diversification experiment 108
monoculture 42 46 99 107 203–4 212–13 294 309 328 334 344 351
Mononchoides, *Mononchus*, predatory nematodes 262

mortality: comparison of steady mortality with effects of sterile release (Knipling's calculations) 164–5; density-dependent 17–22 33 38; killing power, *k*-values 17–25; nematodes, percent kills from soil fumigation 265; weeds 30–1 33 38 47–53; winter moth 16–22
mosquitoes 120 121 128 152–3 163–73 184 189 214
mosses as weeds 179
moths, resistance of *Prunus* spp. 82
Mucuna, L-dopa in seeds 7
mule, characteristics of 169
Musca, hormones and mimics 154–5
Musca domestica 166 168 172 188 218
mushroom flies, chemical control 334
mushroom phorids, vectors of *Verticillium* 334
mustard 272 286
Mustela erminea and *M. nivalis* to control water voles 213
mutual interference constant 18–26
mycoplasma 179
myna bird controlling locusts 214
Myriophyllum spicatum in diet of grass carp 235
myxomatosis 65 215–17 351–2
Myzus cerasi 81

National Vegetable Research Station, Wellesbourne 284 287 288
Myzus persicae 189 270 283 299–300
Nasonovia ribisnigri 79
Nasturtium officinale and grass carp 235
National Institute of Agricultural Botany, Cambridge 284
National Institute of Agricultural Botany, Trawscoed 273
natural enemies (see also 'biological control', 'integrated control', 'parasites', 'pathogens for control', 'predators', and names of these and of pests)
conditions favouring or hindering 203–204
conditions provided by glasshouses 295–6
conservation of, in integrated control 201–2
many in same groups of insects as pests 180
of *Aphis fabae* and timing of aphicide treatment 200–1
of brassica pests, influence of diversity 107–13
of glasshouse pests 296–301 335–9
of nematodes 262–4
qualities relevant to biological control 68
slow increase in tropics 12
supplementing diet of 202
use in biological control (general account) 60–9
nectarines, resistance to pests 81–2
needle nematodes, see '*Longidorus*'
nematodes 66 82 115 179 249–66 279 318
Nematus ribesii 79
Nemeritis, mutual interference constant 18 24
Neodiprion lecontei, spread of virus 123
Neodiprion sertifer: control 126; virus 122–3
Neoplectana carpocapsae, symbiosis 115
net farm income, definition, values 315–316
Netherlands 213 258 261 265 284 296
nettle, annual, virus vector 283
New South Wales, *Dacus tryoni* 166
New Zealand, biological control 63 64 213
Nazara viridula, biological control 61
Nicholson's theory 17–18 23–4
nicotine smokes to assist biological control 299
Noctuidae, pests (see 'cutworms') and predators (see '*Eublemma*')
Nomadacris septemfasciata, control 214
Norfolk, parsnip production 280
Nuphar luteum and grass carp 235
nutgrass control 184

oak 16–22 204
oestrogens, reproduction inhibitors 217–218
OILB (Organisation Internationale de la Lutte Biologique) 357–60
OILB Commission on Identification of Entomophagous Insects, Centre, Geneva 358
IOLB Commission on Integrated Control 358
IOLB Commissions and Working Groups 358

oil palm defoliators, upsurge 22
olive, UNO project on control of pests 358
olive fly, biological control 358
olive moth, hormone insecticides 152
onion smudge, host resistance 88
onion, wild, taint from 99
onions: rotation 280; white rot 281
onions, green 328
Onychiurus armatus Pl. opp. 262
Operophtera brumata, see 'winter moth'
Ophiobolus graminis 241 258
Opuntia spp. (prickly pear cacti) 28 61 97 99 102 103
orchards, pest control (see also 'fruit', 'host-plant resistance', and fruits by name) 203–4 358 360
Organisation Internationale de la Lutte Biologique (OILB) 357–60
organophosphate pesticides: improvements 346; selective 191
oriental fruit fly, control 166
oriental fruit moth 61 82
Oryctolagus cuniculus (rabbit) 65 213 215–217 275 351–2
Ostrinia nubilalis and bat sounds 130
Oxford, conference on pressures upon British farmers 315

Pakistan, parasites of *Aphis gossypii* 336
Panonychus ulmi 80 82 107 120
pansy, field, herbicides 98
Papaver spp., regulation 50
paper factor (juvabione) 148–9 151
paraquat herbicide 187 215
parasite and host: differently affected by temperature 297; generations unsynchronized 297; oscillations 17
parasites
  bacteria: of insects 66 115–18 183 358; of plants 117 239–40 281; of mammals 192
  fungi: of insects 116–18 273 300 335; of plants, see 'plant diseases' and pathogens by name; of nematodes 262–4
  insect parasitoids 8–9 11–12 16–18 23–26 61–4 66–7 183 214 297–300 305 335–6 358
  nematodes: of insects 66 115; of plants 82 179 249–66 279 318; of mammals 153
  of nematodes 253 262–4
  of plants, see 'plant diseases' and 82 117 179 239–40 249–66 279 281 318
  of insects, see 'insect parasitoids' above and 66 115–18 183 273 300 335 358
  of mammals 153 192
  Protozoa: of insects 116; of nematodes 263
  Rickettsiae, of insects 116 117
parasitism: area of discovery 18; mutual interference constant 18 24–6; Nicholson's theory or model 17–18 23–4; quest theory, quest constant 18 23–6; Watt's attack formula 25
parathion 191 298
parsnip canker, host resistance 285
parsnips: Norfolk 280; output value 279
Parus ater feeding on *Thera* 214
Passer domesticus and insect virus 123
pathogens for control of insects
  bacteria: see '*Bacillus*' and also 66 116–118
  fungi: *Cephalosporium* on chrysanthemum aphid 300; *Entomophthora* for control of thrips 335; general account 116–18
  general and miscellaneous: environmental hazards 352; future scope 126–7; general accounts 66 115–27; hazards to vertebrates 117–18; Insect Pathology Section of Glasshouse Crops Research Institute 300; Institute for Biological Control, Darmstadt 357–8; Laboratories in Madrid and Zurich 357; Laboratory for Microbial Control 358; mechanisms of clearance for use need attention 339 341; methods of use 118–19; nematodes and symbionts 66 115; nematode *Deladenus* to sterilize females of *Sirex noctilio* 66; OILB Commission on Insect Pathology 358; Protozoa 116; Rickettsiae 116 117
  viruses: attempts at virus control of *Thaumatopoea* (OILB Working Group) 359; cytoplasmic polyhedrosis viruses (CPV) 120–1; dispersal of 122–4; epizootiology 120–6; general account 117–27; granulosis viruses (GV) 121–2 124 126; hazards to vertebrates 117–18; host specificity 117; inoculum

pathogens for control of insects—*cont.*
potential 124–6; mosquito iridescent virus (MIV) 120–1; nuclear polyhedrosis viruses (NPV) 120–5; occluded and non-occluded 117 120; persistence of latent viruses 121; production of 117; Sigma virus of *Drosphila* 121; survival of 120–2; virulence 124
pathogens for control of weeds 65
pea midge, ADAS warnings 290
pea moth 126 290
peach (*Prunus persicaria*) 76 81–2 182
peach-potato aphid 189 270 283 299–300
pear 80–1 182 316
pear bedstraw aphid 81
pear leaf blister mite 81
pear leaf midge 81
pear sucker 81
pear thrips 81
peas: 279 282 286 290 316 322 325 328
Pegomyia rubivora 78
pellet-seeding to a stand 321
Pemphigus bursarius 284–5
Penicillium rubrum, competing 241
Peniophora gigantea: controlling *Fomes* 244–6; economics 245; inoculation 245
Perillus bioculatus, predator of Colorado beetle 359
permanence of plant population, and feasibility of control measures 203–4
Peronospora farinosa 93–4 271–3
Peronospora parasitica 281
Peronospora schactii 271
pest
avoidance: of bullfinches threatening pears 81; of insect pests on apple 80; of weaver bird by sorghum 76
complexes of specific crops 180–2 184
epidemics, result of conditions created by man 351
management 15 193 196 296–301 334–9
meaning of 178 211 349–50
pesticides (see also 'chemical control', 'fungicides', 'herbicides', 'industry (pesticides)', 'insecticides', and individual compounds by name)
and wildlife 350–1; developing improved chemicals 346; essential for agriculture 274; formulation and recommendations for safe use 346; granular preparations for selective action 346; high plant density hinders application 281; many do little damage 350–1; many spectacular successes 350; minimal use desirable 351; over-dependence on, consequences 277; resistance to, induced 59 73 97 171 199; 'side effects' 15 22 134 339; success on vegetable crops 277; suspicion largely engendered by conservationists 350; 'third generation' 147
Pesticides Safety Precautions Scheme, ruling on soil drenches 339
pests, general aspects
arthropod pests of apple in UK, numbers in major taxa 181
groups of animals that include pests 179
main groups of insect pests of crops 180–1
noxious and beneficial species often closely related 180
number of species 178–9
outbreaks induced by pesticides 22 59 107 132 181 202 276 358
vertebrate, control of 65 211–29
Phaedon cochleariae, per cent kill with menazon 188
Phalacrocorax carbo 213
Phalangium opilio, effect of ground cover 109 111
pheromones, insect: ant trail pheromone 140–1; for selective action against Lepidoptera 183; sex pheromones as baits and confusants 130
Phialophora radicicola inhibiting *Ophiobolus* 241
Philonthus and winter moth 16
Phoma spp., parsnip canker 285
phorid flies on mushrooms 334
Phorodon humuli: resistance of *Prunus* 81; insecticide resistance 189
Phragmites communis eaten by grass carp 235
Phryganidia californica, effect of virus epizootic 124
Phylloxera vastatrix, resistance of vines 74
phytoecdysones 155
Phytomyza syngenesiae 299–300 336
Phytophthora cactorum, collar rot, apple 80
Phytophthora infestans 89
Phytoptus ribis 75 78
Phytoseiulus persimilis 296–7 299 337–8 and Pl. opp. 296

Pieris brassicae cheiranthi 121–2 124
Pieris rapae 63 108–10 112–13 213
pig production in UK 328
pigeon, domestic, and insect virus 123
pigeon, feral, control 218
pigeon, wood 213 222–9
pilot trials as ecological safeguards 44
pine, outbreaks of looper and bark beetle 304–9
pine looper 126 304–9
pine sawflies, chemical control cost 334
pine shoot beetle 304 306–9
pines, sawfly as major pest 126
pirimicarb and control of glasshouse aphids 298 335
pitfall trapping 108 109 111
plant-animal interactions (see also 'host-plant resistance'): a theme of current research 285–6; attack by animals 3–14 47–8 53; co-evolution 6–9; crop density and insect infestation 281 329; different requirements of scolytid species 304; effect of insects on growth of oaks 204; plant resistance assisting control by parasites 299; plant's defences (see also 'host-plant resistance') 3–8 12; tree density and searching insects 4
plant breeding (see also 'host-plant resistance'): for disease resistance in crops 87–95; for pest resistance in fruit 73–83; mutagenic radiation extending variability 87
plant density, see 'crop density' and 'competition'
plant diseases (see also specific diseases and pathogens by name): antagonists 242; biological control 239–46; control through nutrient supply 240 243; control by breeding resistance 73–83 87–95; development of new races of pathogens 89–90; *Fomes annosus* control 243–6; ring barking 242; seed inoculation 240 241; soil amendments 240 241; use of competitors 241–6; viruses, see 'virus diseases of plants'
plant exudates: affecting weeds 100; affecting nematodes 258–9
plant growth regulators, research requirements 347
Plantago lanceolata, population study 50
Plasmodiophora brassicae 279 286
plum, resistance to various pests 81–2
plum and prune, numbers of arthropod pests, USA 182
plum curculio, resistance of *Prunus* 81
Plutella maculipennis: biological control 63; dosages of menazon survived 188
Plutella xylostella, microbiological control 126
Podosphaera leucotricha 80
Poland, grass carp 234
Polygonum amphibium and grass carp 235
Polyporus adustus competing with *Armillaria* 246
Polystictus versicolor competing with *Armillaria* 246
ponasterone, moulting hormone mimic 155
Popillia japonica 81 116 189
population: census 29; density of aphids on sugar beet 274; density of cabbage root fly eggs 111; density, nematodes 254–8 264; density of *Pieris rapae* larvae on Brussels sprouts 108; density of *Ranunculus repens* 30–4; dynamics, long-term studies belong in universities 345–6; dynamics, study to help in weed control 104; equilibrium density, nematodes 257; fluctuation, nematodes 255–9; fluctuation (flux) of weeds 29–39 47–8; fluctuation notable in certain pests 318; increase, nematodes 255–8 264–5; increase, cotton aphid and parasite on cucumbers 297–8; instability, in glasshouses 294 295; life table studies of winter moth 16–17, of forest insects needed 311; models 15–27 30 164–5 278 310 336; mortality, see under 'mortality'; oscillations may be exploited for control 205; oscillations of nematodes 256–7; plant population studies 28–39 42–53; regulation 4 15–27 28–39 47–53 290; replacement preferable to eradication 173; sampling 16 108–12; stability, meaning of 205; stability, *Ranunculus repens* 35; stability and complexity of ecosystem 42–3 106 202–5 212–13; turnover 30
Potamogeton pectinatus and grass carp 235
potato cyst-nematode 252 254–9 261–3 264–6

Potato Marketing Board estimate of increase in processing 325
potato root eelworm, integrated control 201
potato, wart disease 88 286
potatoes 75–6 88 89 182 280 286 316–17 318 321 324 325 328
Pratylenchus penetrans 82 251 252
Pratylenchus sp., increase 264–5
Pratylenchus vulnus, resistance of *Prunus* spp. 82
predation: models of 18–19 21–2 25 336; of winter moth 16–26
predator-prey interactions (see also 'biological control')
  forest birds and prey insect populations 214
  invertebrate predators and brassica pests 109 113
  models of glasshouse mite interactions 336
  vertebrate predators and prey 212–14
  winter moth interaction models 19–22
predators: ants (*Formica rufa*) and forest pests 359; ants to control citrus pests 212; birds 66 123 213–14; Ciliata 262 and Pl. opp.; Collembola 262–3, and Pl. opp. 262; dying from birds poisoned by dressed grain 350; fish 66 98 214 231 234–8 352; fungi 263, and Pl. opp. 262; influence on sugar beet aphids 275–6; insects 16 60 80 109 111–13 180 275 335 358 359; invertebrate, of brassica pests 108–9 111–13; invertebrate, transmitting virus 123; life tables for 16–17; mammals 66 213–14 350; mites 66 296–7 299 337–8; mutual interference 19; nematodes 262 and Pl. opp.; of nematodes 251 262–3, and Pl. opp. 262; of seed-eating beetles 9; of winter moth 16 20–2 26; *Phalangium* 109 111; Proteomyxa Pl. opp. 262; seed 'predators', see 'seed eaters'; staphylinid beetles reducing cabbage root fly 113; soil predators not effective under cultivation and rotation 270; Tardigrada 262 and Pl. opp.; nematodes 262; of nematodes 251 262–3; of seed-eating beetles 9; of winter moth 16 20–2 26; *Phalangium* 109 111; seed 'predators', see 'seed-eaters'; staphylinid beetles reducing cabbage root fly 113; soil predators not effective under cultivation and rotation 270
preservatives in toxic baits 131
prickly pear cacti 28 61 97 99 102 103
progestins, reproduction inhibitors 217
pronamide, pre-emergence herbicide 283
Prospaltella perniciosi controlling San José scale 358
'protected' crops (see also 'glasshouses'), environmental features and feasibility of pest control measures 203–4 294–296
Proteomyxa Pl. opp. 262
protocatechuic acid and onion smudge 88
Protozoa 116 157 262 and Pl. opp. 263
Prunus spp. (stone fruits), resistance to various pests 81–2
Pseudaulacaspis pentagona 82
Pseudomonas aeruginosa 117
Pseudomonas phaseolicola 281
Pseudomonas solanacearum 239–40
Psila rosae 281 288–90 318 347
Psylla pyricola 81
Pteridium aquilinum 100
Pterochlorus persicae 81
Ptychomyia remota controlling *Levuana* 61
Puccinia chondrillina controlling *Chondrilla* 101
Puccinia glumarum f. sp. *tritici* 89
Puccinia graminis 89
Puccinia hordei 94
Puccinia sorghi 91
Puccinia striiformis 89 91–3
Puerto Rico lacks terrestrial vertebrates 11
pygmy beetle on sugar beet 270
Pyrrhocoridae and juvenile hormone mimics 149 151
Pyrus spp., resistance to *Psylla* 86
Pythium debaryanum 240

Queensland fruit fly 166
Quelea (weaver bird), 'resistant' sorghum 76
quest theory 18 23

rabbit 65 213 215–17 275 351–2
radish 286 287

Radophobes similis 82 253
ragwort 50 99
Ramularia defoliating sugar beet 272
Ranunculus acris 29
Ranunculus bulbosus 29 35 100
Ranunculus repens: competition 32; control 38–9; environment 34–5; life cycle 29; life expectancy 32–33; mortality 30–2 36–9; population density 30–4; population fluctuation (flux) 29–35 39; population growth rate 35; population regulation 29–35 38–9; population stability 35; population turnover 30; reproduction 29–39; seeds 30–9 (mortality, dormancy 36–9, predation of 37–9)
rape and cabbage root fly 287
raspberry (*Rubus idaeus*) 75–8 325
raspberry beetle 78
raspberry cane maggot 78
raspberry cane midge 78
raspberry leaf and bud mite 78
raspberry sawfly 78
rats: cats to control 213–14; warfarin bait 220
redbeet 271 280 316
red locust control by myna birds 214
red-necked cane borer 78
red spider mite, see '*Tetranychus urticae*'
red spider mites, progress in control 347
repellents 130
reproduction inhibitors against birds 217–18
research: applied, management of 338; pure, and needs of applied biology 197–8; research workers' viewpoint on integrated control 332–41
Research Institutes, collaborative work 199–201
reservoirs of natural enemies, factor in control 339
resistance, induced, to pesticides 59 73 97 131 152 153 158 171 184 189 198 199 295 299 334 335 346
resistance of plants to pests and pathogens, see 'host-plant resistance'
Rhagoletis cerasi, *R. cingulata*, *R. fausta* 81
Rhinocoris annulatus passing virus 123
Rhodesia, nematicides on tobacco 249
Rhodnius 151 157
Rhopalosiphum insertum 80
Rhopalosiphum nymphaeae 81
rhubarb 283–4 328
Ribes nigrum, see 'blackcurrant(s)'
Ribes nigrum sibiricum 78
Ribes spp., resistance to pests 78–9
rice, number of pests 179
Rickettsiae as biological control agents 116 117
roach, parasitized, eaten by cormorants 213
Rodolia cardinalis controlling *Icerya* 60
root exudates (see also 'plant exudates') 258–9
root-knot nematodes (*Meloidogyne*) 76 81–2 260
root-rot of beans 240
rosy apple aphid 79–80
rosy leaf-curling aphid 79–80 83
rotation of crops: effect on weeds 45–6 270; for pest control 351; restricted rotations, and pests 199; sugar beet, and pests and predators 270; to combat nematodes 257 259–60 265 270 325; usual with arable crops 269; vegetables 279–280
Rothamsted Experimental Station 258
Rothschild Report 345
Rubus fruticosus, blackberry 78
Rubus idaeus, see 'raspberry'
Rubus spp., resistance to pests 75–8
Rumania, grass carp 234
Rumex acetosa, *R. acetosella* 50
rusts, cereal, biological races 74
rutabagas (swedes) and cabbage root fly 286
ryania against codling moth 182–3

St John's Wort, control by *Chrysolina* 97 102
Saissetia oleae, infestation increased by insecticides 358
San José scale: control with *Prospaltella* 358; Working Group of OILB on 358
Sanninoidea spp., resistance of *Prunus* spp. 82
Santa Cruz, biological control 103
Saprobes, nematodes, population increase 264
sawflies, resistance of *Prunus* spp. 82
scale insects, resistance of *Prunus* spp. 82
scarabaeid beetles, control by Rickettsiae 116

scentless mayweed carrying virus 283
Sclerotium cepivorum and crop density 281
Sciurus variegatoides eating toxic seeds 7
Scolytidae 303–4 306–9
Scolytus multistriatus and Dutch elm disease 303
Scotland, blackcurrant and gall mite 78
screw-worm 163 165–6
seakale beet 271
seed bank 30 36–9 45 47–9 283
seed defences 6 7
seed dormancy 36–9 47
seed dressings, organochlorines 190 198 350
seed eaters: birds and dressed grain 350; bruchid beetles 6 8 9 10 13; immigration 10 11; on tropical islands 10–11; parasitization of 8–9; predators of 9; resource limitation 8; satiation 7 8; specificity 6 7 9; vertebrates 6–11 37 350; waiting insects 9–10; weevils 6 8 9
seed predator guild 3–11
seed shadow 5
seedling eaters 4
Seinura, predatory nematode 262
Senecio jacobaea 50 99
Senecio vulgaris carrying virus 283
Serratia marescens, bacterium 117
sesoxane, structural formula 149
shrew, *Sorex minutus* 255
shrews and winter moth 16
Siberia, grass carp 234
Sicily, sterile houseflies released 168
Sigma virus in *Drosophila* 121
siloxane to waterproof ant bait 141
silverfish, chemical control 131
Silvilagus, source of myxoma virus 65 215
Sirex noctilio, nematode used in control 66
Sitophilus granarius and menazon 188
size (area) of unit and feasibility of controls 203–4
slugs: damage to potatoes 316–17; fluctuation 318
smuts, cereal, biological races 74
snails: as biological control agents 66; molluscicides 184
Society of Chemical Industry 355
sodium arsenite, toxin in baits 131
sodium fluoride: as stressor 122; in baits 131
Solanum demissum 89
Solanum tuberosum ssp. *andigena*, nematode resistance 261
Solanum vernei, nematode-resistance 75–6
Solenopsis saevissima, fire ant 131–2
Solenopsis saevissima v *richteri*, imported fire ant 131–4
Sonchus, source of parasites for control 300
Sorex minutus, shrew 255
sorghum, arthropod pests of 182
Sorghum, 'resistance' to *Quelea* 76
sowing to a stand 273 280
soya bean, nematode-resistance 261
speedwells (*Veronica* spp.) as weeds 98
Spicaria, entomophagous fungus 116
Spilopsyllus cuniculi, myxoma vector 216
spinach 271 286
spindle, winter host of *Aphis fabae* 290
Spodoptera 152
spotted alfalfa aphid, control 191
spruce, control of *Fomes* 243–6
spruce budworm, populations, control 197
spruce sawfly, virus in control 120–2 124 126–7
stability: may not help pest control 205; greater in complex ecosystems 106 202 212; instability and low diversity in glasshouses 294 295; instability and pest control 205 294 295; instability under crop rotation 270; of climate, and feasibility of control measures 203–5 294–5; of population, meaning of 205; of population of *Ranunculus repens* 35; supporting by intervention 202
standard man days (SMD) 323–4
staphylinid beetles, predators in brassica plots 109 111–13
starling, chemosterilants for 218
Statistical and Economics Services of Ministry of Agriculture 321
Stellaria media: carrying virus 283; seed population 283
stem nematode 255 259 260
stem rust of wheat 89
Stereum purpureum, competition from *Trichoderma* 241–2
sterile insect release for control 162–70: blowfly 166–7; competitiveness of sterile insects 153 162–3 166–70; fruit flies 166; housefly 166 168; mathematical models 164–5; mosquitoes 167–8; preliminary reduc-

tion with insecticides 168; rearing-factories 166; reasons for use of males only 163; screw-worm campaign 165–6, use of both sexes 163; sex-separation methods 163 169; significance of polygamy and monogamy 163–4; some failures and their causes 166–7; sterile hybrids 169–70; tsetse fly 168–9
sterilization, sexual, of insects: chemosterilants 150–5 167–9 288, hazards to man and animals 167, less effect on females 167, transmission in copulation 150, 153; cytoplasmic incompatibility 170; dominant lethal mutations 162; hybrid sterility 169–70; irradiation 165–167, effects on competitiveness 163; translocations 171; with ecdysones and mimics 154–5; with juvenile hormone and mimics 150 153
sterilization, sexual, of vertebrates 217–220; efficiency models 218–20
Stomoxys, moulting hormones and 154–5
stored products, environmental features and feasibility of control measures 203–4
strawberries in polythene tunnels 321
Sturnus vulgaris, chemosterilants 218
sugar beet (see also 'downy mildew', 'virus yellows'): breeding policy 273; forward contracting 325; husbandry and rotation 107 270–3 281 325; integrated control of pests and diseases 269–76; pathogens and their control 93–4 189 197 198 270–275; pests and their control 107 182 189 197 198 270–6 317 346 347 349; UK production prospects 328
sugar beet mosaic virus, control 271
sugar cane nematodes 179
sugar cane borer, outbreaks after spraying 132
sugar cane moth borer controlled by *Apanteles* 62 71
sulphur (vaporized) against mildew upsetting biological control 298
supernumerary nymphs of Hemiptera 151
survival, see 'mortality'
Sweden: butt-rot in spruce 243; virus of *Neodiprion* 123
swedes (rutabagas) and cabbage root fly 286
symbiotic micro-organisms 115 119
Synanthedon tipuliformis 79
Synchytrium endobioticum 88 286
syrphid flies, predators 107

Taeniothrips inconsequens 81
Taiwan, grass carp 234
take-all of wheat and barley 241 258
Talpa europaea eaten by buzzards 213
Tanzania, pests of coffee 184
tape-seeding to a stand 321
Tardigrada 262 and Pl. opp.
Tasmania, biological control 63
technical services (see also 'ADAS') 278 280 282 289–90
Tenerife, Mediterranean fruit fly 166
tepa, chemosterilant 167–9
terpenoid ethers, juvenile hormone mimics 149
tetrachlorvinphos against pine looper 305
Tetranychus althaeae and *Rubus* spp. 78
Tetranychus desertorum in weed control 101
Tetranychus telarius and menazon 188
Tetranychus urticae 79 296–7 299 317 334 336 337–8
thallium acetate in ant bait 132
thallium sulphate in toxic baits 131
Thames Valley, upper, sugar beet 271
Thaumatopoea pityocampa control 359
Thera spp. eaten by tits 214
Therioaphis maculata, control 63 67
Therioaphis trifolii, control 202–3
thioglucoside in rape 287
thioglycolic acid, stressor 122
thiotepa, chemosterilant 167–8
thiram, maize seed treated 240
Thomasiniana theobaldi on raspberry 78
Thomomys bottae navus and coyotes 213
thrips 81 130 298 335 339
Thrips tabaci, control 298
Tipulidae (leatherjackets) 280 318
tobacco, numbers of pests, USA 182
tobacco budworm and azadirachtin 156
tobacco mosaic virus, new race 90
tobacco rattle virus and nematode vector 265
tobacco wilt disease 239–40
tolerance levels of infestation or damage 289 316
tomato: *Cephalosporium* and *Fusarium* 239; mosaic virus 90

tomato leaf-miner, control 334
tomato wilt and *Cephalosporium* 239
tomatoes: insecticides toxic 334; integrated pest control 301; resistance to *Didymella* 286
Tomicus piniperda 304 306–9
toxic baits against ants, etc. 131–4
translocations, use in genetic control of insects 171–2
trees (see also 'forests'): bullfinches as pests 349; diseases, biological control of 241–6; oak and winter moth 15–16; pine, and insect control 302–311; reproduction 4 5; spacing 5 9 11
Trialeurodes vaporariorum 63 297 298 318
triarimol against mildew and aphids 298
triazine weedkillers 346
Tribolium and moulting hormones 155
Trichinella, gut nematodes 153
trichlorphon against cabbage root fly 289
Trichoderma viride competing with *Fomes* 244 and *Stereum* 242
Trichodorus spp.: crop rotations 259; soil conditions 254; tobacco rattle virus 265
Trichogramma, biological control agent 64
Trichoplusia ni: control 126; virus 123
triethylene-melamine, reproduction inhibitor 218
Trinidad, leaf-cutting ants 134–5
'Triol', moulting hormone mimic 155
Trioxys sinensis, parasite of cotton aphid 298
Tripleurospermum maritimum ssp. *inodorum* 283
Tripyla, predatory nematode 262
tropics: agriculture in 11; exploitation of 13–14; leaf-cutting ants 134–5; rain forest and pest control 203–4; weeds in 179
Trypodendron lineatum in logs 304
tsetse flies: both sexes transmit tryphanosomiasis 163; chromosome translocation 172; sterile release 168–9
Tuberculina maxima displacing *Cronartium* 242
Tubifex eaten by grass carp 234
turnips and cabbage root fly 286
Tylenchorhynchus, population increase 264–5
Typha spp. eaten by grass carp 235
Tyrophagus longior on cucumbers 298

Ulex, fate of seedlings 52
United Kingdom (UK): agricultural production, by commodities, 1968 and 1980 projection 326; arthropod pests of apple, numbers in major taxa 181; number of pests of crops 178; projection 1980 of grain production and land needed 327; subjects for microbiological control 126–7
United States (USA) (passing references omitted): arthropod pests of field and fruit crops, numbers in major taxa 182; eradication of screw-worm 165–6; imported fire ant, and control campaign 131–4; soil-borne pathogens and control 239–43
Universities, collaborative work on control needed 199–200
Upper Volta, sterile hybrid *Anopheles* released 169
Uromyces on sugar beet 271
Urostyla sp. Pl. opp. 262
Urtica urens carrying virus 283
USSR 65 78 234
Ustilago nuda, resistance in barley 93

vectors of human disease, chemical or habitat control 352
vegetable production: change in numbers of holdings 323; changes in rotations 279–80; changes in practice, and pest control 277–82; continuance of 'home' market 322–3; control of pests and diseases 277–90; costs under scrutiny 289; crops seldom free of pest damage 316; demand for continuity of supplies 326; environmental hazards in modern trends 279; forecasts of pest and disease incidence 290; herbicides, key to modern methods 280; host-plant resistance 284–8; identifying factors regulating pests 290; increased percentages of crops grown under contract 325; increasing local concentration of each crop 280; less land used for certain crops, but more produced (UK) 328; mini-cauliflower cropping system 282; modern features 279; need to monitor new cropping systems 282; new crops and cultivars, and

pest and disease control 281–2; new cultural methods affecting pests and diseases 280–1; new methods 321; output value of various crops 279; pests and diseases upset uniformity of produce 280; plant density affects pests and diseases 281; sowing to a stand increases pest hazards 280; survey of infestation, damage and control measures 289–290; systems influenced by economic, social, political factors 279; technical services 289–90
Venturia inaequalis, resistance of apple 80
Veronica spp. as weeds 98
Verticillium on mushrooms; vectors, control 334
Verticillium on olives, soil amendment 240–1
vinyl phosphates, diversification of 346
Viola arvensis, effect of herbicides 98
violet root rot and vegetable rotation 279
virus diseases of crops 77–9 179 197 198 270–5 281 283–4 286 289 340
virus yellows of mangold, aphid vector 270
virus yellows of sugar beet: control 197 198 269–76; growing virus-free seed crops 270–1; plant density effects 272 281; sources of infestation and aphid vectors 271 (and Pl. opp.) 274–5
viruses and mycoplasma of plants, over 300 forms 179
viruses for pest control, see 'pathogens for control . . .'
Vitis 74 182 347
voles: and winter moth 16; eating seeds 37
Voria ruralis parasite and virus vector 123

Wales, virus of *Gilpinia* 126
warfarin in rat baits 220
wart disease of potato 88 286
waterproofing agents in baits 131
weaver birds (*Quelea*) 76
Weed Research Organization and *Agropyron* 100
weeds
aquatic 233–7 352
chief problem, control of grasses in cereals 183 194
control: and ecology 42–53; basis of agricultural revolution in Britain 199; biological 29 44–5 49 51 61 64–6 97–104 233–7; by management 45 49 99; chemical, see 'herbicides'; cultural 45–6 99 270; direct and indirect 49; economic aspects 98–9
ecology 28–39 42–53
effects of crop and other plants on weeds 100–1 281
fungicides and insecticides affecting weeds 50
grazing, effects of 98 103 233–7 352
groups of plants that include weeds 179 180 183
hosts of pests and pathogens 43 271 283
immigration, invasion 28 43–4 48
in monoculture 99 183 194
life cycles 29 47–9
numbers in tropical crops 179
of grassland, population studies 28–39
populations, population dynamics 29–39 44–53 100–1 283
removal, and effects of 43–4 49 107 199
seed bank 30 36–9 45 47–9 283
seed dormancy 36–9 47
tolerance by crops 43 108 110 112–13
Wellesbourne, National Vegetable Research Station 284 287 288
West Indies, control of houseflies 168
West Palaearctic Regional Section (WPRS) of IOBC 357–60
wheat: resistance to pathogens 88–93; growth regulators to reduce lodging and eyespot 347; increase in size of farms 324; insecticides, selective placement 189–90
wheat bulb fly (*Hylemyia coarctata*)
control: integrated 198; insecticides, selective placement 190; seed dressings 350; sowing dates 201
major pest, abundant every year 318
white amur (grass carp, *Ctenopharyngodon*) 98 214 234–8 352
white pine blister rust controlled by fungus 242
white rot of onions and crop density 281
wild oats 47 49–50 100–1 103
willow-carrot aphid, and crop density 281
winter moth: in Canada 25 63; methods of population study 16; models of

winter moth—*cont.*
population change, mortality, parasitism, predation 17–22 26; parasites of 18; population dynamics 16–20; predation of 16–22 26; quest theory, Nicholson's theory 17–24
wireworms (*Agriotes* spp.) 190 280 316–317
Wisconsin, cabbage resistance to aphid 286
woodmice and winter moth 16
wood-pigeon: eaten by goshawks 213; experiments on behavioural scarecrow 222–9
woolly apple aphid, host resistance 74–5 79 80 318
woolly pear aphid, host resistance 81
Wykeham Forest, Yorkshire, pine looper outbreak 304–6
Wytham, near Oxford, winter moth studies 22

Xiphinema diversicaudatum, soil relations 251
Xiphinema spp., generation time 255
Xyleborus saxesini, host resistance 81

yellow rust of wheat, host resistance 89 91–3
Yugoslavia, use for host resistance to *Aphis idaeae* 78

Zeiraphera diniana: microbiological control? 126; changing susceptibility to virus 124
Zemun, Yugoslavia, laboratory for biological control 357 359
Zurich, laboratory for insect pathology 357